Prentice Hall Endocrinology Series

Mac E. Hadley
Series Editor

The
Adrenal
Cortex

The Adrenal Cortex

Gavin P. Vinson

Department of Biochemistry
Queen Mary and Westfield College
London

Barbara Whitehouse

Biomedical Sciences Division
King's College London

Joy Hinson

Department of Biochemistry
Queen Mary and Westfield College
London

PRENTICE HALL, Englewood Cliffs, New Jersey 07632

Library of Congress Cataloging-in-Publication Data

Vinson, Gavin P.
   The adrenal cortex / Gavin P. Vinson, Barbara Whitehouse, Joy
Hinson.
      p.   cm.
   Includes bibliographical references and index.
   ISBN 0-13-395153-7
   1. Adrenal cortex—Physiology.   2. Adrenal cortex—Diseases.
   3. Adrenocortical hormones.   I. Whitehouse, Barbara   II. Hinson,
   Joy   III. Title.
   QP188.A28V55   1992
   612.4'5—dc20   91-29388
                                                                     CIP

Editorial/production supervision
   and interior design: Laura A. Huber
Production Assistant: Jane Bonnell
Acquisitions Editor: Michael Hays/Betty Sun
Editorial Assistant: Dana L. Mercure/Maureen Diana
Cover Design: Wanda Lubelska Design
Prepress Buyer: Mary Elizabeth McCartney
Manufacturing Buyer: Susan Brunke
Marketing Manager: Alicia Aurichio

Prentice Hall Endocrinology Series

© 1992 by Prentice-Hall, Inc
A Simon & Schuster Company
Englewood Cliffs, New Jesey 07632

Printed in the United States of America

10   9   8   7   6   5   4   3   2   1

ISBN   0-13-395153-7

PRENTICE-HALL INTERNATIONAL (UK) LIMITED, London
PRENTICE-HALL OF AUSTRALIA PTY. LIMITED, Sydney
PRENTICE-HALL CANADA INC., Toronto
PRENTICE-HALL HISPANOAMERICANA, S.A., Mexico
PRENTICE-HALL OF INDIA PRIVATE LIMITED, New Delhi
PRENTICE-HALL OF JAPAN INC., Tokyo
SIMON & SCHUSTER ASIA PTE. LTD., Singapore
EDITORA PRENTICE-HALL DO BRASIL, LTDA., Rio de Janeiro

To Ben, Bronwen, Jenny, Kate, Mary, Michael, Michaela,
Peter and William

Contents

Acknowledgments

A very large number of colleagues and friends have assisted us in the preparation of this book in very many ways. In particular, we would like to thank the following for permission to reproduce their previously published figures, and for many helpful discussions: S. Barker, J.P. Baxter, G.M. Besser, J.P. Bonvalet, E.L. Bravo, P.F. Bruning, C.W. Burke, P. Chambon, I. Chester Jones, P. Coates, B. Cook, M.C. Cornwall, J. Chronshaw, I. Doniach, R. Flower, M.G. Forest, R. Fraser, C.E. Gomez-Sanchez, J.E. Hall, W. Hanke, M.A. Holzwarth, V.H.T. James, S. Kominami, J.J. Lambert, L.W. Levine, A. Levitzki, S. Malamed, R. Mauret, J.C. Melby, W. Mosley, A. Munck, D.W. Nebert, M.I. New, G. Nussdorfer, M. Oelkers, M. O'Hare, M.G. Parker, M. Peach, A.J. Pettit, I.R. Phillips, J.R. Puddefoot, J. Pudney, P.W. Raven, D. Schulster, E.M. Smith, C. Wanless, G.H. Williams, P. White, D.B. Young.

In addition, we would like to thank Katherine Macdonald, Clementina Obinkwo and Joy Smith for willing word processing. Special thanks too to Bridget Landon and staff of the Medical Illustration and Photographic Departments at St. Bartholomew's Hospital for their excellent work which has contributed so much to the book's final appearance.

Finally, thank you to the series editor, Mac Hadley, for inviting us to write the book, and for all his help and encouragement.

Introduction

The Adrenal Cortex

A brief historical perspective

The adrenals were described by several authors in the sixteenth and seventeenth centuries, and were identified by their shape, size, and proximity to the kidney. Although sometimes classified rather loosely as glands, meaning, according to the usage of the time, simply soft tissue [1]—hence *glandulis quae renibus incumbent* (Eustachius, ca. 1501–1574: pre 19th C references are given in full refs. 2 and 3)—the concept of internal secretion of course arose much later (see below). Authors of this period also used the terms *capsulae renales* (Spigelius, 1578–1625), *capsulae suprarenales* (Jean Riolan the Younger, 1580–1657), *capsulae atrabilaria* (Bartholinus, 1585–1629) and others [1–3]. Thomas Wharton (1614–1673) introduced a more rigorous concept of a gland as an organ of secretion, and clearly identified the adrenals (*glandulae ad plexum nerveum sitae*, also *glandulae renales* and *capsulae atrabilariae*) as secretory organs together with lymph glands, pancreas, salivary glands, mammary glands and thymus, as well as the endocrine organs, thyroid, pituitary, and gonads. His concepts of the classification of secretions into excretory (e.g., from the testis) and nutritional (e.g., from the thymus) and those which have some other useful function (such as the adrenal products) of course only partly coincide with ours! There followed among various authors a degree of speculation, not firmly based, about the possible functions of such secretions, and their routes of egress from the gland, [1,4]. It is, however, worth noting that Riolan the Younger thought that the glands were functional only in fetal life, because of their relatively large size at that time. Nevertheless, although several essays were submitted in response to the challenge issued in 1716 by the Adademie des Sciences in Bordeaux: "Quel est l'usage des glandes surrénales?" none was judged by Charles de Montesquieu (1689–1755) to be worthy of the prize.

It was in the nineteenth century that clear cut observations paved the way for the identification of the adrenal as a gland of internal secretion, in the sense that this term is now used. The general concept that the ductless glands might secrete their products directly into the blood stream had a complex origin [1,4], but the term "sécretion interne" was first used by Claude Bernard [5] to describe the production of glucose into the blood stream by the liver, and he later invoked the concept in relation to the spleen, thyroid, and adrenals, though according to Young [4] he was giving voice to ideas which were already in vogue. In the adrenal, Vulpian [6] and Virchow [7] demonstrated that dilute aqueous ferric chloride gave a strong green color in the adrenal medulla, and showed too that this reaction occurred with the blood from the adrenal vein. Nevertheless, it was Thomas Addison who gave the field its crucial impetus in his description of the syndrome which now bears his name, and his correct interpretation of the significance of the diseased adrenals of his patients [8]. A few years later, the histological structure of the gland was first described by Arnold, who introduced the terms *zona glomerulosa, zona fasciculata*, and *zona reticularis* [9]. Brown-Séquard [10] first demonstrated experimentally that the presence of the adrenals was essential to life, although his interpretation of their function, as a type of specialized excretory organ, deviates considerably from the concept of a secretory gland. His statements were regarded as highly controversial for some time. Perhaps this scepticism was later sustained in part by doubts about his subsequent activities, especially the vigorous, and even more highly controversial, advocacy of "organotherapy," the treatment of disease by administration of tissue extracts. In particular, the claims of rejuvenation achieved by treatment with testicular extracts raised serious scientific objections, as well as the moral and philosophical reactions characteristic of the period. This later approach, however, clearly reflected the strengthening view that the glands may provide materials required for essential physiological functions. The concept was molded into a generally accepted form following the demonstration by Oliver and Schaeffer [11] of the presence of a pressor substance in adrenal extracts. Adrenaline was then isolated by Aldrich [12] and synthesized by Stolz [13]. Adrenaline was thus the first hormone, although the word "hormone" itself was not used until it was applied to secretin, which was isolated by Bayliss and Starling shortly afterwards [14].

Evidence that the adrenal cortex might have a function separate from that of the medulla emerged more slowly, owing to the difficulties of preparing extracts which were free of adrenaline. However, partial success was achieved by Rogoff and Stewart [15], and a completely adrenaline free extract was obtained by Hartman, Macarthur, and Hartman [16] and they and Pfiffner and Swingle [17] first showed that such an extract could preserve the life of an adrenalectomized cat. Later, the effects of the extract "cortin" in normal subjects and Addisonians, with particular emphasis on electrolyte balance were definitively summarized [18].

At about this time too, the chemical nature of the active principles involved was becoming clear. Oestrone was the first steroid hormone to be isolated in crystalline form [19,20] and, from studies initiated in 1935, three groups, those of Kendall, Wintersteiner and Pfiffner, and Reichstein led within three years to the isolation from adrenal tissue of a series of 21 substances in crystalline form. Six of these were able to prolong life in adrenalectomized animals: today their trivial names are deoxycorticosterone, corticosterone, 11-dehydrocorticosterone, cortisol, 11-deoxycortisol, and cortisone, though they were then known by a series of letters of the alphabet, confusingly different in each group (see ref. 21 for a fuller description). Their structures were elucidated in a long series of painstaking studies by these and other workers [21]. It was, however, only with the development of paper chromatographic methods in the 1950s that it was possible to show that either cortisol or corticosterone was the major steroid product secreted into the adrenal vein in a variety of mammalian species [22,23].

The study of the physiological role of the adrenocortical secretion was given great impetus by the work of Selye [24], who emphasized the responses of the system to "stress," and the concept of the "General Adaptation Syndrome" through which chronic stress would lead to the onset of various diseases. This concept, however, proved perhaps too simple, and many aspects of the relationship between stress and disease remain to be fully explained [30].

It was recognized that adrenal steroids were involved in the regulation of both carbohydrate and electrolyte metabolism, and both are perturbed in Addisonian patients (e.g., ref. 25). Indeed one early effective treatment for Addisonian patients was to provide saline [26] although with a false rationale: it was thought to dilute toxins. The nature of these separate functions became clearer in the 1930s as several workers characterized the electrolyte and carbohydrate responses to adrenalectomy [23,27,28]. However, it came as a great surprise to find that adrenal steroids also had anti-inflammatory actions, to the extent that they could be used successfully in the treatment of rheumatoid arthritis [29,30].

The concept that different types of hormones might be involved in these effects was not fully accepted until Simpson and Tait and their co-workers isolated and characterized aldosterone as the principal mineralocorticoid, essentially lacking in glucocorticoid activity, [31,32]. It was the technical problems associated with crystallizing and purifying the relatively minute amounts of this extremely potent steroid present in adrenal glands which delayed the characterization of this, the last metabolically important steroid hormone to be identified. Indeed the difference in experimental approach compared with what had gone before is interesting, and whereas in earlier work crystalline steroids were isolated in yields which could be handled using classical chemical methods, the isolation of aldosterone depended heavily on the application of the new sophisticated paper chromatographic methods, as well as on a particularly sensitive bioassay (see Tait & Tait, [23]). The other, less active, 18-oxygenated steroids were first characterized by Birmingham and Ward, and Péron (18-hydroxydeoxycorticosterone) [33,34], and Ulick and his co-workers (18-hydroxycorticosterone) [35]. It was also Ulick's group who some 20 years later then characterized 18-hydroxycortisol and 18-oxocortisol [36,37].

The relationship between the pituitary and adrenocortical function in clinical conditions was explored extensively from 1912 onwards by Harvey Cushing [38,39] among others, and the dependence of the adrenal on the presence of the pituitary was clearly shown once satisfactory methods for hypophysectomy had been devised for experimental animals [40]. ACTH was first isolated by the groups of Li and Sayers [41,42]. That the glomerulosa is the source of mineralocorticoid activity and is hardly affected by hypophysectomy, whereas the glucocorticoids emanate from the inner adrenocortical zones under pituitary control, was first suggested by Swann in 1940 [43]. This was, of course, many years before the separate identity of mineralocorticoids and glucocorticoids was universally accepted (see ref. 23). However, the theory gained in experimental support, for example, through the morphometric studies on the rat adrenal by Deane & Greep, [44]. Following the characterization of aldosterone, however, direct functional evidence, obtained by incubating separated adrenocortical zones from bovine or rat glands, showed that aldosterone was produced only in the glomerulosa, and that the fasciculata-reticularis produced cortisol or corticosterone [45,46]. At practically the same time as the characterization of aldosterone, there appeared the classical descriptions of primary aldosteronism caused by adenoma of (presumably) glomerulosa type cells in humans [47]. The mechanisms for the regulation of aldosterone secretion remained an enigma for many years, and indeed still cannot be said to be completely resolved (see Chapter 4). However, a decisive advance was made in 1958 when Gross drew attention to the possible importance of the renin-angiotensin system [48].

Elucidation of the mechanism for biosynthesis of adrenal steroids first became a practical possibility with the advent of readily available isotopically labeled precursors: first acetate and cholesterol, and later the steroids such as progesterone. The initial approach was that of Hechter, Pincus, and co-workers [49] who perfused calf adrenals with these substances, and thus laid the basis for our understanding of the biosynthetic pathways by which the steroid hormones are formed. Stone and Hechter [50] identified cholesterol side-chain cleavage as the ACTH sensitive site in the biosynthetic pathway using similar techniques. Subsequently, in vitro incubation of adrenocortical preparations was the preferred method. First introduced by Tepperman

[51], adrenocortical incubations have had a productive history and were used by Saffran and Bayliss and Birmingham et al., for ACTH bioassays [52, 53]. Some 15 years later, the introduction of isolated cell suspensions, in general, supplanted the use of the quartered glands [54]. Prior to this, ascorbic acid depletion tests were widely used for ACTH bioassays (e.g., ref. 55), and indeed in the 1970s there was a resurgence of this approach, refined to be performed at the cytochemical level [56].

Incubation of subcellular fractions of adrenal glands for the study of the biosynthetic enzymes for corticosteroid biosynthesis by several groups [57–59], showed that 11β-hydroxylation, 20- and 22-hydroxylation, and 18-hydroxylation were mitochondrial, and Ryan and Engel [60] demonstrated the microsomal location of steroid 21-hydroxylase. Hayano and Sweat and their co-workers showed the requirement for molecular oxygen [61,62], and Brownie and Grant [63] that Krebs cycle metabolites may support steroidogenesis, thus initiating the study of cofactor requirements. Characterization of the monooxygenases involved studies by many workers (see refs. 64–67 for full summaries), however, it was Kimura and Omura and their co-workers [68,69] who characterized the three mitochondrial hydroxylase components as a flavoprotein dehydrogenase, a non-haem iron protein and cytochrome P-450.

Much of the work on intracellular messengers in the adrenal cortex is recent, and is dealt with fully in Chapter 4, but it is interesting to note that the requirement for calcium in the response to ACTH was in fact established many years ago, by Birmingham and co-workers [53], sometime before calcium's role as a ubiquitous intracellular messenger was recognized. It is also worth noting that although the importance of cAMP as an intracellular mediator had been shown in the liver, the first endocrine tissue in which this was shown was the adrenal cortex [70,71].

Although the main thrust of adrenocortical research has been in the clinical field, and, experimentally, in a limited number of mammalian species, it is interesting to note how much comparative work was done at an early stage. Thus Perrault [72] apparently described the adrenals of birds and reptiles in 1671, and Swammerdam [73] those of the frog in 1714. In fish, elasmobranch adrenal tissue was first described by Nagel [74], (who also made extensive observations in other vertebrates, including birds, amphibia, and reptiles), Leydig [75], and Stannius [76]. This work was drawn together by Balfour [77] who showed the homology between the unpaired interrenal body and the cortex, and the paired bodies along the dorsal aorta with the mammalian adrenal medulla. Giacomini [78], similarly, dealt with the identification and disposition of the chromaffin tissue (the equivalent of the mammalian medullary tissue) and the interrenal (the adrenocortical homologue) in the teleost. Swale Vincent [79] showed the paired bodies in elasmobranchs contained pressor substance(s) similar to those in the mammalian medulla, but one may speculate that Cleghorn [80] was puzzled by his finding that extracts of the elasmobranch interrenal failed to support adrenalectomized cats. We now know why: the elasmobranch interrenal is the only adrenocortical homologue among the vertebrates which does not produce corticosterone, cortisol, or aldosterone! (see Chap. 9).

This brief account is of necessity selective, and much has been omitted. For a fuller historical introduction to the study of the adrenal cortex, the reader is referred again to the reviews by Medvei [1], Rolleston [2], Schumacker [3], Tait and Tait [23], and Gaunt [30].

References

1. Medvei, V.C. 1982. *A history of endocrinology.* Lancaster: MTP Press.

2. Rolleston, H.D. 1936. *The endocrine organs in health and disease.* Oxford: University Press.

3. Schumacker, H.B. 1936. The early history of the adrenal glands. *Bull. Inst. Hist. Med.* 4:39–56.

4. Young, F.G. 1970. The evolution of ideas about animal hormones. In: J. Needham, ed. *The chemistry of life.* Cambridge: University Press, pp. 125–55.

5. Bernard, C. 1855–56. Lecons de physiologie experimentale appliquee a la medecine. 2 vols. Paris: J.B. Bailliere.

6. Vulpian, E.F.A. 1856. Note sur quelques reactions propres a la substance des capsules surrenales. *C.R. Acad. Sci. Paris.* 43:663–65.

7. Virchow, R. 1857. Zur Chemie der Nebennieren. *Arch. Path. Anat.* 12:481–83.

8. Addison, T. 1855. *On the constitutional and local effects of disease of the suprarenal capsules.* London: S. Highley.

9. Arnold, J. 1866. Ein beitrag zu der feineren structur und dem chemismus der nebennieren. *Virchows Arch. Path. Anat. Physiol.* 39:64–117.

10. Brown-Séquard, C.E. 1856. Recherches experimentales sur la physiologie et la pathologie des capsules surrenales. *Arch. Gen. Med. Paris* 5 s VIII, 385:372.

11. Oliver, G., and Schaefer, E.A. 1895. The physiological effects of extracts of the suprarenal capsules. *J. Physiol. Lond.* 18:230–76.

12. Aldrich, T.B. 1901. A preliminary report on the active principle of the suprarenal gland. *Am. J. Physiol.* 5:457–61.

13. Stolz, F. 1904. Ueber Adrenalin und Alkylaminoacetobrenzcatechin. *Ber. Dtsch. Chem. Ges.* 37:4149–54.

14. Bayliss, W. M., and Starling, E.H. 1902. The mechanism of pancreatic secretion. *J. Physiol., Lond.* 28:325–53.

15. Rogoff, J.M., and Stewart, G.N. 1927. The influence of adrenal extracts on the survival period of adrenalectomised dogs. *Science,* 66:327–28.

16. Hartman, F.A., Macarthur, C.G., and Hartman, W.E. 1927. A substance which prolongs the life of adrenalectomized cats. *Proc. Soc. Exp. Biol. Med.* N.Y.: 25:69–70.

17. Pfiffner, J.J., and Swingle, W.W. 1929. The preparation of an active extract of the suprarenal cortex. *Anat. Rec.* 44:225–26.

18. Thorn, G.W., Garbut, H.R., Hitchcock, F.A., and Hartman, F.A. 1937. The effect of cortin upon the renal excretion of sodium, potassium, chloride, inorganic phosphorus and total nitrogen in normal subjects and patients with Addison's disease. *Endocrinology.* 21:202–12.

19. Doisy, E.A., Thayer, S., and Veler, C.D. 1930. The preparation of the crystalline ovarian hormone from the urine of pregnant women. *J. Biol. Chem.* 86:499–509.

20. Butenandt, A. 1932. Chemical constitution of the follicular and testicular hormones. *Nature, Lond.* 130:238.

21. Fieser, L.F., and Fieser, M. 1949. Natural products related to phenanthrene, 3rd ed. New York: Reinhold.

22. Bush, I.E. 1953. The paper chromatography of steroids and its application to assay problems. *Ciba Fndn. Colloq. Endocrin.* 5:203–15.

23. Tait, J.F., and Tait, S.A.S. 1979. Recent perspectives on the history of the adrenal cortex. *J. Endocr.* 83:3P–24P.

24. Selye, H. 1936. Stress syndrome. A syndrome produced by diverse nocuous agents. *Nature, Lond.* 138:32.

25. Harrop, G.A., and Weinstein, A. 1932. Addison's disease treated with suprarenalcortical hormone. *J. Am. Med. Assoc.* 98:1525–31.

26. Stewart, G.N., and Rogoff, J.M. 1925. Studies on adrenocortical insufficiency. *Proc. Soc. Exp. Biol. Med., N.Y.* 22:394–97.

27. Britton, S.W., and Silvette, H. 1937. The adrenal cortex and carbohydrate metabolism. *Cold Spring Harbor Symp. Quant. Biol.* 5:357–61.

28. Hartman, F.A., and Brownell, K.A. 1934. Relation of adrenals to diabetes. *Proc. Soc. Exp. Biol. Med., N.Y.* 31:834–35.

29. Hench, P.S., Kendall, E.C., Slocumb, C.H., and Polley, H.F. 1949. The effect of a hormone of the adrenal cortex (17-hydroxy-11-dehydro-corticosterone: compound E) and of pituitary adrenocorticotropic hormone on rheumatoid arthritis. *Proc. Mayo Clinic.* 24:181–97.

30. Gaunt, R. 1974. History of the adrenal cortex. In: Blaschko, H., Sayers, G. and Smith, A.D. (Eds.) Handbook of Physiology sect. 7. American Physiological Society, Washington.

31. Grundy, H.M., Simpson, S.A., and Tait, J.F. 1952. Isolation of a highly active mineralocorticoid from beef adrenal extract. *Nature, Lond.* 169:795–96.

32. Simpson, S.A., and Tait, J.F. 1955. Recent progress in methods of isolation, chemistry and physiology of aldosterone. *Recent Progr. Hormone Res.* 11:183–210.

33. Birmingham, M.K., and Ward, P.J. 1961. The identification of the Porter-Silber chromogen secreted by the rat adrenal. *J. Biol. Chem.* 236:1661–67.

34. Péron, F.C. 1961. Isolation of 18-hydroxydeoxycorticosterone from rat adrenals. *Endocrinology,* 69:39–45.

35. Ulick, S., and Kusch, K. 1960. A new C-18 oxygenated steroid from bullfrog adrenals. *J. Am. Chem. Soc.* 82:6421–22.

36. Ulick, S., and Chu, M.D. 1982. Hypersecretion of a new corticosteroid, 18-hydroxycortisol in two types of adrenocortical hypertension. *Clin. Exp. Hypertension.* (A) 4:1771–77.

37. Ulick, S., Land, M., and Chu, M.D. 1983. 18-oxocortisol, a naturally occurring mineralocorticoid agonist. *Endocrinology.* 113:2320–22.

38. Cushing, H.W. 1908. *The pituitary body and its disorders.* Philadelphia: J.B. Lippincott.

39. Cushing, H.W. 1938. The basophil adenomas of the pituitary body and their clinical manifestations (pituitary basophilism). *Bull. Johns Hopkins Hosp.* 50:137–95.

40. Smith, P.E. 1930. Hypophysectomy and replacement therapy in the rat. *Am. J. Anat.* 45:205–74.

41. Li, C.H., Simpson, M.E., and Evans, H.M. 1943. Adrenocorticotropic hormone. *J. Biol. Chem.* 149:413–24.

42. Sayers, G., White, A., and Long, C.N.H. 1943. Preparation and properties of pituitary adrenocorticotropic hormones. *J. Biol. Chem.* 149:425–36.

43. Swann, G. 1940. The pituitary-adrenocortical relationship. *Phys. Rev.* 20:493–521.

44. Deane, H.W., and Greep, R.O. 1946. A morphological and histochemical study of the rat's adrenal cortex after hypophysectomy, with comments on the liver. *Am. J. Anat.* 79:117–46.

45. Ayres, P.J., Gould, R.P., Simpson, S.A., and Tait, J.F. 1956. The *in vitro* demonstration of differential corticosteroid production within the ox adrenal gland. *Biochem. J.* 63:19P.

46. Giroud, C.J.P., Stachenko, J., and Venning, E.H. 1956. Production of aldosterone by rat adrenal glands *in vitro*. *Proc. Soc. Exp. Biol. Med., N.Y.* 92:855–59.

47. Conn, J.W., and Louis, L.H. 1956. Primary aldosteronism, a new clinical entity. *Ann. Intern. Med.* 44:1–15.

48. Gross, F. 1958. Renin und hypertensin, physiologische oder pathologische wirkstoffe? *Klin. Wschr.* 36:693–706.

49. Hechter, O., and Pincus, G. 1954. Genesis of the adrenocortical secretion. *Physiol. Rev.* 34:459–96.

50. Stone, D., and Hechter, O. 1954. Studies on ACTH action in perfused bovine adrenals: The site of action of ACTH in corticosteroidogenesis. *Arch. Biochem. Biophys.* 51:457–69.

51. Tepperman, J. 1950. Effects of purified ACTH added *in vitro* on the oxygen consumption and ascorbic acid content of surviving dog adrenal slices. *Endocrinology.* 47:384–85.

52. Saffran, M., and Bayliss, M.J. 1953. *In vitro* bioassay of corticotrophin. *Endocrinology.* 52:140–48.

53. Birmingham, M.K., Elliott, F.H., and Valere, P.H.L. 1953. The need for the presence of calcium for the stimulation *in vitro* of rat adrenal glands by adrenocorticotropic hormone. *Endocrinology.* 53:687–89.

54. Kloppenborg, P.W.C., Island, D.P., Liddle, G.W., Michelakis, A.M., and Nicholson, W.E. 1969. A method of preparing adrenal cell suspension and its applicability to the *in vitro* study of adrenal metabolism. *Endocrinology.* 82:1053–58.

55. Sayers, M.A., Sayers, G., and Woodbury, L.A. 1948. The assay of adrenocorticotrophic hormone by the ascorbic acid depletion method. *Endocrinology.* 42:379–93.

56. Daly, J.R., Loveridge, N., Bitensky, L., and Chayen, J. 1974. The cytochemical bioassay of corticotrophin. *Clin. Endocrinol., Oxf.* 3:311–18.

57. Sweat, M.L. 1951. Enzymatic synthesis of 17-hydroxycorticosterone. *J. Am. Chem. Soc.* 73:4056–57.

58. Halkerston, I.D.K., Eichhorn, J., and Hechter, O. 1959. TPNH requirement for cholesterol side chain cleavage in adrenal cortex. *Arch. Biochem.* 85:287–89.

59. Nakamura, Y., and Tamaoki, B.I. 1964. Intracellular distribution and properties

of steroid 11B-hydroxylase and steroid 18-hydroxylase in rat adrenal. *Biochem. Biophys. Acta.* 85:350–52.

60. Ryan, K.J., and Engel, L. 1957. Hydroxylations of steroids at carbon 21. *J. Biol. Chem.* 225:103–14.

61. Hayano, M., Lindberg, M.C., Dorfman, R.I., Hancock, J.E.H., and von Doering, W. von E. 1955. On the mechanisms of the C-11β-hydroxylation of steroids. A study with H_2O^{18} and O_2^{18}. *Arch Biochem.* 59:529–32.

62. Sweat, M.L., Aldrich, R.A., de Bruin, C.H., Fowlks, W.L., Heiselt, L.R., and Mason, H.S. 1956. Incorporation of molecular oxygen into the 11β-position of corticosteroids. *Fed. Proc.* 15:367.

63. Brownie, A.C., and Grant, J.K. 1954. The *in vitro* enzymic hydroxylation of steroid hormones. 1. Factors influencing the enzymic 11β-hydroxylation of 11-deoxycorticosterone. *Biochem. J.* 57:255–63.

64. Harding, B.W., Bell, J., Oldham, S.B., and Wilson, L.D. 1968. Corticosteroid biosynthesis in adrenal cortical mitochondria. In: K.W. McKerns, ed. *Functions of the adrenal cortex.* vol. II, Amsterdam: North Holland Publishing Co., pp. 831–96.

65. Cooper, D.Y., Narasimhulu, S., Rosenthal, O., and Estabrook, R.W. 1968. Studies on the mechanism of C-21-hydroxylation of steroids by the adrenal cortex. In K.W. McKerns, ed. *Functions of the adrenal cortex.* vol. II, Amsterdam: North Holland Publishing Co., pp. 897–942.

66. Cammer, W., Cooper, D.Y., and Estabrook, R.W. 1968. Electron-transport reactions for steroid hydroxylation by adrenal cortex mitochondria. In K.W. McKerns, ed. *Functions of the adrenal cortex.* vol. II, Amsterdam: North Holland Publishing Co., pp. 943–92.

67. Kimura, T. 1968. Electron transfer system of steroid hydroxylases in adrenal mitochondria. In K.W. McKerns, ed. *Functions of the adrenal cortex.* vol. II, Amsterdam: North Holland Publishing Co., pp. 993–1006.

68. Kimura, T., and Suzuki, K. 1965. Enzymatic reduction of non-heme iron protein (adrenodoxin) by reduced nicotinamide adenine dinucleotide phosphate. *Biochem. Biophys. Res. Comm.* 20:373–79.

69. Omura, T., Sato, R., Cooper, D.Y., Rosenthal, O., and Estabrook, R.W. 1965. Function of cytochrome P450 of microsomes. *Fed. Proc.* 24:1181.

70. Haynes, R.C. 1958. The activation of adrenal phosphorylase by the adrenocorticotrophic hormone. *J. Biol. Chem.* 233:1220–22.

71. Grahame-Smith, D.G., Butcher, R.W., Ney, R.L., and Sutherland, E.W. 1967. Adenosine 3',5'-monophosphate as the intracellular mediator of the action of adrenocorticotropic hormone on the adrenal cortex. *J. Biol. Chem.* 242:5535–41.

72. Perrault, C. 1671. Suite de memoires pour servir de l'Histoire naturelle des animaux. 1:155. Paris.

73. Swammerdam, J. 1714, English trans. 1758. *The book of nature or the history of insects.* trans: T. Flloyd, revised J. Hill, London, C.G. Seyffert, reprinted 1978, New York: Arno Press.

74. Nagel, M. 1836. Ueber die Struktur der Nebennieren. *Arch. Anat. Physiol. Lpz.*, pp. 365–83.

75. Leydig, F. 1853. *Anatomische-histologische untersuchen uber Fische und reptilien.* Berlin: G. Reimer.

76. Stannius, H. 1846. *Lehrbuch der vergleichenden Anatomie der Wirbeltiere.* Berlin.

References

77. Balfour, F.M. 1878. *Monograph on the development of elasmobranch fishes.* London: Macmillan.

78. Giacomini, E. 1908. Il sistema interrenale e il sistema cromaffine (sistema feocromo) nelle anguille adulte, nelle cieche e nei leptocefali. *Mem. R. Accad. Bologna* (sezione delle scienze naturali). Serie VI, 5:113–28.

79. Vincent, S. 1896. The suprarenal capsules in the lower vertebrates. *Proc. Birm. Nat. Hist. Phil. Soc.* 10:1–25.

80. Cleghorn, R.A. 1932. Observations on extracts of beef adrenal cortex and elasmobranch interrenal body. *J. Physiol. Lond.* 75:413–27.

Introduction

1

Morphology

1.1 General morphology

In the mammal, the adrenals are paired glands lying anterior to the kidneys. In some species, such as the human, they are closely adpressed to the cephalic pole of the kidney, while in others such as the rat or rabbit, the association is not so intimate. Their combined wet weight is approximately 0.01 to 0.02 percent of the total body weight, thus weighing in the human species about 8 g in the adult. In the adult female rat, the combined weight is about 55 mg, but in the male only about 40 mg. Such sexual dimorphism is found in a number of species (mostly rodent), but not in others, including dog, sheep, bovine, and human [1–4]. In general, the gland receives arterial blood through a number of small arteries which originate directly from the dorsal aorta and, in some species (including the human) from branches of the renal and inferior phrenic arteries. From these, blood is dispersed in a network of arterioles in the connective tissue capsule. The venous drainage is through a single vein which discharges either into the renal vein (as in the left gland in the rat or human) or directly into the vena cava (as in the right gland in these species) [Fig. 1.1(a) and 1.1(b)].

1.2 Microscopic detail

There are many descriptions of the structure of the adrenal cortex as seen in light microscopy, in a number of species (for reviews see refs. [1–5]). The

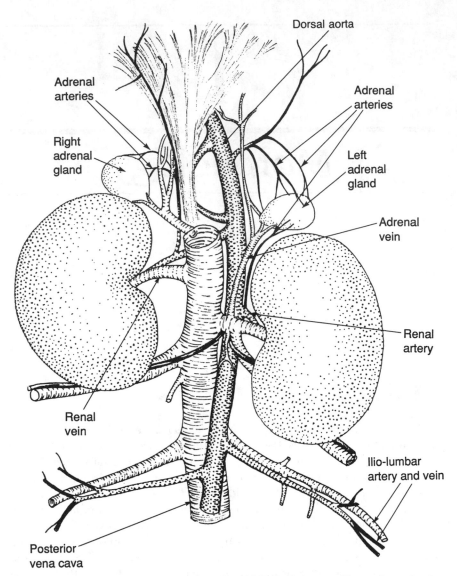

Dorsal aorta

Adrenal
arteries

Right
adrenal
gland

Adrenal
arteries

Left
adrenal
gland

Adrenal
vein

Renal
artery

Renal
vein

Ilio-lumbar
artery and vein

Posterior
vena cava

FIGURE 1.1(a) Vascular supply to the adrenal gland of the rat. Figure drawn by B. Landon, after an illustration in ref. 4.

classical view is that there are at least three layers of cells arranged as concentric shells. From the outside, these are the zona glomerulosa, the zona fasciculata, and the zona reticularis [Fig. 1.2(a)]. The cells of the different zones are generally distinguished according to their shape and size, and their arrangement and position within the gland. The zona glomerulosa lies just below the connective tissue capsule and consists of an arrangement of cells described variously as whorls, loops, or baskets. Cells of the zona

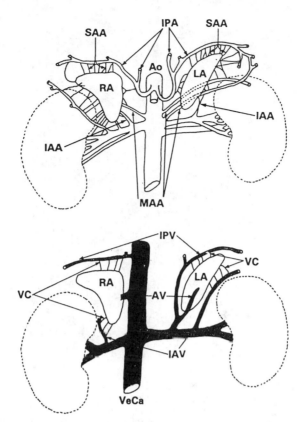

FIGURE 1.1(b) Arterial (top) and venous (bottom) vascular system of the human adrenal gland. Abbreviations: LA, RA = left and right adrenal, IPA = inferior phrenic arteries, SAA, MAA, IAA = superior, medial, and inferior adrenal arteries, Ao = aorta, VeCa = vena cava, IPV = inferior phrenic veins, AV = adrenal veins, IAV = inferior adrenal veins, VC = venae comitantes. (Figure reproduced, with permission, from ref. 1.)

fasciculata are invariably arranged as a series of centripetally orientated cords. In the zona reticularis these become less clearly organized and the reticularis takes its name from the network which the cells are seen to assume. Further zones are recognized, though their appearance may vary considerably from species to species. The zona intermedia lies between the glomerulosa and the fasciculata and is at most two- or three-cell layers deep. There is considerable species variation in the appearance of this zone (and of certain of the additional inner adrenocortical zones described below). Since most descriptions of morphology of the gland deal with either the rat or human species, perhaps the easiest way is to consider these first and then to deal in a more general way with the appearance of the gland in other species.

In the rat, the zona glomerulosa consists of small cells with a relatively large nuclear/cytoplasmic ratio [Fig. 1.2(b)]. In collagenase dispersed cell suspensions, these cells assume a spherical form, with diameter approxi-

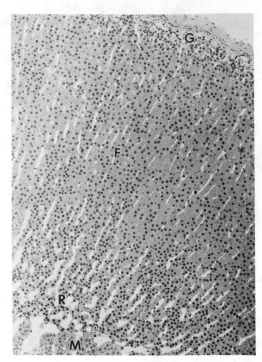

FIGURE 1.2(a) Section through the adrenal cortex of an adult Sprague–Dawley strain male rat, showing the arrangement of adrenocortical cells and zones. G = zona glomerulosa, F = zona fasciculata, R = zona reticularis, M = medulla. This gland was perfusion fixed under conditions of operative stress, which gives dilatation of the vasculature. Note that sinusoids passing through the zona reticularis are continuous with those of the medulla. × 92. (Micrograph prepared by J. Pudney, and reproduced with permission from ref. 13.)

mately 10 microns. There are between four and six layers of glomerulosa cells comprising about 38 percent of the total volume of the cortex. There are approximately one million glomerulosa cells per gland although this will vary depending on the physiological status of the animal (see Chap. 4).

The mitochondria of the zona glomerulosa cells are larger than those found in other zones and in further contrast to the inner adrenal cortical zones, the cristae are of shelflike or lamelliform appearance. Smooth endoplasmic reticulum, Golgi apparatus, lysosomes, and other structures are also clear but are not especially characteristic of the glomerulosa cell. As in other adrenocortical cells, lipid droplets may be abundant, sparse, or absent altogether. The plasma membrane is irregular, and depending on the degree of stimulation, may develop numerous filopodia or microvilli which project into the intercellular spaces [Figs. 1.2(b), 1.3]. The cells may nevertheless form tight junctional complexes with adjacent glomerulosa cells (Fig. 1.3). Coated pits are not abundant, and secretory granules, and pinocytotic vesicles, though described by some authors, are not prominent.

The cells of the zona fasciculata are larger than glomerulosa cells and,

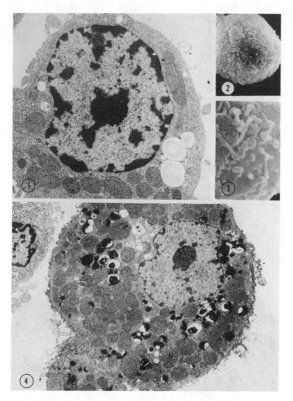

FIGURE 1.2(b) Electron micrograph of adrenocortical cells from a Wistar strain rat. (1) Transmission EM of an isolated glomerulosa cell, demonstrating general characteristics of the cell, with high nuclear–cytoplasm ratio, lipid droplet, mitochondria with lamelliform cristae. × 6240. (2) and (3) Scanning EM of zona glomerulosa cell, showing the surface covered with many filopodia. (2) × 1300 and (3) × 10400. (4) Transmission EM of isolated zona fasciculata cell showing general cytological features: low nuclear-cytoplasm ratio, lipid droplets, numerous mitochondria containing vesicular cristae. × 3328. (Micrograph prepared by J. Pudney, and reproduced with permission from ref. 12.)

when isolated, are approximately 20 microns in diameter [Fig. 1.2(a),(b)]. The nuclear/cytoplasmic ratio is thus smaller than in the zona glomerulosa. According to Idelman [4], the cells of this zone comprise about 54 percent of the total volume of the cortex, with rather more than a million cells per gland. One of the striking features of the zona fasciculata cell is the number of mitochondria with which the cytoplasm is packed. These are seemingly smaller in diameter than those of the glomerulosa, and the cristae are, in contrast with the glomerulosa cells, tubulovesicular in appearance [Fig. 1.2(b)]. The smooth endoplasmic reticulum and Golgi apparatus are prominent. Lipid droplets may again be present and it is possible that they show variations depending on the sex and age of the animal. The plasma membrane is irregular, as in the glomerulosa cell and this is again related to the state of stimulation, the stimulated cell characteristically showing a greater development of filopodia or microvilli (Fig. 1.4).

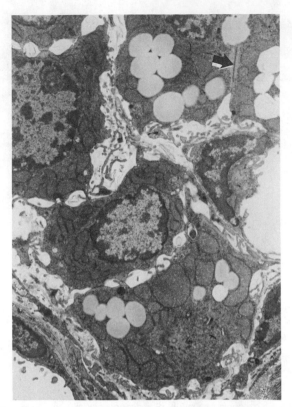

FIGURE 1.3 Transmission EM of zona glomerulosa cells in stimulated perfusion-fixed adrenal of Sprague-Dawley rat. Typical cytological features are seen, cf. Figure 1.2(b), and the cells are connected by junctions (arrow) and develop numerous filopodia which project into the enlarged intercellular and subendothelial spaces. × 6360. (Micrograph prepared by J. Pudney, and reproduced with permission from ref. 12.)

The zona reticularis has fewer cells, which are smaller than those of the fasciculata and have a diameter of about 10 μm. The mitochondria are rather more sparse than in the fasciculata, while still showing tubulovesicular cristae. Other features resemble those of fasciculata cells, with the exception that lysosomes are sometimes relatively abundant (Fig. 1.5).

A good deal of the earlier literature, using the light microscope, was devoted to various forms of Sudan staining, the study of lipid droplets, and the differences they may show between different zones, and between different treatments. Since the lipid droplets play an important part in the production of steroid hormones (see Chap. 2) it is to be expected that their abundance and size may vary greatly in different animals and, indeed, in different cells of the same animal. For this reason it is difficult, on the basis of such data, to reach general conclusions about the zonal distribution of lipid.

In the human adrenal cortex, the zona glomerulosa is relatively much smaller than in the rat: Neville and O'Hare [1] say that in the normal gland,

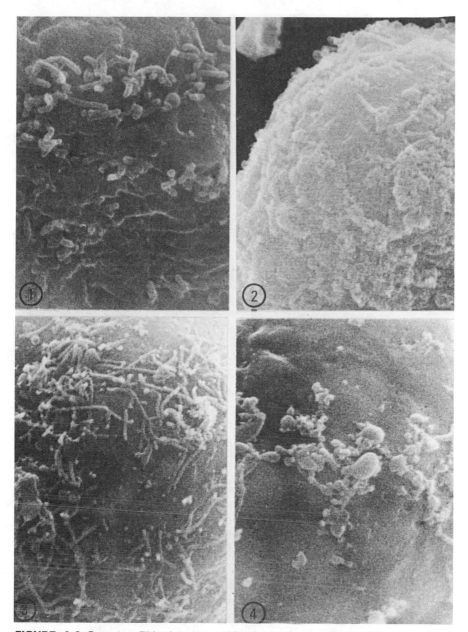

FIGURE 1.4 Scanning EM of isolated Wistar strain rat adrenal cells. (1) Rats stimulated with ACTH in vivo; cells exhibit extensive filopodia ($\times$ 11760). (2) Rats treated with cortisol in vivo; cells show poor development of filopodia ($\times$ 11760). (3) Cells incubated with ACTH in vitro, which induces a profuse development of filopodia ($\times$ 5460). (4) Cells incubated in the absence of ACTH, which do not possess many filopodia ($\times$ 17640). (Micrograph prepared by J. Pudney, and reproduced with permission from ref. 12.)

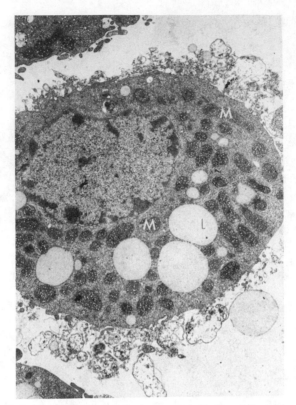

FIGURE 1.5 Transmission EM of typical zona reticularis cell from the Wistar strain rat adrenal. Note the high nuclear–cytoplasm ratio, lipid droplets (L) and relatively small number of mitochondria, containing tubulovesicular cristae (M). × 7200. (Micrograph prepared by A.J. Pettit.)

it seldom occupies more than 5 percent of the total cortex. This is partly because the zone does not occupy a continuous shell in the outer part of the gland, but occurs in the form of more or less isolated islets (Fig. 1.6). As in the rat, however, the cells are small and round with a high nuclear/ cytoplasm ratio, and the mitochondria are once again characterized by their lamelliform or shelflike cristae (Fig. 1.7). The smooth endoplasmic reticulum is sparse, and ribosomes and polysomes are visible throughout the cytoplasm. Fasciculata cells are larger, with abundant cytoplasm and mitochondria which contain tubulo-vesicular cristae (Fig. 1.8), although this may vary in appearance between the outer and inner parts of the zone [1]. In conventional stains, these cells are lighter in appearance than reticularis cells and hence have been termed "clear cells" as opposed to the "compact cells" of the reticularis, which stain more deeply [1]. As in other species, the clear cells of the fasciculata are arranged in centripetally orientated cords. The zona reticularis comprises one third of the gland with cells intermediate in size between those of the glomerulosa and fasciculata. Mitochondria are elongated

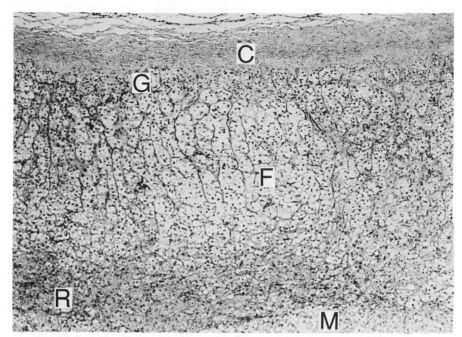

FIGURE 1.6 Normal human adrenal cortex. The outer glomerulosa (G) has a focal distribution beneath the capsule (C), while the "compact" cells of the zona reticularis (R) occupy the inner quarter of the cortex. The "clear" cells of the zona fasciculata (F) comprise the remainder. A small area of adrenal medulla (M) is present on the innermost aspect. × 68. (Micrograph reproduced with permission from ref. 1.)

with tubulovesicular cristae (Fig. 1.9). Microbodies (lipofuchsin granules) are also present; lipid droplets occur as well.

It is important to notice that although the descriptions of the three zones of the definitive cortex that have been given here apply to the most widely studied species, there is a good deal of species variation in the zonation of the gland. One possibly important feature, which is, however, rarely discussed, is that the zona glomerulosa in some species has a quite different appearance from the glomerulosa of the rat or the human species. In the Carnivora (dogs, cats, bears, etc.), the zona glomerulosa cells are large, palely staining, flattened structures strikingly stacked in big loops and whirls. The glomerulosa in these species also appears to occupy a larger percentage of the total gland volume (Fig. 1.10). It is not clear whether this appearance of cells in these species has particular functional correlation.

1.3 The circulatory system

In general, conventional histological preparation does not reveal a great deal of the structure of the vascular system of the gland. Nevertheless, in both

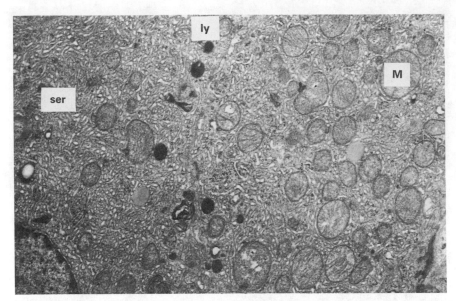

FIGURE 1.7 Transmission EM of human adrenal zona glomerulosa cells, showing mitochondria with tubulolaminar cristae (M), abundant smooth endoplasmic reticulum (ser), and lysosomes (ly). × 10080. (Micrograph reproduced with permission from ref. 82.)

rat and human glands, endothelial cell nuclei can be discerned, but not the shape, size, and arrangement of the vessels which they form, unless the gland is perfusion fixed [Fig. 1.2(a)]. Several descriptions of the vascular system of the gland have been given however, and mostly these depend on the injection of plastics or other substances which remain after the rest of the gland is digested away. Microscopic detail of the structure of these vessels is therefore rarely given.

The earliest description of the vascular system of an adrenal gland, that of the dog, was given by Flint [6] and this paper is still widely quoted. A variable number of adrenal arteries supply the gland in the dog and Flint found that they first formed a poorly defined plexus in the capsule where, after some subdivision, they gave rise to three types of arteries, arteriae capsullae, arteriae corticis, and arteriae medullae, supplying the connective tissue capsule, cortex, and medulla, respectively. After breaking up into a capillary network, blood is collected by a separate venous system in each of the three regions which thus have entirely separate vascular systems [Fig. 1.11(a)]. This interpretation of separate vasculature for cortex and medulla has been followed by some subsequent authors although in general it is not held that the capsule has a separate vasculature. Most authors describe the medullary and cortical arteries as arising from the capsular arteriolar network. The medullary arteries run through the cortex, and only in the medulla do they divide into arterioles and eventually into capillaries. Gersh and Grollman [7] follow Flint [6] in holding that the circulation of blood in the

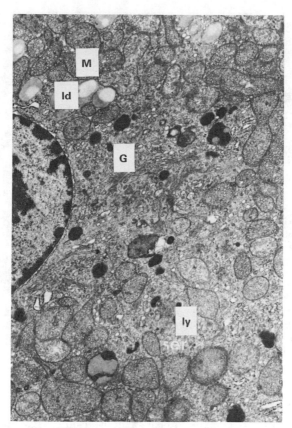

FIGURE 1.8 Transmission EM of human adrenal zona fasciculata cells. Mitochondria (M) showing vesicular cristae and a well-developed juxta nuclear Golgi apparatus (G). ly = lysosome, ld = lipid droplet. × 11400. (Micrograph reproduced with permission from ref. 82.)

capillaries of the cortex and medulla is related only in that of blood from both regions enters the same veins, and that they are almost entirely independent in other ways. Subsequently, some authors have followed this rigid line of separate vasculatures for the cortex and medulla, while others, although viewing the medullary arteries as a major source of blood supply to the medulla, nevertheless consider that blood from the cortex also supplies the sinusoids of the medullary region [1,8,9], see Figure 1.11(b).

Micrographs obtained using perfusion fixed rat adrenals [6] certainly uphold the view that the cortical and medullary circulations are continuous [Fig. 1.2(a), 1.12]. It is agreed that the medullary arteries, whatever their function, are few in number, and in the rat gland, for example, there are only four to six arteries.

Histological detail of the structure of these blood vessels has only been described rarely. Some authors (e.g., ref. 10) nevertheless produced remarkably clear drawings based on conventional light microscopic tech-

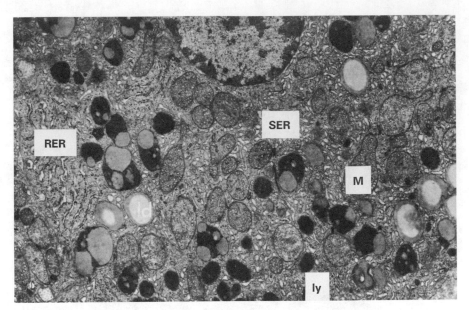

FIGURE 1.9 Transmission EM of human adrenal zona reticularis cells. Mitochondria (M) with tubulovesicular cristae, and a rather well-developed smooth endoplasmic reticulum (SER). RER = rough endoplasmic reticulum, ly = lysosomes, ld = lipid droplet. × 9280. (Micrograph reproduced with permission from ref. 82.)

niques, which closely reflect the detail revealed in later electron microscope (EM) studies. In the rat, the subcapsular arterioles have muscular walls suggesting they may be a site for control of blood flow through the gland (Fig. 1.12). Sinuses, which run through the cortex, on the other hand, are lined only by a single sheet of extremely attenuated endothelial cells and the same is true of the lining of the medullary sinusoids. The wall of the adrenal vein is also very thin as it traverses the gland, although it becomes thicker as it is exteriorized (Fig. 1.13). Because they are difficult to find, the histological detail of the structure of the medullary arteries is also rarely reported. Light microscopy suggests that the walls of these vessels are slightly more complex than those of the sinusoids (Fig. 1.13) [11–13].

The vasculature of the human gland is rather more complex in two respects [1,14]. One is that there exist arteriovenous loops in which blood is apparently carried from the cortical arterioles in a vessel which sweeps down through the zona glomerulosa and the zona fasciculata but then loops back to the exterior of the gland in a site adjacent to its origin (Fig. 1.14). A further complexity is that as the adrenal vein leaves the gland, a section of cortex is inverted and wraps around the vessel in the form of the cortical cuff (Fig. 1.14). The three zones of the cortex are recognizable in the cuff region which is supplied with blood derived from the *arteriae comitantes* which run in the wall of the large veins, and the drainage of this region is through vessels transporting blood from the medulla through the cortex to

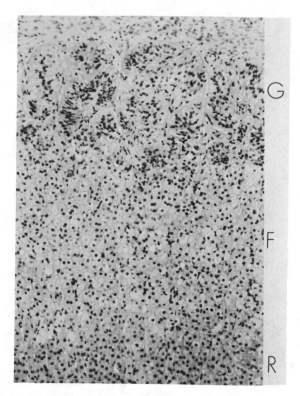

FIGURE 1.10 Light micrograph of dog adrenal cortex, showing extensive zona glomerulosa consisting of elaborate whorls of relatively large flattened cells (G), zona fasciculata (F), and part of the zona reticularis (R). × 71. (Micrograph generously provided by I. Doniach.)

the central vein i.e. from the reticularis towards the glomerulosa, in contrast to the situation in other parts of the cortex.

1.4 Additional adrenocortical zones

In the human species, the fetal adrenal is a large gland which during the second trimester of pregnancy reaches a maximum relative weight of about 0.3 percent of body weight. This declines slightly in the third trimester and precipitously in the neonate. The large size of the fetal adrenal is mostly attributable to the presence of an additional adrenocortical zone called the fetal zone, lying between the definitive cortex and the medulla (Fig. 1.15). In contrast to the relatively small basophilic cells of the outer definitive zone, the larger cells (20–50 μm) in the fetal zone are eosinophilic. Neville and O'Hare [1] describe the definitive zone mitochondria as having lamelliform cristae whereas the fetal zone mitochondria have tubulovesicular internal structures. There appears to be an intermediate zone between the

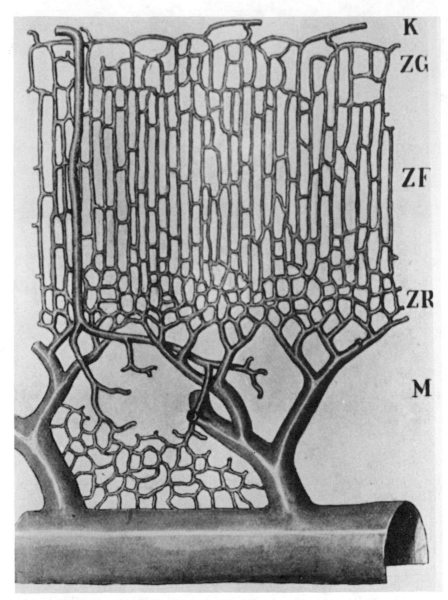

FIGURE 1.11(a) Separate arterial supply to the cortex and medulla of the dog adrenal, as envisaged by Flint [6]. K = capsule, ZG = zona glomerulosa, ZF = zona fasciculata, ZR = zona reticularis, M = medulla. (Figure reproduced with permission from ref. 6.)

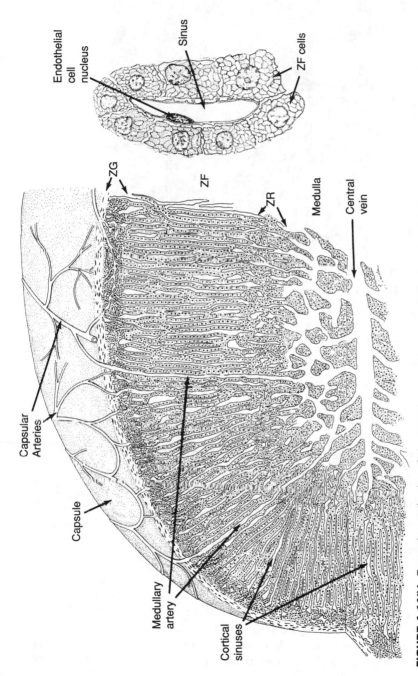

FIGURE 1.11(b) General organization of the vasculature of the mammalian adrenal gland. Note that, contrary to the views of Flint [see Fig. 1.11(a)], the medulla in fact receives blood both via the cortex, and also directly through the medullary arteries. Figure drawn by B. Landon, modified from illustration in ref. 8. ZG = zona glomerulosa, ZF = zona fasciculata, ZR = zona reticularis. The inset figure illustrates the relationship between a cortical sinus and zona fasciculata cells.

1.4 Additional adrenocortical zones

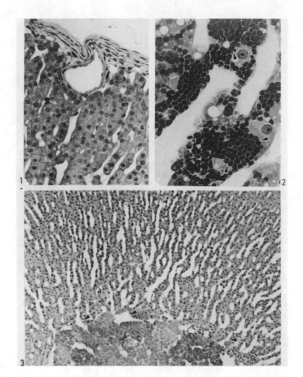

FIGURE 1.12 (1) Detail of the outer adrenal cortex of the rat, showing arteriole beneath the capsule. × 144. (2) Micrograph of rat adrenal after chronic ACTH treatment, which causes extravasation of red blood cells from the sinusoids into the extracellular space around the cortical cells. × 332. (3) Micrograph of the inner cortex and medulla, showing where the cortical sinusoids (arrows) are confluent with those of the medulla. × 77. (Micrograph prepared by J. Pudney, and reproduced with permission from ref. 11.)

definitive cortex and the fetal cortex in which the mitochondria are of the fetal zone type but the endoplasmic reticulum is less abundant. The fetal cortex is also seen in other primates but does not appear to be present in the fetal adrenal of other species [2,3].

Further variants on inner adrenocortical zones are, however, present in other species. In the mouse, the X zone, a densely staining zone, lies between the reticularis and the medulla in the adult nulliparous female, and in the prepuberal animals of both sexes. This zone involutes at puberty in the male, and at first pregnancy in the female, leaving a connective tissue capsule dividing the reticularis from the medulla. Cells of the X zone are smaller than those of the zona fasciculata, and it has been suggested that their maintenance is caused by the action of luteinizing hormone from the pituitary [3]. Involution of the X zone is a result of the production of gonadal androgens to which it is exquisitely sensitive. There are considerable strain differences in the appearance of the X zone in mice, however, and in some the zone does not disappear completely during pregnancy in the female

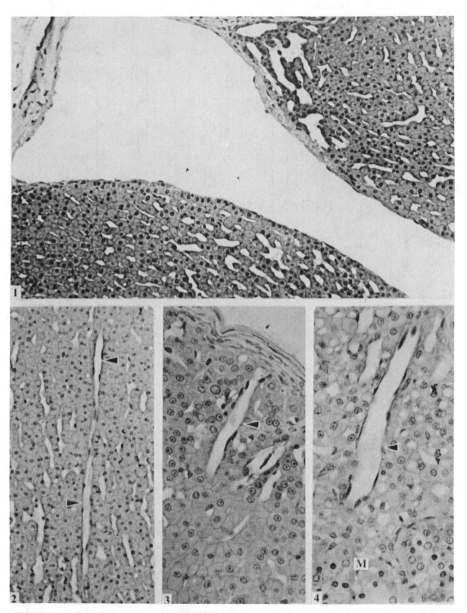

FIGURE 1.13 (1) Appearance of the adrenal vein of the rat, as it passes through the cortex. The walls of this vessel lack muscle coats, and its structure is scarcely more complex than that of the surrounding sinusoids. × 132. (2), (3), and (4) Medullary arteries (arrows) present in the cortex of the rat adrenal, in the zona fasciculata (2), arising from the vessels in the subcapsular plexus (3) and (4) entering medullary tissue (M). (× 156, × 166, × 290, respectively). These arterioles have slightly thicker muscular walls than the surrounding sinusoids. (Micrograph prepared by J. Pudney, and reproduced with permission from ref. 11.)

1.4 Additional adrenocortical zones

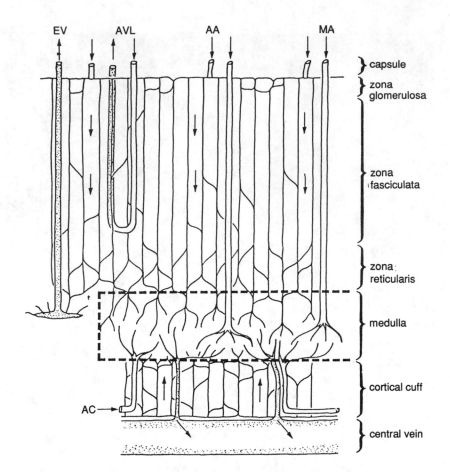

EV   AVL   AA   MA

capsule

zona
glomerulosa

zona
fasciculata

zona
reticularis

medulla

cortical cuff

AC →

central vein

Intraglandular Vascularization of the human adrenal gland

FIGURE 1.14 Intraglandular vascularization of the human adrenal cortex. Details of the vascular pattern on the left illustrate the regions of the gland without interposed medulla, and on the right, the patterns found in the head of the gland, where there is a medulla. The arrows represent the direction of normal blood flow. EV = emissary vein, AVL = arteriovenous loop, AA = adrenal artery, MA = medullary artery, AC = arteriae comitantes. (Reproduced with permission from ref. 1.)

while in others it may be scarcely discernible at any stage [15]. An additional inner adrenocortical zone of tissue is also found in the marsupial, the brush-tailed possum [16]. This large zone, which has been termed the "special zone," consists of large palely staining cells and is found only in the adult female. It is still not clear what relationship these additional adrenocortical tissues have to the inner adrenocortical zone of the normal gland, that is, the zona reticularis. Structures apparently related to the X zone of the mouse have been described in a number of other related species [15].

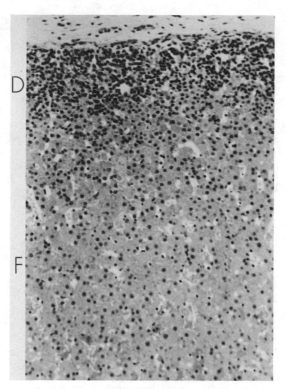

FIGURE 1.15 Light micrograph of human fetal adrenal (18 weeks) showing relatively small cells of the definitive zone (D) surrounding the larger cells of the voluminous fetal zone (F). × 85. (Micrograph generously provided by I. Doniach.)

1.5 Innervation of the adrenal cortex

Most early studies reported that all nerve fibers entering the adrenal gland passed straight through to the adrenal medulla without branching and with no evidence of nerve endings in the cortex. It was, therefore, concluded that the adrenal cortex received no direct innervation [10,17]. One early study, however, observed a "rich, intimate nerve supply to the cortex" of human adrenal glands and described small, unmyelinated fibers passing between the adrenocortical cells and anastomosing to form networks enclosing the cells with nerve endings within the cortical cells [18]. More recently, the presence of nerve fibers in the adrenal cortex, apparently synapsing with the adrenocortical cells, has been reported in the rat, pig, and mouse [19,20]. Robinson et al., [21] have described unmyelinated nerve axons terminating in close proximity to endocrine cells in the sheep adrenal cortex. Catecholaminergic innervation in the zonae glomerulosa and fasciculata in the rat adrenal gland, and vasoactive intestinal peptide (VIP) immunoreactive nerve fibers have also been described (Fig. 1.16), some of which may not

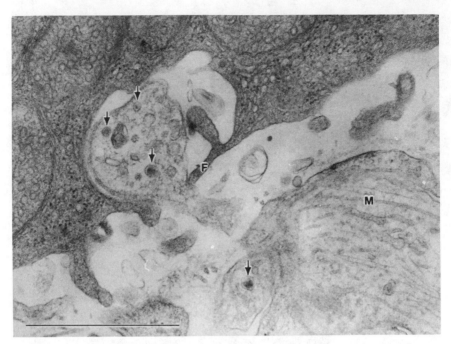

FIGURE 1.16 Transmission EM of rat adrenal, showing a nerve ending on a zona fasciculata cell. The ending, encapsulated within the adrenocortical cell, is a branch of the nerve process in the lower right of the figure. Arrows = dense core vesicles, 60–80 nm in diameter, suggesting that the ending is adrenergic. F = filopodium of fasciculata cell, M = microtubule in nerve process. Marker is 1 μm. × 41000. (Micrograph generously provided by S. Malamed.)

arise from the splanchnic nerve [22–24]. There is also evidence suggesting sensory innervation of the adrenal gland. Niijima and Winter [25] found afferent nerves in the adrenal gland of the rabbit and cat, and reported that the unit activity of these nerves could be modified by mechanical distention of the capsule or by alterations in blood pressure in the adrenal vein, suggesting a possible neural control of adrenal blood flow. Adrenocortical innervation also has been proposed as a part of the mechanism for the control of compensatory adrenal hypertrophy [26] and circadian adrenocortical rhythmicity in the rat [27].

1.6 Experimental modification of adrenocortical morphology

1.6.1 Hypophysectomy

The decline of adrenal weight after hypophysectomy by at least 50 percent in the long term has been recorded on many occasions [2,3]. This is largely

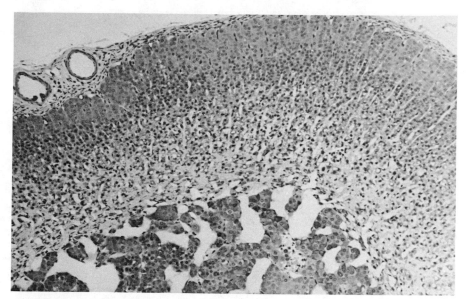

FIGURE 1.17(a) Light micrograph of rat adrenal, 28 days after hypophysectomy. Note healthy glomerulosa, but greatly reduced and disorganized inner adrenocortical zones. × 126.

attributable to the degeneration of the zona fasciculata and the reticularis, while in the rat, the zona glomerulosa may become wider [2], (Fig. 1.17). With the advent of morphometric techniques applied at the electron microscope level, it has been possible to measure more subtle changes. The volume of the zona glomerulosa, and its subcellular components, especially nuclei and mitochondria may actually decrease, despite the increase in width of the zone. This effect is transitory, however, and by 30 days, the volume is restored through further increases in width [28,29].

These effects are marginal compared with those seen in the inner zones, however, and in the rat there is overall loss of cells, which also decrease in size [2,30]. The volume of the mitochondrial compartment decreases dramatically on hypophysectomy, and the number of mitochondria may decrease by half within 3 days. In combination, these effects produce sparse populations of large mitochondria with few vesicles which may be transformed in appearance. Smooth endoplasmic reticulum is also decreased [4,30]. These changes are consistent with the studies on circulating hormone levels which suggest that aldosterone, uniquely a product of the zona glomerulosa, is only moderately affected by hypophysectomy under otherwise controlled conditions, whereas secretion of glucocorticoids (i.e., corticosterone in the rat) rapidly decays (see Chap. 3). These effects are attributable to the loss of secretion by the anterior pituitary of adrenocorticotropic hormone (ACTH, corticotropin). Alternatively, it has recently been proposed

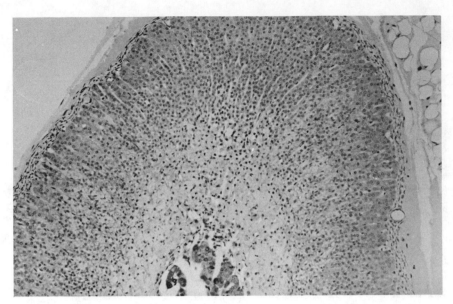

FIGURE 1.17(b) Hypophysectomized rat adrenal, as for Figure 1.17(a), but after subsequent 3 days treatment with $ACTH_{1-24}$ (50 μg per day). Note partial restoration of zona fasciculata, when compared with Figure 1.17(a). × 126. (Preparations by G. Price, micrographs by J. Pudney.)

that other pro-opiomelanocortin (POMC) derived peptides, including non-γ-MSH containing N-terminal peptides, specifically control adrenocortical cell division [31].

1.6.2 Effects of ACTH on adrenocortical morphology

Acute actions   Acutely, the most marked morphological effects of ACTH (which can occur within an hour of administration to the intact animal or to cells in culture) are a change in shape of the adrenocortical cell, and an increase in its surface area. A number of in vitro studies have been made using the transplantable adrenocortical tumor cell line of the mouse (Y1). When these cells are incubated with ACTH or cyclic-AMP, they "round up," with concomitant loss of stress fibers and an increase in the number of filamentous microvilli and filopodia [32,33]. Internally, there are increases in endosomes and Golgi apparatus and it is possible that in part these changes are correlated with the internalization of low-density lipoprotein or with the hydrolysis of this material by lysosomes to yield free cholesterol [34] (cf Chap. 2). It is possible that microfilaments are involved since cytochalasin B, a microfilament inhibitor, inhibits ACTH-induced steroido-genesis and accumulation of mitochondrial cholesterol [33,34]. Some of these changes parallel events visible in normal adrenocortical cells, most commonly studied in the rat. Thus, Rhodin [35] published evidence for a novel mechanism of secretion which he termed endoplasmocrine secretion, a modified form of exocytosis. Further evidence for exocytotic mechanisms have been reported [36–38]; however the putative hormone storage granules described by these authors are not, in general, particularly abundant.

The cell surface events described in the Y1 cell line are, however, readily visible in rat adrenocortical cells acutely stimulated with ACTH [12,38,39]. In the perfusion-fixed rat adrenal cortex, unstimulated cells show uncomplicated surfaces, and the plasma membranes of adjacent cells are closely adpressed to one another. On ACTH stimulation, the cells retract from one another, and filopodia project into the intercellular spaces thus formed, and into the subendothelial space. This occurs in all of the adrenocortical zones [12]. Similar actions of ACTH on the cell surface may occur in the human adrenal cortex [40]. These effects can also be elicited acutely in rat adrenal cells incubated with ACTH in vitro [12], (Fig. 1.4).

In vivo administration of ACTH also causes another major change, seen in some histological preparations, in that the gland very quickly becomes hyperemic. This has been shown by various authors (e.g., ref. 41) by measuring the blood content of the gland after various treatments. In particular, they drew attention to the contribution which blood content makes to adrenal weight following ACTH treatment. In the clinical situation, Selye and Stone [42] and Symington [5], drew attention to a number of conditions, perhaps related to disease-induced stress, in which the blood content of the human gland is greatly increased, in many cases apparently leading to hemorrhagic infarction. In other work, direct study of flow rates has been made and clear stimulatory effects of ACTH on the rate of blood flow rate through the gland have been found in a number of species including the human (see [12] for references). Usually, the increase in flow rate is correlated directly with the increase in steroid secretion. In vivo, there are clearly a number of ways in which ACTH treatment, or stress, might increase flow through the gland. Raised arterial blood pressure for example, increases adrenal blood flow, but in conditions of low-arterial blood pressure, the adrenal is protected [43]. Flow may also be increased, however, in perfusion systems, which isolate the gland from the systemic circulation. It therefore seems that, in addition to systemic actions, there may also be an intraglandular mechanism for this effect. It is possible that this is mediated by the action of ACTH on the capsular and subcapsular arterioles, and reduction of tone in these vessels may allow a greater flow of blood through the gland. The morphological consequence of these events is seen as expansion of intraglandular vascular space, with dilatation of the sinusoids throughout the gland in perfusion-fixed specimens [12,13], (Fig 1.18). In conventionally preserved specimens there is often an increase in the number of erythrocytes, particularly in the inner layers of the reticularis. This accumulation of blood in the inner reticularis led Symington [5] to propose the existence of a vascular dam between the reticularis and the medulla in the human gland.

Chronic effect of ACTH on adrenocortical morphology Chronically, the

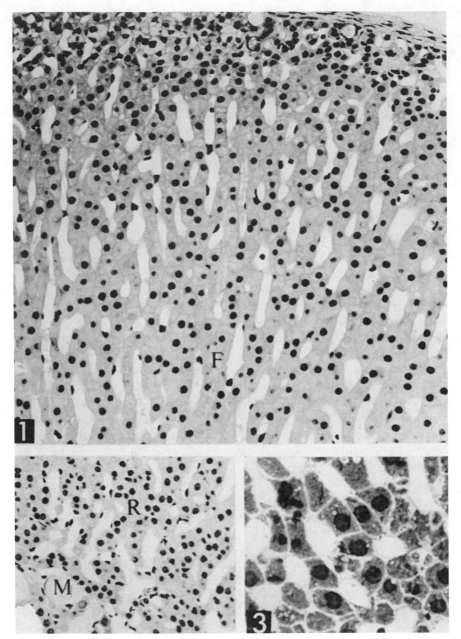

FIGURE 1.18 (1) Endogenous stimulation of rat adrenal glands results in extreme dilatation of the vascular channels (sinusoids) in the zona glomerulosa (G), and (2) zona fasciculata (F), zona reticularis (R), and medulla (M). × 294. (3) There is also an increase in intercellular space among the fasciculata cells. × 650. (Micrograph prepared by J. Pudney, and reproduced with permission from ref. 12.)

actions of ACTH are more complex. Administration of ACTH in sufficient doses to hypophysectomized animals maintains cell number and content and prevents the decline in adrenal weight [2,3,30,44,45]. It similarly prevents the decline of steroid concentrations in circulating blood (see Chap. 3). In ACTH treated intact animals, adrenal weight is increased, and the contribution to this of increased blood content has already been noted. In addition, however, there is both cell hypertrophy and proliferation. There are increased mitoses and incorporation of amino acids, and both the volume of the mitochondrial compartment and the number of mitochondria per cell also increase. Furthermore, there is a characteristic response of the smooth endoplasmic reticulum (SER) which shows increased surface area and dilation of tubular elements, together with development of Golgi apparatus and decrease in lysosomes (for reviews, see refs. 4,45). At the biochemical level, many of these effects would appear to be correlated with increased steroidogenesis, the metabolism of cholesterol esters and the induction of the various species of cytochrome P450 concerned with steroid hydroxylation both in the endoplasmic reticulum and in the mitochondria (e.g., refs. 46,47).

In tissue culture of fetal rat adrenals, Kahri [48], showed that the characteristic structure of the cells is changed and the cells of the zona intermedia are transformed into the fasciculata type, as judged by the transformation of the cristae in the mitochondria to the tubulovesicular variety. In vivo, several authors have shown that with ACTH treatment the shelflike cristae characteristic of zona glomerulosa mitochondria may be lost with the development either of an intermediate [49] or fasciculata type of cell [13,47], (Fig. 1.19). This is correlated with the loss of certain glomerulosa functions, such as the capacity of the cells to produce aldosterone or to respond to the specific zona glomerulosa stimulants (e.g., refs. 49–52). At the EM-level, it is evident that ACTH may produce inclusions in the cell which are not generally seen in normal animals, for example, dense membranous whorls both in mitochondria and in endoplasmic reticulum (ER). Chronically, ACTH induces changes which can be interpreted as a continuation of the acute effect. The vascular space becomes very much enlarged and the endothelial lining of the sinusoids is attenuated, to the extent that eventually it breaks down and erythrocytes may be extravasated. These then pack densely around the cells of the inner adrenal cortex [13], (Fig. 1.12). Such events, readily seen in the rat, are reminiscent of changes seen in the human adrenal cortex when subjects die under stressful conditions [5].

1.6.3 Morphological consequences of zona glomerulosa stimulation

Physiologically, the most important chronic stimulant of the zona glomerulosa is dietary sodium restriction. This produces an increase in the width of the zona glomerulosa, resulting from an increase both in the number and in the volume of cells [4,53,54]. These effects are the reverse of those seen with

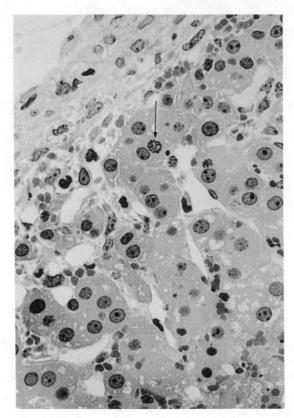

FIGURE 1.19 Light micrograph of rat adrenal after chronic ACTH treatment, showing fasciculata-like cells extending to the capsule and the absence of clearly identifiable zona glomerulosa cells. Also present are mitotic figures (arrow) which are restricted to cells in the outermost region. × 400. (Micrograph prepared by J. Pudney, and reproduced with permission from ref. 13.)

aldosterone treatment, or in sodium loading [54–57]. In dietary sodium restriction, mitochondrial number is increased in zona glomerulosa cells, and there is an increase in the number of mitochondria with straight tubular cristae [53,55,57]. These take the form of stacks of parallel lamelli which differ strikingly in dimension and other morphological features from normal mitochondrial cristae. The smooth endoplasmic reticulum, coated vesicles, coated pits, and microvilli are also increased [4,58]. The actions of potassium loading are similar to those of sodium depletion [57], while sodium loading causes a decrease in the width of the zona glomerulosa, together with further morphological consequences of hypoactivity [56,57]. In all of these conditions, and after treatment with angiotensin II, changes in lipid content may reflect mobilization of stored cholesterol as a steroid precursor [59]. It has been suggested that the increased surface area of the mitochondrial cristae is related to increased 18-hydroxylase activity.

Manipulation of the renin-angiotensin system (see Chap. 3) has produced evidence that angiotensin II can stimulate growth and maintenance of the structure of the zona glomerulosa [59–61, 62–64]. Further substances which have a trophic effect on the zona glomerulosa of the rat adrenal include a methionine enkephalin analogue, prolactin, vasoactive intestinal peptide, and α-MSH [65–68]. Of these effectors, both angiotensin II and met-enkephalin also may have some trophic effects on the rat adrenal zona fasciculata [69,70]. Somatostatin, which inhibits the adrenocortical response to angiotensin II also inhibits the growth and steroidogenic capacity of the rat adrenal and the zona glomerulosa [71]. Although, in general, these results may suggest the close coupling of the morphology of the zona glomerulosa with aldosterone secretion, Riondel et al. [72], found that angiotensin II, administered simultaneously with ACTH, prevents the mitochondrial transformation from the glomerulosa into the fasciculata type, but is unable to prevent the fall in secretion of aldosterone caused by chronic ACTH treatment. Doubtless, the interaction between all these glomerulosa stimulants is complex [62,63].

1.6.4 Sexual dimorphism

A feature of the rat adrenal which has attracted attention over the years is the sexual dimorphism in the adult, to which reference has been made. The adrenal of the female is approximately 50 percent larger than that of the male, and ovariectomy results in some adrenal atrophy. Orchidectomy on the other hand, results in adrenal hyperplasia [73–75]. Volumes of subcellular components, including mitochondria and smooth endoplasmic reticulum, are all higher in the fasciculata and reticularis cells of female animals. Such observations suggest that sexual dimorphism of the rat adrenal cortex may depend on the inhibitory action of testosterone, and the stimulatory effect of estrogen on the hypothalamo-adrenal axis. The nature of this interaction is still obscure and may occur at the level of the hypothalamus, the liver, the adrenal itself, in the plasma binding of steroids, or because of the hypocholesterolemia produced by estradiol [74,76]. Actions of other hypolip-idemic drugs support the view that reduction of circulating lipoprotein may result in stimulation of the hypothalamo-pituitary axis [77,78].

1.7 Embryology

Adrenocortical tissue arises from the mesoderm in the region of the epithelial cells which line the coelom. As Chester Jones [3] points out, development of the adrenocortical tissue is intimately bound with that of the kidney, gonads, and the posterior venous system. The cortical anlagen arise from this tissue in the region of the angle between the dorsal mesentery and the germinal ridge (Fig. 1.20). In mammals, the cortical primordia appear early in development (in the third or fourth week in the human embryo), and the cells tend to become arranged into cords, eventually forming an oval

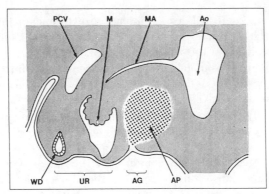

FIGURE 1.20 Cross section through the trunk region of a 4-week human embryo. The adrenal primordium (AP) is situated above the adrenal groove (AG), next to the urinogenital ridge (UR) and is bounded by the mesonephros (M), mesonephric artery (MA), and aorta (AO). Other adjacent structures include the posterior cardinal vein (PCV) and the Wolffian duct (WD). (Reproduced with permission from ref. 1.)

mass protruding a little into the dorsal coelomic cavity. Following this, sympathochromaffin cells migrate into the cortex and eventually take up their central position.

In humans, early studies suggested that there are two separate proliferations of the cortical primordia, the first giving rise to the fetal cortex, and the second to the definitive cortex which surrounds it [79]. An alternative view of the origins of the fetal cortex was given by Crowder [80] who found no evidence of separate origins of the fetal and definitive cortex. Instead, during the third month of gestation, there is a large relative increase in adrenal weight. Active proliferation is still confined to the outer definitive cortex and the fetal zone now comprises the majority of the overall cortical width. This leads to the conclusion that the rapid enlargement of the fetal cortex during the first trimester stems from the centripetal contributions of the definitive cortex.

In contrast, Sucheston and Cannon [81] consider that the development of the adrenal cortex may be divided into five phases, and reaffirmed the early view that the initial condensation of coelomic epithelium (first phase) is followed by a second proliferation [79] (second phase), giving rise to the permanent definitive cortex which thus caps the foetal zone. This occurs between 6 and 8 weeks of gestation. For 3 months there is only irregular separation of fetal and permanent cortex, and medulla. The third phase is characterized by the appearance of periodic acid-Schiff (PAS) positive material within the fetal zone, which is seen in 18-week fetuses and persists until the time of birth. The fourth phase is the decline and disappearance of the fetal cortex. In the last phase the zona glomerulosa, fasciculata and medulla clearly begin to delineate by the twenty-eighth and thirtieth week, although the increase in medullary material makes it difficult to distinguish between it and the fetal cortex. All adult zones of the adrenal gland are present by 6 months after birth and at this time the term definitive cortex can most appropriately be used. These authors consider that there is a process of maturation of the fetal cortex into an inner fasciculata and reticularis. Thus, the establishment of definitive zones occurs through proliferation of the permanent cortex, maturation of the fetal cortex, and the growth of the medulla. They therefore dispute the concept of massive involution of the fetal cortex at birth (see refs. [2,3]) and suggest that there is little evidence of cell loss and that few cells show mitotic activity. In most species however, the chromaffin and cortical tissues remain intermingled at the center of the gland until after birth [2].

References

1. Neville, A.M., and O'Hare, M. 1982. *The Human Adrenal Cortex*. Berlin: Springer-Verlag.

2. Deane, H.W. The anatomy, chemistry, and physiology of adrenocortical tissue. *Handbuch der experimentallische pharmakologie*. 14:1–185.

3. Chester Jones, I. 1957, *The Adrenal Cortex*. Cambridge: The University Press.

4. Idelman, S. 1978. The structure of the mammalian adrenal cortex. In: I Chester Jones and I.W. Henderson, ed. *General, Comparative and Clinical Endocrinology of the Adrenal Cortex*. vol. II, pp. 1–199. London: Academic Press.

5. Symington, T. 1969. *Functional Pathology of the Human Adrenal Gland*. Edinburgh: Livingstone.

6. Flint, J.M. 1900. The blood vessels, angiogenesis, organogenesis, reticulum and histology of the adrenal. *Johns Hopkins Hospital Reports* 9:153–229.

7. Gersh, I., and Grollman, A. 1941. The vascular pattern of the adrenal gland of the mouse and rat and its physiological response to changes in vascular activity. *Contributions to embryology of the Carnegie Institute*, 29:112–25.

8. Harrison, R.C., and Hoey, M.J. 1960. *The Adrenal Circulation*. Oxford: Blackwell.

9. Lazorthes, G., Gaubert, J., Poulhes, J., Roulleau, J., and Martinez-Cobo, C. 1959. Note sur la vascularisation de la surrenale. *Comptes Rendus, Assoc. d'anat.* 45:489.

10. Bennett, H.S. 1940. The life history and the secretory cycle of the cells of the adrenal cortex of the cat. *Am. J. Anat.* 67:151–227.

11. Vinson, G.P., Pudney, J., and Whitehouse, B.J. 1985. The mammalian adrenal circulation and the relationship between adrenal blood flow and steroidogenesis. *J. Endocr.* 105:285–94.

12. Pudney, J., Sweet, P.R., Vinson, G.P., and Whitehouse, B.J., 1981. Morphological correlates of hormone secretion in the rat adrenal cortex and the role of filopodia. *Anat. Rec.* 201:537–51.

13. Pudney, J., Price, G.M., Whitehouse, B.J., and Vinson, G.P., 1984. Effects of chronic ACTH stimulation on the morphology of the rat adrenal cortex. *Anat. Rec.* 210:603–15.

14. Dobbie, J.W., and Symington, T. 1966. The human adrenal gland with special reference to the vasculature. *J. Endocr.* 34:479–89.

15. Deacon, C.F., Mosley, W., and Chester Jones, I. 1986. The X zone of the mouse adrenal cortex of the Swiss albino strain. *Gen. Comp. Endocr.* 61:87–99.

16. Call, R.N., and Janssens, P.A. 1984. Hypertrophic adrenocortical tissue of the Australian brush-tailed possum (*Trichosurus vulpecula*): uniformity during reproduction. *J. Endocr.* 101:263–67.

17. Elliott, T.R. 1913. The innervation of the adrenal glands. *J. Physiol.* 46:285–90.

18. Alpert, L.K. 1931. The innervation of the suprarenal glands. *Anat. Rec.* 50:221–23.

19. Unsicker, K. 1971. On the innervation of the rat and pig adrenal cortex. *Z. Zellforsch.* 116:151–56.

20. Migally, N. 1979. The innervation of the mouse adrenal cortex. *Anat. Rec.* 194:105–12.

21. Robinson, P.M., Perry, R.A., Hardy, K.J., Coghlan, J.P., and Scoggins, B.A. 1977. The innervation of the adrenal cortex in the sheep. *J. Anat.* 124:117–29.

22. Kleitman, N., and Holzwarth, M.A. 1985. Catecholaminergic innervation of the rat adrenal cortex. *Cell Tiss. Res.* 241:139–47.

23. Holzwarth, M.A. 1984. The distribution of vasoactive intestinal peptide in the rat adrenal cortex and medulla. *J. Autonom. Nerv. Sys.* 11:269–83.

24. Hokfelt, T., Lundberg, J.M., Schultzberg, M., and Fahrenkrug, J. 1981. Immunohistochemical evidence for a local VIP-ergic neuron system in the adrenal gland of the rat. *Acta Physiol. Scand.* 113:575–76.

25. Niijima, A., and Winter, D.L. 1968. The effect of catecholamines on unit activity in afferent nerves from the adrenal glands. *J. Physiol.* 196:647–56.

26. Dallman, M.F., Engeland, W.C., and Holzwarth, M.A. 1977. The neural regulation of compensatory adrenal growth. *Ann. N.Y. Acad. Sci.* 297:373–92.

27. Ottenweiler, I.E., and Meier, A.H. 1982. Adrenal innervation may be an extra pituitary mechanism able to regulate adrenocortical rhythmicity in rats. *Endocrinology.* 111:1334–38.

28. Nussdorfer, G.G., Mazzochi, G., and Rebuffat, P. 1973. An ultrastructural stereologic study of the effects of ACTH and adenosine 3'5'-cyclic monophosphate on the zona glomerulosa of the rat adrenal cortex. *Endocrinology.* 92:141–51.

29. Nickerson, P.A., and Brownie, A.C. 1975. Effect of hypophysectomy on the volume and ultrastructure of zona glomerulosa in rat adrenal. *Endocrinologie experimentale (Bratislava).* 9:187.

30. Canick, J.A., and Purvis, J.L. 1972. The maintenance of mitochondrial size in the rat adrenal cortex zona fasciculata. *Exp. Mol. Pathol.* 16:79–93.

31. Estivariz, F.E., Carino, M., Lowry, P., and Jackson, S. 1988. Further evidence that N-terminal pro-opiomelanocortin peptides are involved in adrenal mitogenesis. *J. Endocr.* 116:201–06.

32. Mattson, P., and Kowal, J. 1980. Acute steroidogenic stimulation of cultured adrenocortical tumor cells; an electron microscope study. *Tissue and Cell.* 12:685–701.

33. Mattson, P., and Kowal, J. 1983. The effects of ACTH on microtubules, GERL and lysosomes of adrenal tumor cells. *J. Cell Biol.* 97:104a.

34. Lorenz, S., and Mattson, P. 1986. Cinemicrographic observations of cultured adrenocortical cells. *Virchows Archiv.* (cell pathology). 52:221–36.

35. Rhodin, J.A.G. 1971. The ultrastructure of the adrenal cortex of the rat under normal and experimental conditions. *J. Ultrastr. Res.* 34:23–71.

36. Gemmel, P.T., Laychock, S.G., and Rubin, P.P. 1977. Ultrastructural and biochemical evidence for a steroid-containing secretory granule in the perfused cat adrenal gland. *J. Cell Biol.* 72:209–15.

37. Mazzocchi, G., Belloni, S.A., Rebuffat, P., Robba, C., Neri, G., and Nussdorfer, G.G. 1979. Fine structure of the rabbit adrenal cortex, and the effects of short-term ACTH administration. *Cell Tiss. Res.* 201:165–79.

38. Setoguti, T., Inoue, Y., and Shin, M. 1987. Freeze-fracture replica studies of effects of ACTH treatment and hypophysectomy on the cell surface of the rat adrenal inner cortex. *Acta Anat.* 128:124–28.

39. Loesser, K.E., and Malamed, S. 1987. A morphometric analysis of adrenocortical actin localized by immunoelectron microscopy: The effect of adrenocorticotropin. *Endocrinology.* 121:1400–04.

40. Setoguti T., and Inoue, Y. 1981. Freeze-fracture replica study of the human adrenal cortex, with special reference to microvillous projections. *Acta Anat.* 111:207–21.

41. Payet, N., Lehoux, J.G., and Isler, H. 1980. Effects of ACTH on the proliferative and secretory activities of the adrenal glomerulosa. *Acta Endocr.* 93:365–74.

42. Selye, H., and Stone, H. 1950. *On the Experimental Morphology of the Adrenal Glands.* Springfield, Ill.: Charles C. Thomas.

43. Frank, H A., Frank, E D., Korman, H., Macchi, I.A., and Hechter, O. 1955. Corticosteroid output and adrenal blood flow during hemorrhagic shock in the dog. *Am. J. Physiol.* 182:24–28.

44. Sabatini, D.D., de Robertis, E.D.P., and Bleichmar, H.B. 1962. Submicroscopic study of the pituitary action on the adrenocortex of the rat. *Endocrinology.* 70:390–406.

45. Nussdorfer, G.G., Mazzochi, G., and Meneghelli, V. 1978. Cytophysiology of the adrenal zona fasciculata. *Int. Rev. Cytol.* 55:291–365.

46. Boyd, G.S., and Trzeciak, W.H. 1973. Cholesterol metabolism in the adrenal cortex: Studies on the mode of action of ACTH. *Ann. N.Y. Acad Sci.* 212:361–77.

47. Waterman, M.R., and Simpson, E.R. 1985. Regulation of the biosynthesis of cytochromes P450 involved in steroid hormone synthesis. *Mol. Cell Endocr.* 39:81–89.

48. Kahri, A.I. 1966. Histochemical and electron microscopic studies on the cells of the rat adrenal cortex in tissue culture. *Acta Endocr.* 52:sup. 108, pp. 1–96.

49. McDougall, J.G., Butkus, A., Coghlan, J.P., Denton, D.A., Muller, J., Oddie, C.J., Robinson, P.M., and Scoggins, B.A. 1980. Biosynthetic and morphological evidence for inhibition of aldosterone production following administration of ACTH to sheep. *Acta Endocr.* 94:559–70.

50. Komor, J., and Muller, J. 1979. Effects of prolonged infusions of potassium chloride, adrenocorticotrophin or angiotensin II upon serum aldosterone concentration and the conversion of corticosterone to aldosterone in rats. *Acta Endocr.* 90:680–91.

51. Vazir, H., Whitehouse, B.J., Vinson, G.P., and McCredie, E. 1981. Effects of prolonged ACTH treatment on adrenal steroidogenesis and blood pressure in rats. *Acta Endocr.* 97:533–42.

52. Aguilera, G., Fujita, K., and Catt, K.J. 1981. Mechanism of inhibition of aldosterone secretion by adrenocorticotrophin. *Endocrinology.* 108:522–28.

53. Giacomelli, F., Wiener, J., and Spiro, D. 1965. Cytological alterations related

to stimulation of the zona glomerulosa of the adrenal gland. *J. Cell Biol.* 26:499–517.

54. Shelton, J.M., and Jones, A.L. 1971. The fine struture of the mouse adrenal cortex and the ultrastructural changes in the zona glomerulosa with low and high sodium diets. *Anat. Rec.* 170:147–82.

55. Fisher, E.R., and Horvat, B. 1971. Ultrastructural features of aldosterone production. *Arch. Pathol.* 92:172–79.

56. Nickerson, P.A., and Molteni, A. 1972. Reexamination of the relationship between high sodium and the adrenal zona glomerulosa in the rat. *Cytobiologie,* 5:125–38.

57. Nussdorfer, G.G. 1980. Cytophysiology of the adrenal zona glomerulosa, *Int. rev. Cytol.* 64:307–68.

58. Palacios, G., and Lafarga, M. 1976. Coated vesicles in the rat adrenal glomerular zone after a low-sodium diet. *Experientia.* 32:381–83.

59. Rebuffat, P., Belloni, A.S., Mazzocchi, G., Vassanelli, P., and Nussdorfer, G.G. 1979. A stereological study of the trophic effects of the renin-angiotensin system on the rat adrenal zona glomerulosa. *J. Anat.* 129:561–70.

60. Mazzocchi, G., Robba, C., Rebuffat, P., and Nussdorfer, G.G. 1982. Effects of sodium repletion and timolol maleate administration on the zona glomerulosa of the rat adrenal cortex: An electron microscope study. *Endokrinologie.* 79:81–88.

61. Mazzocchi, G., and Nussdorfer, G.G. 1984. Long term effect of captopril on the morphology of normal rat adrenal zona glomerulosa. *Exp. Clin. Endocr.* 84:148–52.

62. Mazzocchi, G., Meneghilli, V., and Nussdorfer, G.G. 1983. Effects of angiotensin II on the zona glomerulosa of sodium-loaded dexamethasone-treated rats administered or not with maintenance doses of ACTH: Stereology and plasma hormone concentrations. *Acta Endocr.* 102:129–35.

63. Mazzocchi, G., Rebuffat, P., Robba, C., Malendowicz, L.K., and Nussdorfer, G.G. 1985. Trophic effects of potassium loading on the rat zona glomerulosa: Permissive role of ACTH and angiotensin II. *Acta Endocr.* 108:98–103.

64. Kasemri, S., and Nickerson, P. 1976. Quantitative ultrastructural study of the rat adrenal cortex in renal encapsulation. *Am. J. Pathol.* 82:143–51.

65. Robba, C., Mazzocchi, G., and Nussdorfer, G.G. 1986. Effects of chronic administration of a methionine-enkaphalin analogue on the zona glomerulosa of the rat adrenal cortex. *Res. Exp. Med.* 186:173–78.

66. Mazzocchi, G., Robba, C., Malendowicz, L.K., and Nussdorfer, G.G. 1987. Stimulatory effect of vasoactive intestinal peptide (VIP) on the growth and steroidogenic capacity of rat adrenal zona glomerulosa. *Biomed. Res.* 1:19–23.

67. Rebuffat, P., Robba, C., Mazzocchi, G., and Nussdorfer, G.G. 1986. Further studies on the effects of prolonged prolactin administration on the zona glomerulosa of the rat adrenal cortex. *Res. Exp. Med.* 186:387–15.

68. Robba, C., Rebuffat, P., Mazzocchi, G., and Nussdorfer, G.G. 1986. Long-term trophic action of α-melanocyte-stimulating hormone on the zona glomerulosa of the rat adrenal cortex. *Acta Endocr.* 112:404–08.

69. Nussdorfer, G.G., Robba, C., Mazzocchi, G., and Rebuffat, P. 1981. Effects of angiotensin II on the zona fasciculata of the rat adrenal cortex: An ultrastructural stereologic study. *J. Anat.* 132:235–42.

70. Robba, C., Mazzocchi, G., and Nussdorfer, G.G. 1986. Evidence that long-term

methionine-enkephaline administration stimulates rat adrenal zona fasciculata. *J. Steroid Biochem.* 24:917–19.

71. Robba, C., Mazzocchi, G., and Nussdorfer, G.G. 1986. Further studies on the inhibitory effects of somatostatin on the growth and steroidogenic capacity of rat adrenal zona glomerulosa. *Experimentale pathologie*, 29:77–82.

72. Riondel, A.M., Rebuffat, P., Mazzocchi, G., Nussdorfer, G.G., Gaillard, R.L., Bockhorn, L., Nussberger, J., Vallotton, M.B., and Muller, A.F. 1987. Long-term effects of ACTH combined with angiotensin II on steroidogenesis and adrenal glomerulosa morphology in the rat. *Acta Endocr.*, 114:47–54.

73. Kitay, J.I. 1968. Effects of estrogen and androgen on the adrenal cortex of the rat. In: K.W. McKerns, ed. *Functions of the Adrenal Cortex.* vol. II, Amsterdam: North Holland Publishing Co. pp. 775–811.

74. Kime, D., Vinson, G.P., Major, P., and Kilpatrick, R. 1979. Adrenal-gonad relationships. In: I. Chester Jones and I.W. Henderson, eds. *General, Comparative and Clinical Endocrinology of the Adrenal Cortex.* vol. III, London: Academic Press, pp. 183–347.

75. Malendowicz, L.K., Robba, C., and Nussdorfer, G.G. 1986. Sex differences in adrenocortical structure and function. *Cell Tiss. Res.* 244:141–45.

76. Magalhaes, M.M., Magalhaes, M.C., Gomes, M.L., Hipolito-Reis, C., and Serra, T.A.M. 1987. A correlated morphological and biochemical study on the rat adrenal steroidogenesis. *Eur. J. Cell Biol.* 43:247–52.

77. Mazzocchi, G., Robba, C., Meneghelli, V., and Nussdorfer, G.G. 1986. Long-term effects of the hypocholesterolaemic drug 4-aminopyrazolo-pyrimidine on the zona fasciculata of the rat adrenal cortex. *J. Anat.* 149:1–9.

78. Robba, C., Mazzocchi, G., Gottardo, G., and Nussdorfer, G.G. 1986. Effects of the hypolipidemic drug Nafenopin on the zona glomerulosa of the rat adrenal cortex: Morphological counterparts of functional alterations. *Anatomische anzeiger*, Jena. 161:35–41.

79. Uotila, U.U. 1940. The early embryological development of the fetal and permanent adrenal cortex in man. *Anat. Rec.* 76:183–204.

80. Crowder, R.E. 1957. The development of the adrenal gland in man, with special reference to origin and ultimate location of cell types and evidence in favor of the "cell migration" theory. *Contrib. Embryol.* 251:193–210.

81. Sucheston, M.E., and Cannon, M.S. 1968. Development of zonular patterns in the human adrenal gland. *J. Morph.* 126:477–92.

82. Belloni, A.S., Mazzocchi, G., Mantero, F., and Nussdorfer, G. 1987. The human adrenal cortex: Ultrastructure and baseline morphometric data. *J. submicrosc. cytol.* 9:657–68.

References

2

Nature, Biosynthesis, and Metabolism of the Adrenocortical Secretion

2.1 Hormone structure

The chemistry of the steroid hormones is best understood by considering the nature of the four parent hydrocarbon molecules in turn, together with the major types of steroid hormone to which they are related (detailed reviews are available in refs. [1–3]).

1. Cholestane (C_{27}) This may be considered to be the stem structure from which all steroid hormones are derived. It consists of four carbon atom rings, three of six carbon atoms (rings A, B, C) and one of five (ring D) linked together as illustrated (Fig. 2.1). In addition there is an eight-carbon atom side chain attached to the D ring, with two further carbon atoms in the form of angular methyl groups between rings A and B, and C and D as shown. The numbering system of the carbon atoms is also illustrated. Quantitatively, the most important compound of this class is cholesterol, which has the structure illustrated (Fig. 2.2), and which differs from the parent hydrocarbon by having a hydroxyl group at C-3 and a double bond between C-5 and C-6. Note that the convention for illustrating the orientation of substituents on the ring structure is that (with ring A to the left, and ring D to the right) substituents projecting from the plane of the ring towards the observer (the orientation of the 3-hydroxyl group in cholesterol), is designated β and is shown by a continuous line. Substituents projecting

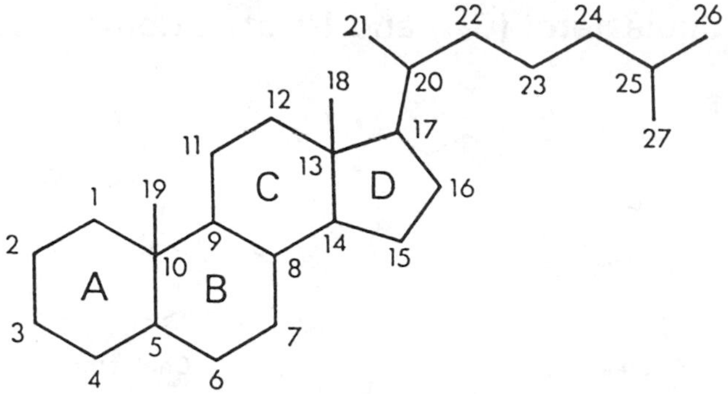

Cholestane

FIGURE 2.1 The basic steroid structure. Cholestane, which has 27 carbon atoms in the positions shown, is not formed in nature, but may be considered chemically as the parent hydrocarbon for cholesterol and related molecules. In this structure, as in other figures in this chapter, the convention is adopted that, unless otherwise indicated, each straight line joins two carbon atoms, and that hydrogen atoms saturate remaining valencies.

from the plane of the ring away from the observer, the α configuration, is designated with a dashed line. Substituents on the side chain are designated by the R (rectus) and S (sinister) system. For a full account of this system and other matters relating to steroid hormone nomenclature, see ref. 3. Also illustrated in Figure 2.2 are the C_{24} bile acids, which are among the main cholesterol metabolites.

2. Pregnane (C_{21}) This parent hydrocarbon, illustrated in Figure 2.3, has only two carbon atoms in the side chain. The major adrenal steroids which are related to this hydrocarbon are also illustrated. Biologically active compounds have ketone groups at C-3 and C-20, and a double bond between C-4 and C-5. They may also have hydroxyl groups at various loci, and one of the striking features of steroid endocrinology is that it is often difficult to predict biological activity on the basis of structure. For example, deoxycorticosterone, with a single hydroxyl group at C-21 is a mineralocorticoid (see Chap. 4), whereas corticosterone, which in addition has an 11β-hydroxyl group, is a glucocorticoid which possesses relatively little mineralocorticoid activity. On the other hand, aldosterone, with its oxygen function at C-18 linked in a hemiacetal configuration with C-11, is a potent mineralocorticoid. Both of these activities are of course very different from that of progesterone, the hormone of pregnancy and of the luteal phase in the ovarian cycle, which has no hydroxyl groups.

3. Androstane (C_{19}) has no side chain on ring D. Major compounds in this group are illustrated (Fig. 2.4). The most important are testosterone, androstenedione, and dehydroepiandrosterone (DHEA). Notice that both testosterone and androstenedione share with progesterone, and

Cholesterol (C$_{27}$) and Related Compounds

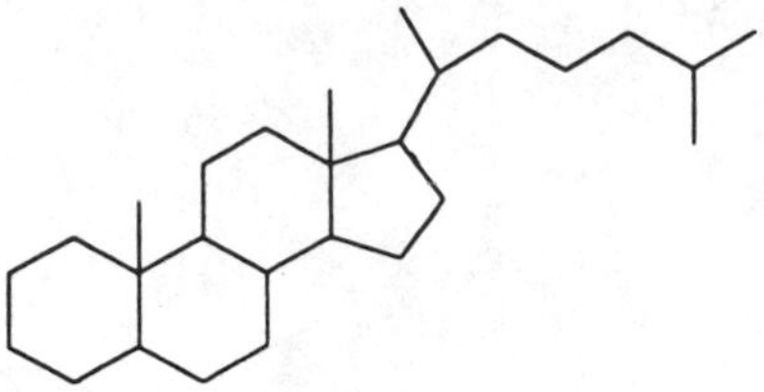

Cholestane

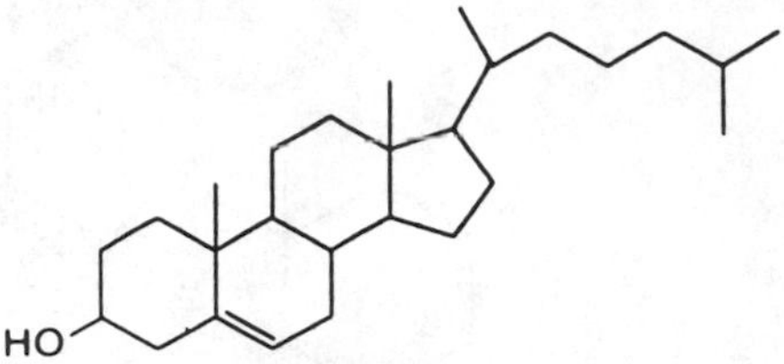

Cholesterol

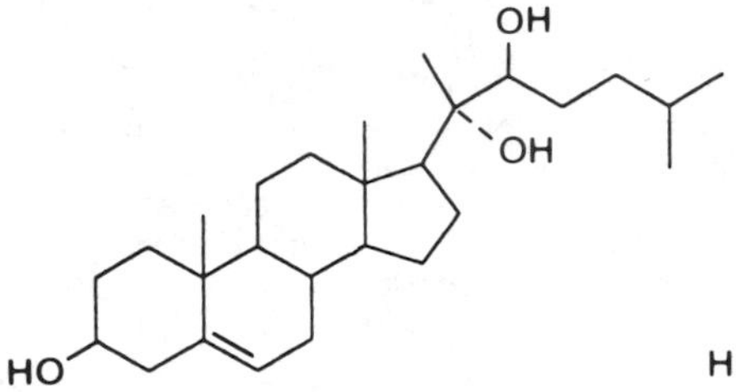

20α, 22R-dihydroxycholesterol

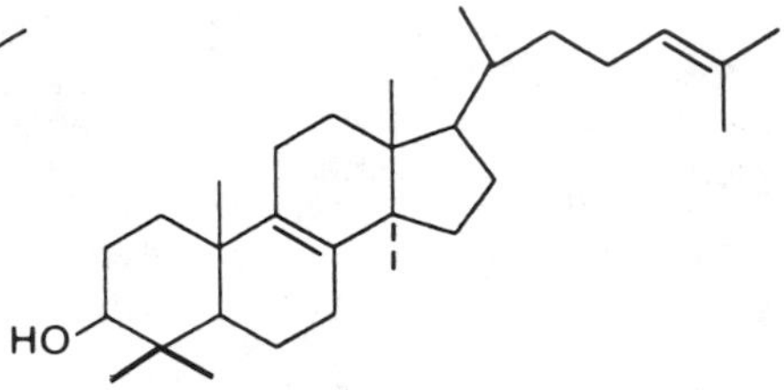

Lanosterol (C$_{30}$)

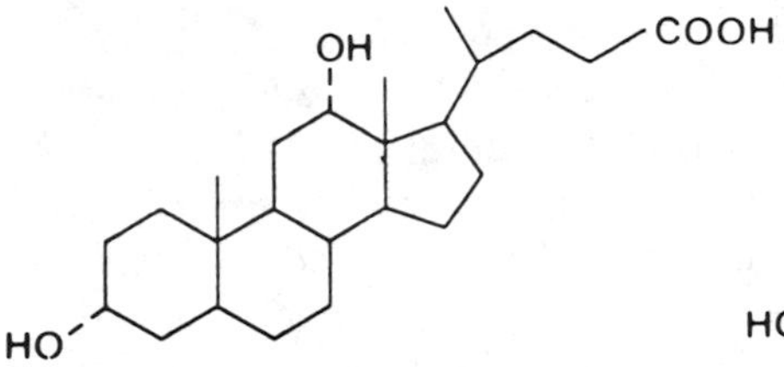

Deoxycholic acid (C$_{24}$)

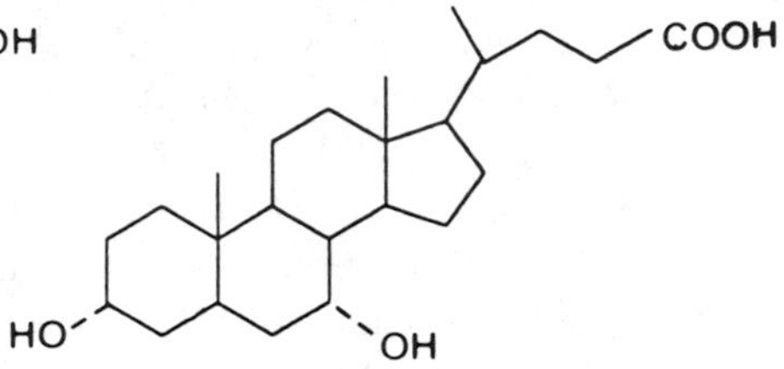

Chenocholic acid (C$_{24}$)

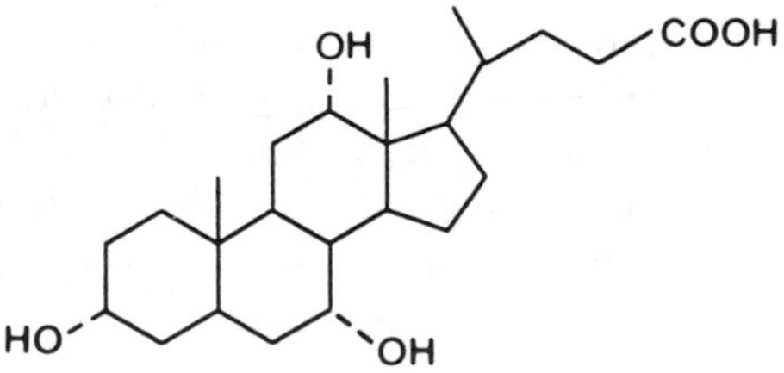

Cholic acid (C$_{24}$)

FIGURE 2.2 Relationship between the structures of cholestane, the naturally occurring C$_{27}$ compounds, and the C$_{24}$ bile acids, which are metabolites of cholesterol.

C$_{21}$ Steroids

Pregnane

Progesterone

CH$_2$OH

Deoxycorticosterone

CH$_2$OH

HO

Corticosterone

CH$_2$OH
HO

Aldosterone

CH$_2$OH
HO   OH

Cortisol

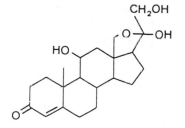

CH$_2$OH
O   OH

18-hydroxydeoxycorticosterone

CH$_2$OH
O   OH
HO

18-hydroxycorticosterone

FIGURE 2.3 C$_{21}$ compounds. Structures of pregnane, the "parent" hydrocarbon, and related steroids.

C$_{19}$ Steroids

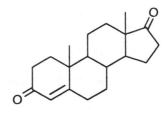

Androstane

Testosterone

Dehydroepiandrosterone

Androstenedione

FIGURE 2.4 C$_{19}$ compounds. Structures of androstane and related steroids.

the other biologically active C$_{21}$ steroids, the ketone group at C-3 and the double bond between C-4 and C-5. DHEA, on the other hand, has a 5-ene-3β-hydroxyl configuration, the same as cholesterol. In circulating plasma, DHEA mostly occurs as the sulfate (DHEAS) [see p. 58, and Fig. 2.16(b)]

4. Estrane (C$_{18}$) This compound differs from androstane in the absence of the angular methyl group, C-19. Associated with this, the biologically active estrogens have a phenolic ring A, with a hydroxyl group at C-3. Biologically, the most active estrogen is estradiol-17β, which has a less active companion, estrone. The other classical estrogen is estriol, which is excreted in increasing amounts in pregnancy in the human species (Fig. 2.5).

2.2 Biosynthesis of adrenal steroids

2.2.1 Origin of cholesterol

There are two sources of cholesterol in the adrenal cortex. It may be biosynthesized de novo from the C$_2$ fragment, acetate [4], or it may be

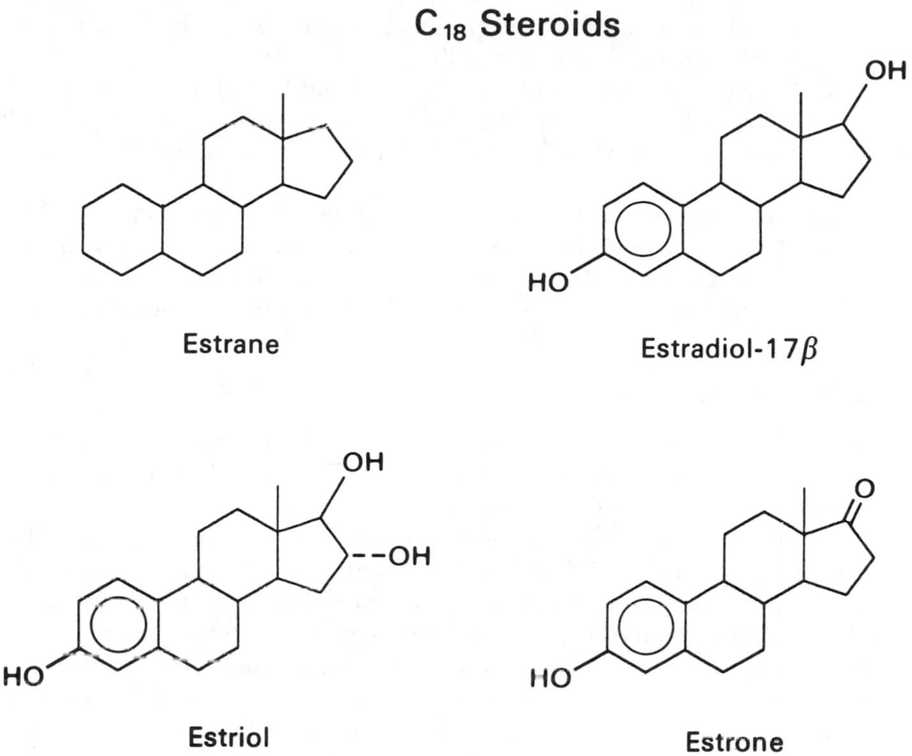

C$_{18}$ Steroids

Estrane

Estradiol-17β

Estriol

Estrone

FIGURE 2.5 C$_{18}$ compounds. Structures of estrane and related steroids.

transported from the plasma into the cell with the circulating plasma lipoproteins, low-density lipoprotein (LDL) in the human and bovine species [5–7], or high-density lipoprotein (HDL) in the rat [8,9]. Internalization is thought to occur by a process of receptor mediated endocytosis, equivalent to the process which is known to occur in the hepatocyte [10,11]. It is likely that these two sources of cholesterol have varying importance in different species. In some species such as the rat, which have "fatty adrenals" [12], circulating cholesterol may be the most important source of adrenal cholesterol, whereas in other species it is known that de novo synthesis of cholesterol within the adrenocortical cell is extremely important. This appears, for example, to be true of the hamster (a "nonfatty" adrenal) [12–13]. In "fatty" adrenals, cholesterol is stored within the adrenal cell in lipid droplets (see Chap. 1), largely in the form of cholesterol esters which may comprise 80 percent of total cholesterol [14].

There are several groups of enzymes which are concerned with the biosynthesis of adrenal steroids. These are

1. Enzymes concerned with cholesterol biosynthesis from acetate. Of these, HMG CoA reductase appears to have a central importance in

the adrenal as a site for control of cholesterol biosynthesis, as it does in the liver [4,13–15], (Fig. 2.6).

2. Enzymes associated with esterification and hydrolysis of cholesterol esters in lipid droplets. Principally, these are acyl CoA: cholesterol acyltransferase (ACAT), and cholesterol ester hydrolase (CEH) [16,17] (see Fig. 2.11).

3. Hydroxylases, which are all of the cytochrome-P450 type [14,18,19,20].

4. Dehydrogenases. These include the 3β-hydroxysteroid-dehydrogenase/isomerase complex, and 11β-hydroxysteroid-dehydrogenase. The 17α-hydroxysteroid-dehydrogenase is relatively inactive in the adrenal.

2.2.2 Cytochrome-P450

The cytochrome-P450 group comprise a family of heme-containing monooxygenases found in a number of tissues (Fig. 2.7), but perhaps most studied in the liver, where these enzymes have a particular interest because of their involvement in detoxification of drugs and other substances [22]. The molecular weight of all of them is about 50,000, but, although clearly related, their immunoreactive properties display a good deal of heterogeneity, as do the sequencing studies which have been performed [23,24] (see Fig. 2.7). The active site of P450 contains a single iron protoporphyrin prosthetic group. Dioxygen is bound, reduced, and activated at this site. The current concept of the sequence of the catalytic cycle is as follows [25–27], (Fig. 2.8):

1. Binding of substrate to low-spin P450 (Fe^{3+}), to give a high-spin ferric complex.
2. A one-electron reduction to give the Fe^{2+} state.
3. Binding of molecular oxygen.
4. A second one-electron reduction.
5. Addition of $2H^+$ and elimination of water, leaving proposed activated atomic oxygen species.
6. Oxidation of substrate and its release leaving ferric P450.

In the adrenal cortex, the mitochondrial cytochrome P450s are $P450_{SCC}$, which catalyzes hydroxylation at 20α and 22R and cholesterol side-chain scission, and $P450_{11\beta}$ which catalyzes 11β and 18-hydroxylation. Reducing equivalents for these hydroxylases are derived from NADPH, via an electron transfer system localized in the mitochondrial matrix. The electrons from NADPH are transferred to the FAD-containing flavoprotein, NADPH-adrenodoxin reductase. This transfers electrons to the iron sulfoprotein, adrenodoxin, which in turn transfers electrons to P450. These components are arranged in association with the membrane-bound P450 as illustrated (Fig. 2.9) [28,29]. Bovine adrenodoxin reductase has a molecular weight of 52,000, while adrenodoxin has a molecular weight of 12,000 [20]. Cytochrome-$P450_{11\beta}$, which also catalyzes 18-hydroxylation, may further catalyze

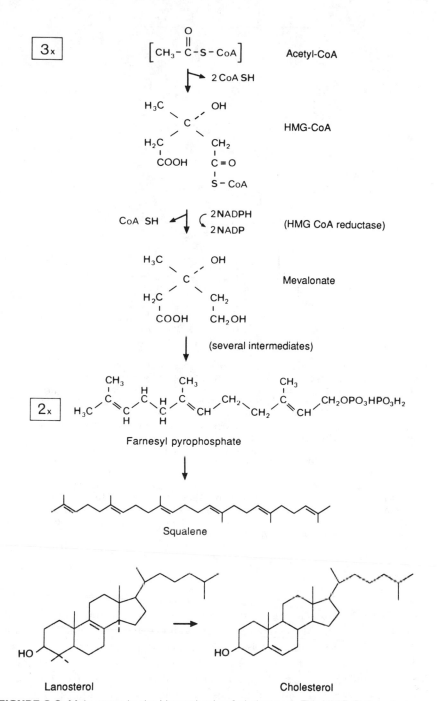

FIGURE 2.6 Main steps in the biosynthesis of cholesterol. The HMG-CoA reductase catalyzed reaction is the major site at which synthesis is regulated.

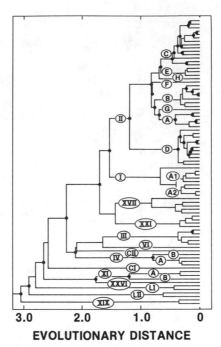

3.0　　　**2.0**　　　**1.0**　　　**0**

EVOLUTIONARY DISTANCE

FIGURE 2.7 The family of cytochrome P450 genes. This diagram shows the probable sequence and time scale for the evolution of the major families, which are indicated by roman numerals. The homology between the primary structures of existing families is generally no greater than 35 percent. Subfamilies, in which homology is between 35 and 65 percent are indicated by roman capital numerals. The letters on the x-axis designate the proteins for which the genes code. The P450s which catalyze steroid hydroxylation in mammalian systems are indicated here by 11β (which catalyzes 11β and 18-hydroxylations, c21A and c21B (c21B catalyzes 21 hydroxylation, c21A is a pseudogene: see Chap. 6), scc (catalyzing 20 and 22 hydroxylations, and cholesterol side-chain cleavage), 17α (catalyzing 17α-hydroxylation, and the 17α-21 lyase reaction) and arom (catalyzing 19 hydroxylation and the conversion of C_{19} steroids to the aromatic C_{18} compounds). The LI family exists in yeast, and the CI family is present in Pseudomonas: the remaining families are mammalian, and most abundant in the liver. Other P450s (e.g., placental 16α-hydroxylase, and the kidney 25-hydroxycholecalciferol 1α-hydroxylase) await sequencing, and are therefore, not included. The C family may be responsible for hepatic steroid hydroxylations. (Figure reproduced, with permission, from ref. [21].)

the formation of aldosterone from 18-hydroxycorticosterone according to some authors [30]. More recently, the evidence has accumulated for the existence of different species of cytochrome $P450_{11β/18}$ in the fasciculata and glomerulosa zones in rats [31]. A cDNA coding for the glomerulosa form, termed "aldosterone synthase" cytochrome P450, has been cloned. The protein for which it codes catalyzes the formation of aldosterone from deoxycorticosterone, whereas the closely related $P450_{11β}$ form, which occurs throughout the gland, catalyzes the formation of corticosterone, but not aldosterone [32].

The microsomal P450s are 21-hydroxylase and, in many species including

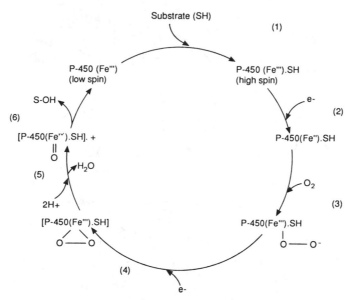

FIGURE 2.8 Probable events in P450 catalyzed steroid hydroxylation. (1) Interaction of steroid with P450, to give the ferric complex. (2) One electron reduction. (3) Formation of P450-substrate-oxygen complex. (4) One electron reduction. (5) Addition of $2H^+$ and elimination of water, leaving proposed activated atomic oxygen species. [ ] = hypothetical intermediate. (6) Spatial juxtaposition of active oxygen and substrate→hydroxysteroid and ferric P450 in concerted action (see ref. 25).

the human and bovine, 17α-hydroxylase. Although the latter enzyme exists in the rat adrenal, its activity is relatively minor. In the case of the microsomal hydroxylases, electrons are transferred from NADPH to cytochrome-P450 by means of a single flavoprotein, NADPH cytochrome-P450 reductase (MW 78000) (Fig. 2.9) [33,34]. It appears to be identical with the FAD/FMN-containing P450 reductase which has been purified from the liver. In both testis and adrenal, a single protein, cytochrome $P450_{17\alpha}$ catalyzes both the 17α-hydroxylation and the 17–20 lyase reactions [35–37].

In the cytosol, the main reaction which generates the NADPH required to support the various hydroxylations, is the glucose 6-phosphate dehydrogenase reaction [38]. In the mitochondrion, there are several possibilities. NADPH is formed in the malic enzyme catalyzed formation of pyruvate from malate [39] and, since the reverse reaction may be favored in the cytosol, it has been proposed that a "malate shuttle" may operate through transporting malate thus formed from the cytosol into the mitochondria. This thereby transfers reducing equivalents from the cytosol to the mitochondrial hydroxylases. Within the mitochondrion, too, it has been suggested that reducing equivalents may be supplied to NAD^+ from succinate, via reversed electron transport, flowing from ubiquinone to NAD^+, then to the energy-linked transhydrogenase system [40,41]. In the rat and in humans, however, NADPH is generated by isocitrate dehydrogenase, which is $NADP^+$ linked in the adrenal of these species (Fig. 2.10) [41].

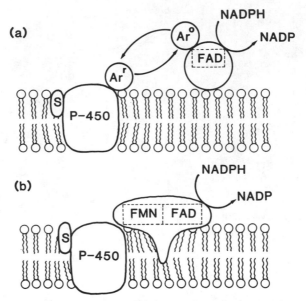

FIGURE 2.9 Schematic representation of the steroidogenic electron transfer system of mitochondria (a) and microsomes (b). In the mitochondria, the FAD-containing adrenodoxin reductase binds to the smaller iron-sulfur protein, adrenodoxin. Electrons derived from NADPH are transferred to the reductase, and then to adrenodoxin. The reductase/adrenodoxin complex then dissociates, and reduced adrenodoxin (Ad^r) forms a new complex with cytochrome P450, to which the steroid substrate (S) is bound. After reduction of P450, the oxidized adrenodoxin (Ar^o) returns to the reductase, and the cycle may start again. In the microsomes, a single FMN and FAD containing reductase transfers electrons between NADPH and cytochrome P450. (Reproduced with permission from ref. 33.)

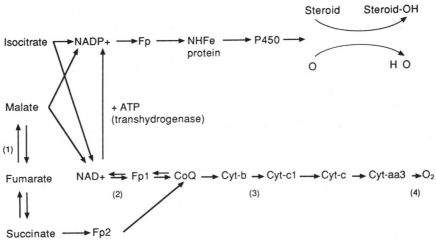

FIGURE 2.10 Electron transport and steroid hydroxylation, and the origins of NADPH utilized by steroid hydroxylases. This diagram also shows the sites of action of the inhibitors which were used in characterization of the system, 1, arsenite; 2, rotenone, amytal; 3, antimycin A; 4, cyanide.

2.2.3 The steroid dehydrogenases

3β-hydroxysteroid-dehydrogenase/isomerase There are two closely linked components in this system. These are (1) the 5-ene-3β-hydroxysteroid: NAD^+ oxidoreductase and (2) 5-ene-3-oxosteroid-4,5-isomerase. The complex is located in the microsomal fraction of the cell, and appears to be similar in its general properties in adrenal, gonads, and placenta. Studies using bovine adrenal preparations have shown that $NADP^+$ is the preferred cofactor, and that DHEA is the preferred steroid substrate, followed by pregnenolone [42,43]. DHEA thus competitively inhibits the oxidation of C_{21} steroids. DHEA-sulfate is not utilized as a substrate, nor does it inhibit steroid oxidation. Cholesterol is only weakly converted.

The isomerase activity requires no cofactor, although the presence of phospholipids appears essential [44]. There is some evidence that different isomerases may be involved in the conversion of C_{19} and C_{27} compounds [45], although kinetic data indicates the existence of a single complex acting on both DHA and pregnenolone in bovine and human adrenals [46–48].

Purification of the complex from ovine and from rat adrenals has been achieved, without separation of the two activities, and in rat adrenals, SDS-polyacrylimide gel electrophoresis gave one protein band, indicating therefore, a single protein with two activities. Its molecular weight was estimated to be 46,500 [49,50].

11β-hydroxysteroid dehydrogenase This enzyme is present in the adrenals of most species which have been examined. It is microsomal and requires $NADP^+$ or NAD^+, and appears to be specific for compounds with the 4-en-3-one configuration. The activity is relatively weak in the adrenal, although this may depend on the conditions [51], but stronger in liver and kidney, in which it has been more closely studied. Studies using rat liver preparations and in human subjects suggest that the 11-oxidase and the 11β-hydroxysteroid dehydrogenase components are in fact different entities [51,52]. It is unclear whether this is the case in the adrenal, although in certain subjects with hypertensive disorders, alterations in interconversion between cortisone and cortisol in the kidney indicate that this may be true of this tissue [53] (see Chap. 4).

17β-hydroxysteroid dehydrogenase This enzyme has been more closely studied in the gonads than in the adrenal, in which its activity is relatively weak. The adrenal gland, therefore, produces considerably more androstenedione than testosterone, and more oestrone than oestradiol-17β [54]. In the gonads, the enzyme is microsomal and requires $NADP^+$. There may be several isozymes with different specificities [55].

"18-hydroxysteroid dehydrogenase" The relationship between 18-hydroxycorticosterone and aldosterone, the former with a hydroxyl group at C_{18} and the latter with an aldehyde function, has suggested to many authors that an

18-hydroxysteroid dehydrogenase may catalyze the interconversion of these two steroids. There is little strong evidence to support this view and the difficulty has arisen because in many in vitro situations yields of aldosterone from 18-hydroxycorticosterone are quite small [54]. It is possible that the relationship between the two compounds does not depend on the activity of a dehydrogenase at all, and that, instead, aldosterone arises through a second hydroxylation on C-18 which results in a spontaneous loss of water, thus producing the aldehyde [56,57].

2.2.4 Extra mitochondrial metabolism of cholesterol

As stated earlier, it is possible that cholesterol may, at least in the adrenals of some species, be synthesized de novo. In many cases, however, including the human species and in the rat, circulating cholesterol derived from low-density lipoprotein (LDL) in the former or high-density lipoprotein (HDL) in the latter is utilized by the gland. The lipoprotein moiety is translocated to intracellular sites through receptor mediated endocytosis and this appears to be a control site for ACTH action as discussed in Chap. 3. In the cytosol, two enzymes are concerned with the subsequent metabolism of cholesterol. These are acyl-CoA:cholesterol acyltransferase (ACAT) which catalyzes esterification of cholesterol, and cholesterol ester hydrolase (CEH) which catalyzes the hydrolysis of the esters (Fig. 2.11). Cholesterol is stored in esterified form within the lipid droplets in the cells and is mobilized following activation of cholesterol ester hydrolase so that it is made available to the mitochondria for subsequent steroidogenesis. Both of these events, the receptor mediated internalization of the lipoprotein, and the metabolism of cholesterol esters may be sites for control of steroidogenesis by ACTH or other factors (see Chap. 3).

2.2.5 The biosynthesis of steroid hormones within the adrenal cell

The actions of all of these components of the adrenocortical cell can account for steroidogenesis (Fig. 2.11). The process is as follows:

Cholesterol, either derived from circulating lipoprotein, or synthesized de novo, is stored in esterified form in the lipid droplets. Following hydrolysis of the cholesterol esters, the sterol is transferred to the inner mitochondrial membrane. This transfer may be facilitated by the so-called sterol carrier protein-2 (SCP_2) [58,59]. Mitochondrial cholesterol is the immediate substrate for steroid production and, in isolated mitochondrial systems, its depletion can be seen to be related stoichiometrically with steroid output. Hydroxylation at 22R then 20α is followed by the loss of the side chain in the form of isocaproic acid. The C_{21} steroid which this produces, pregnenolone, then emerges from the mitochondria and passes to the smooth endoplasmic reticulum. It may then be utilized to form cortisol or corticosterone (Fig. 2.12). In the first case, 17α-hydroxylation of pregnenolone, in the smooth endoplasmic reticulum, now yields 17α-hydroxypregnen-

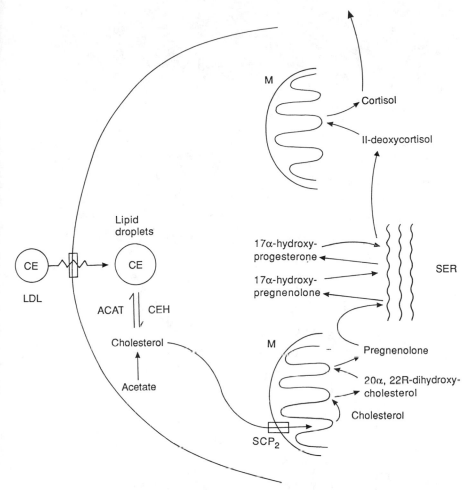

FIGURE 2.11 Cellular organization for cortisol biosynthesis. LDL = low-density lipoprotein (carrying cholesterol esters), CE = cholesterol esters, ACAT = acyl-CoA: cholesterol acyl-transferase, CEH = cholesterol ester hydrolase, SCP_2 = sterol carrier protein$_2$, SER = smooth endoplasmic reticulum, M = mitochondrion. All mitochondria in a cell probably carry out all of the hydroxylases: here shown on different mitochondria for clarity. As well as cortisol, any of the intermediates from pregnenolone onwards may also be secreted from the cell.

olone. This compound is then acted upon by the 3β-hydroxysteroid-dehydrogenase/isomerase system. In the formation of corticosterone, the major pathway involves the action of the 3β-hydroxysteroid-dehydrogenase/isomerase system on pregnenolone, without prior hydroxylation.

The next events are common to the production both of cortisol and corticosterone. The steroid is first hydroxylated at C-21, and then retranslocated back to the mitochondrial site. Here 11β-hydroxylation then yields corticosterone or cortisol. Through the action of the same enzyme, 18-hydroxycorticosterone or 18-hydroxycortisol may also be produced, but in

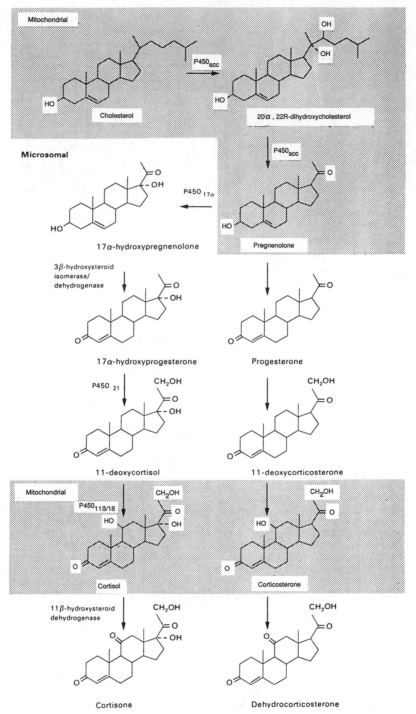

FIGURE 2.12 Major pathway for formation of glucocorticoids from cholesterol. After the formation of 17α-hydroxypregnenolone and pregnenolone, the 17α-hydroxysteroids and the 17-deoxysteroids are formed by the actions of enzymes with similar activities, which are named on the 17α-hydroxysteroid pathway only.

very much smaller yields. No specific transport mechanisms have been clearly identified as being responsible for the translocation of steroid between these sites.

Biosynthesis of aldosterone is more problematical, as previously noted. Its production is related to that of 18-hydroxycorticosterone, which is formed from corticosterone (Fig. 2.13). There is another possibility however, and evidence exists for supposing that aldosterone may be formed from a special pool of 18-hydroxylated intermediates, including especially 18-hydroxydeoxycorticosterone, which are sequestered in the cell membrane of the zona glomerulosa by association with a lipoprotein complex [60,61].

The $C_{17\alpha\text{-}20}$ lyase reaction is also catalyzed by $P450_{17\alpha}$, and may act on 17α-hydroxyprogesterone (to yield androstenedione) or 17α-hydroxypregnenolone (to yield dehydroepiandrosterone) (Fig. 2.14). In the liver, this reaction may also yield steroids with an 11-oxygen function as metabolites of the 17α-hydroxycorticosteroids (Fig. 2.15).

C_{19} steroids may be converted to corresponding phenolic steroids through the action of $P450_{arom}$ (Fig. 2.16), although this reaction is only weak in the adrenal.

2.3 Metabolism of corticosteroids

The major site for metabolism of circulating corticosteroids is the liver, although conversion can also take place at other sites including kidney and intestine. Metabolites are mostly eliminated in the urine; the biliary/fecal route is relatively minor in humans, though more important in the rat. The major types of reactions involved are (see Figs. 2.15, 2.17)

1. Reduction, of ring A to give, first, C_5 saturated (5α-H or 5β-H) configurations, and then a 3α or 3β hydroxyl group (mostly 3α in humans). Reduction can also take place at C-20 giving either α or β hydroxyl groups, and side-chain cleavage, producing C_{19} steroids from 17α-hydroxypregnanes can also occur. The ring A reductases require NADPH.

2. Hydroxylation, at several sites including 6α and 16α and β. Enzymes involved in these transformations have not been studied extensively. It is probable that steroid hydroxylations in the liver are, like adrenal hydroxylations, cytochrome P450 dependent.

3. Conjugation, as sulfates or glucuronides. The sulfate moiety is transferred from 3-phosphoadenosin-5-phosphosulfate via specific sulfotransferases to the 3 locus (if occupied by a hydroxyl group) or the 21 position in 21-hydroxypregnanes. 17β-hydroxy-C_{19} steroids may be sulfoconjugated at C_{17}. Glucuronides are formed through the action of the UDP glucuronyl transferase system at the 3 (or 18) position, and metabolites of cortisol and aldosterone are excreted as glucuronides. Among the C_{19} steroids, only DHEA is excreted predominantly as a

Formation of Aldosterone

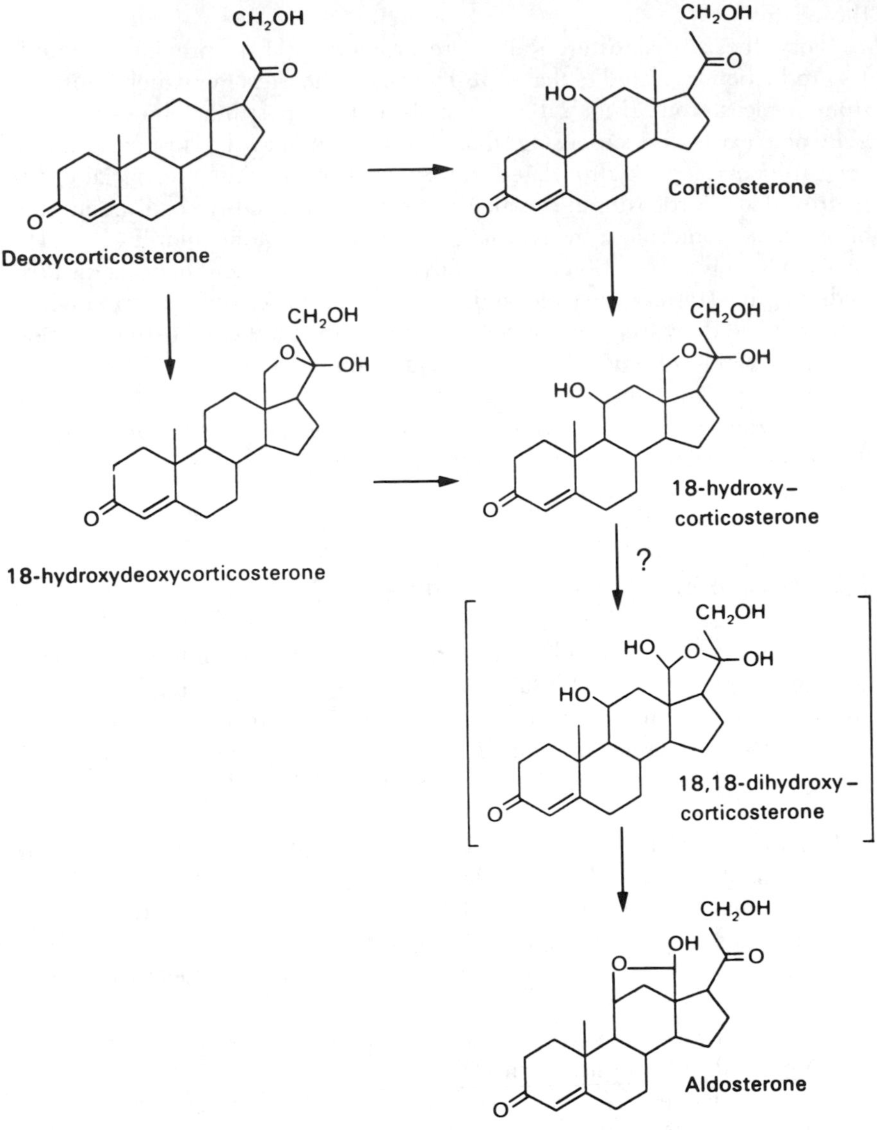

FIGURE 2.13 Possible pathways for the formation of aldosterone from deoxycortico-sterone.

Formation of Androgens

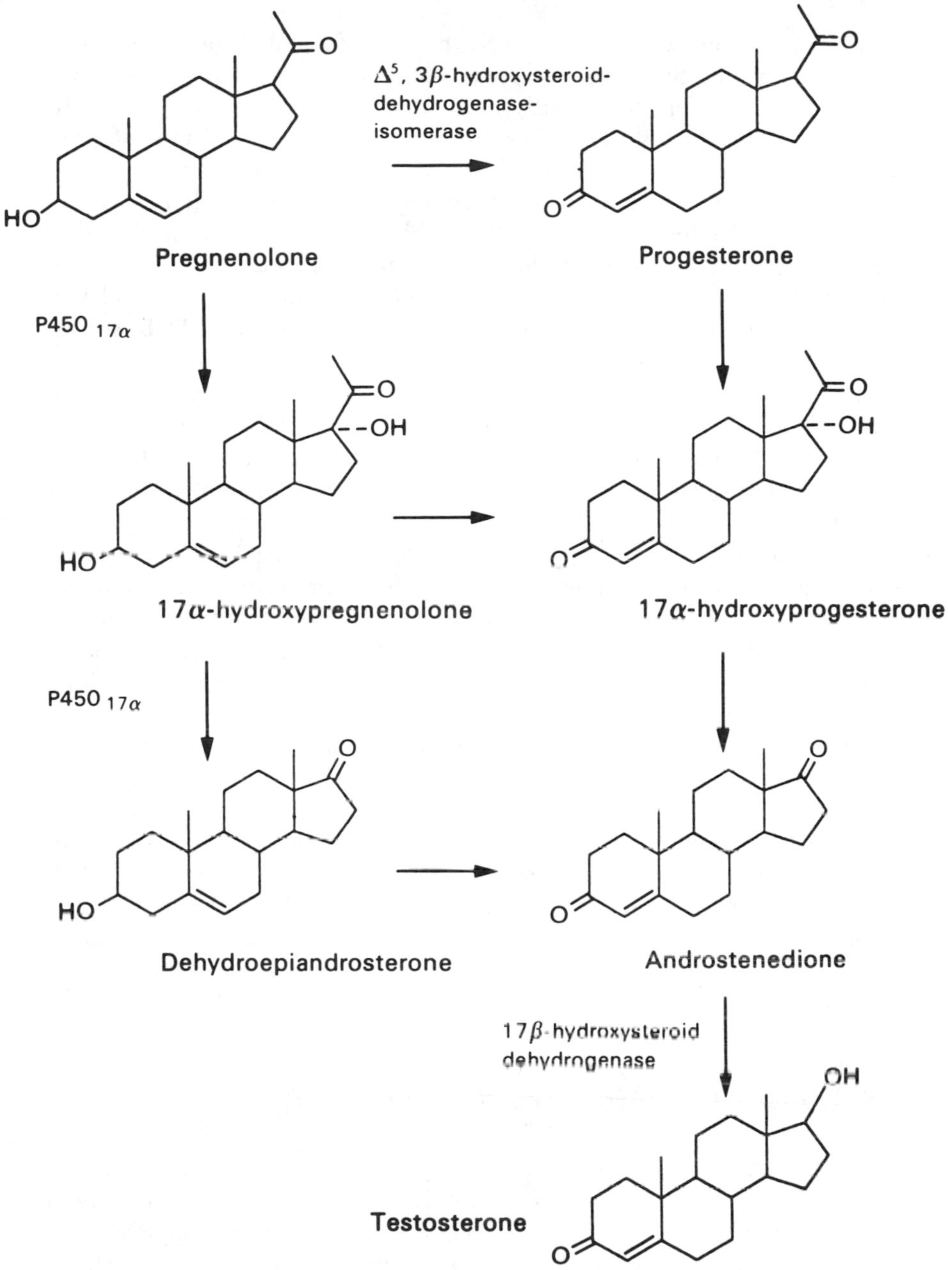

FIGURE 2.14 Pathway for the formation of the androgens. Δ^5, 3β-hydroxysteroids (on the left) may all be transformed into the Δ^4, 3-ketone equivalents by the action of the Δ^5, 3β-hydroxysteroid-dehydrogenase-isomerase system, androstenedione is converted into testosterone by the 17β-hydroxysteroid dehydrogenase, while the remaining reactions are all catalyzed by $P450_{17\alpha}$.

2.3 Metabolism of corticosteroids **51**

sulfate, and others, including testosterone and the 17-oxosteroids are excreted as glucuronides for the most part (see refs. [1,2] for reviews).

It is conventionally assumed that the process of metabolism and conjugation not only deactivates the steroids but also makes them more water soluble, and therefore, apparently more readily excreted in the urine. In the case of the sulfates, this is clearly not true. Dehydroepiandrosterone sulfate is a major circulating steroid component in the human species, and occurs in peripheral plasma in concentrations a good deal higher (up to 5–6 µmol/1, similar to that in adrenal vein blood) than those of cortisol (0.4 µmol/1) [62]. Cortisol is, nevertheless, the major secretory product of the adrenal with (under surgical stress) a concentration of up to 10 µmol/1 in adrenal vein blood. It is probable that the high concentration of DHA sulfate in peripheral plasma is attributable to its long half life, presumably reflecting the relatively poor renal extraction of steroid conjugates in the presence of plasma proteins [63]. The biological significance of sulfoconjugation, therefore, is somewhat obscure, except in so far as it is known that DHEA sulfate, after 16α-hydroxylation is a substrate for estriol formation in pregnant women (Fig. 2.16 and see Chap. 9).

Although steroid assay in plasma and salivary samples is widely used to assess adrenal function, following the advent of simple radioimmunoassay (RIA) and other assay systems, it is interesting to note that urinary steroid analysis of adrenal steroid metabolites can still be used to give valuable information. The particular advantages for this method are that the technique is noninvasive and that, in the collection of a 24-hour urine, the values obtained are an index of integrated hormone output over the whole period, whereas plasma and salivary sampling is inevitably subject to episodic variations. Use of capillary column gas chromatography is now standard in many analytical laboratories and some striking examples of the chromatographic profiles of adrenal steroid metabolites are shown in Figure 2.18 [64]. Another method, not at present widely applied, is the use of fast atom bombardment (FAB) mass spectrometry which can give extremely rapid profiles of steroid conjugates after the introduction of crude urine samples into the machine.

2.4 Transport of corticosteroids in plasma

The glucocorticoids, corticosterone and cortisol, are transported in plasma in a protein bound form. At normal circulating concentrations, probably in excess of 90 percent of these steroids is bound to a specific transport protein called transcortin or corticosteroid binding globulin (CBG) which has a single binding site for corticosteroids with a dissociation constant of approximately 10^{-7} to 10^{-8} mols/l. There is also a lower affinity binding (dissociation constant approximately 10^{-3} mols/l) to plasma albumin. Transcortin is present in the plasma in a concentration of approximately 550 nmols/l.

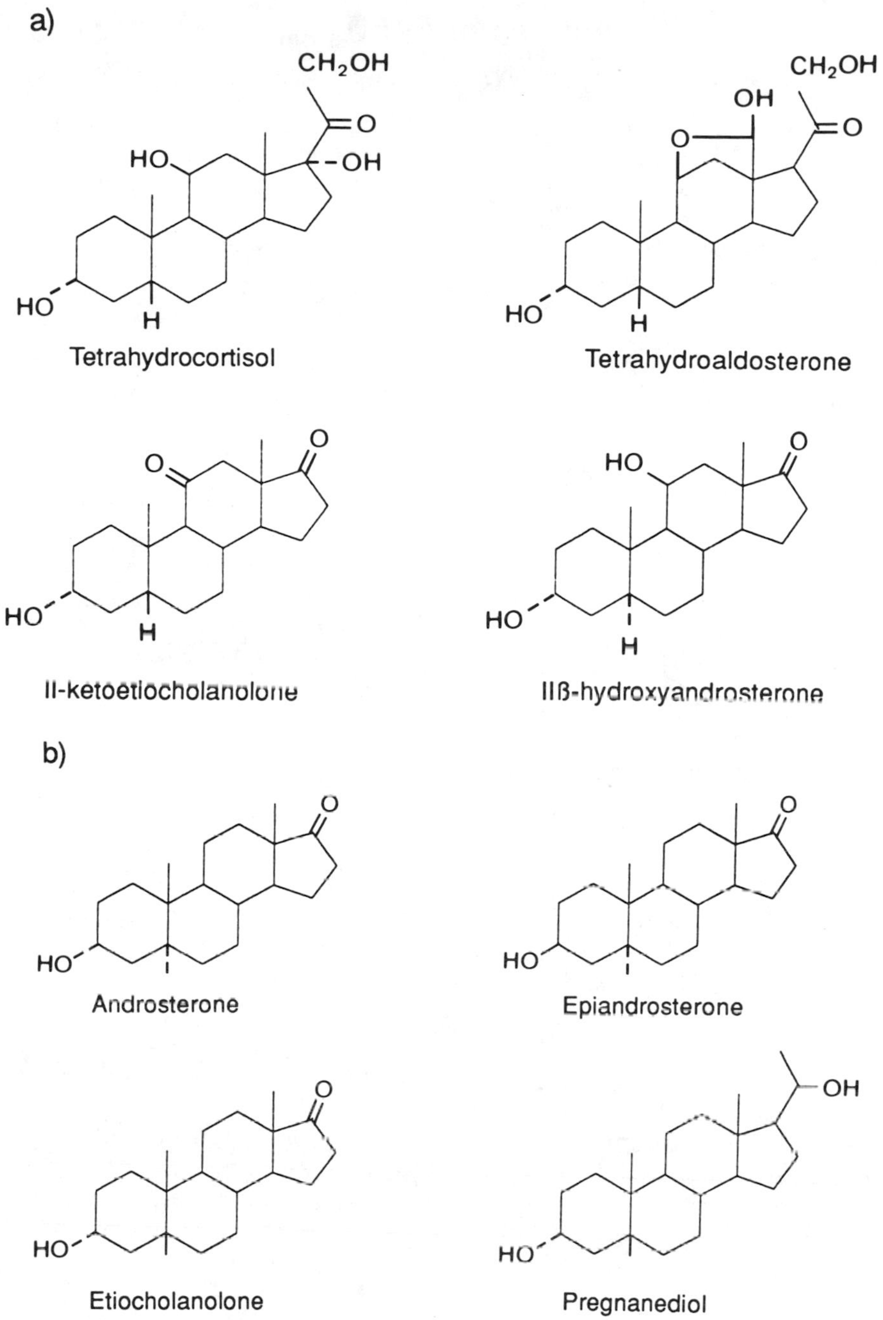

a)

Tetrahydrocortisol

Tetrahydroaldosterone

Il-ketoetiocholanolone

IIβ-hydroxyandrosterone

b)

Androsterone

Epiandrosterone

Etiocholanolone

Pregnanediol

FIGURE 2.15 Steroid metabolites of (a) corticosteroids and (b) C_{19} steroids and progesterone.

Actions of P450 arom

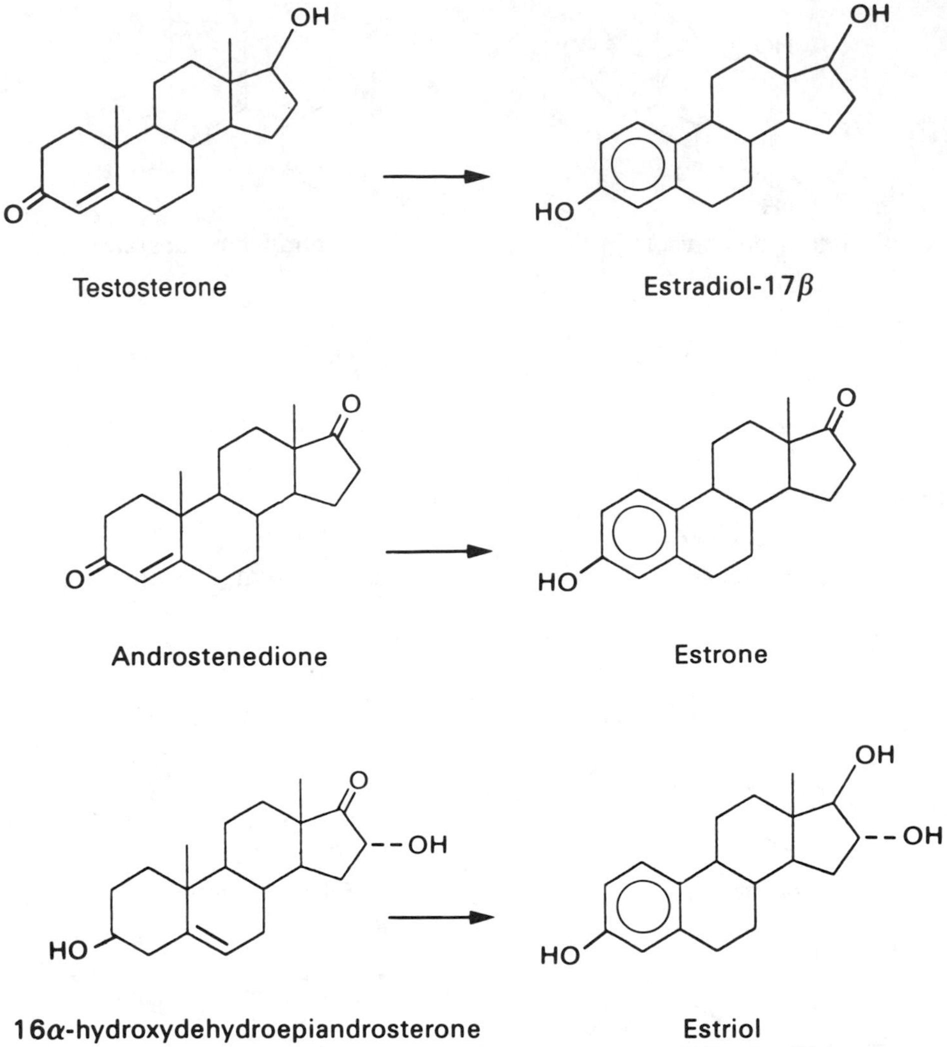

Testosterone → Estradiol-17β

Androstenedione → Estrone

16α-hydroxydehydroepiandrosterone → Estriol

FIGURE 2.16 Formation of C_{18} phenolic steroids, the estrogens, from C_{19} precursors, through the action of P450$_{arom}$. The formation of estriol from 16α-hydroxydehydroepiandrosterone, which occurs in the human placenta, also involves dehydrogenase action.

Consequently, glucocorticoid at concentrations higher than this can only be bound by the lower affinity sites of albumin. The concentration of transcortin is itself variable however, and notably is greatly increased during pregnancy because of the actions of oestrogens [64,65].

The consequences of plasma binding are still somewhat conjectural

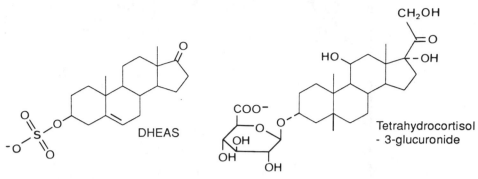

FIGURE 2.17 Examples of steroid conjugates.

although it is widely assumed that only the nontranscortin bound fraction of the glucocorticoids is biologically active. It is for this reason that attention has been paid to the assay of the free fraction alone in tests of adrenocortical function. In particular, this was seen as a specific advantage of assaying steroids in saliva rather than plasma since it is thought that the salivary content represents only the free (filtrable) fraction of the plasma content [64].

In recent years however, the concept of the bound fraction as a circulating pool of biologically inactive steroid has been open to some challenge since the extraction of steroid from plasma, and its transport across microcirculatory barriers, is not limited in vivo to the free fraction alone. These findings have been taken to suggest that a mechanism of enhanced dissociation of hormone from the plasma protein caused by transient conformational changes may facilitate the entry of the steroid into the target organ [66]. Specific binding sites for corticosteroid binding globulin and sex hormone binding globulin in potential target organs would seem to support this hypothesis [67,68]. However, since free hormone is in a state of equilibrium with bound hormone, removal of part of the free fraction, by its retention in target tissue, does not necessarily lead to a loss of free circulating hormone for any significant period. Consequently, while it may be true that only free hormone can leave the circulation, it does not necessarily follow that the bound hormone is inactive because unavailable. Any such unavailability is, in essence, reversible. The significance that plasma binding of hormones has for their biological activity remains enigmatic [69,70].

It certainly seems to be the case that the plasma kinetics of steroids are greatly affected by their plasma binding. In Table 2.1, data is given which was obtained from human subjects by observing the transport and excretion of labeled corticosteroids. In calculating this data, protein binding of the steroids was not explicitly considered and the data were derived by considering the distribution as if it occurred into a single homogeneous body fluid compartment and was removed by a single inactivating excretion process [65]. Although because of the simple assumptions which have been

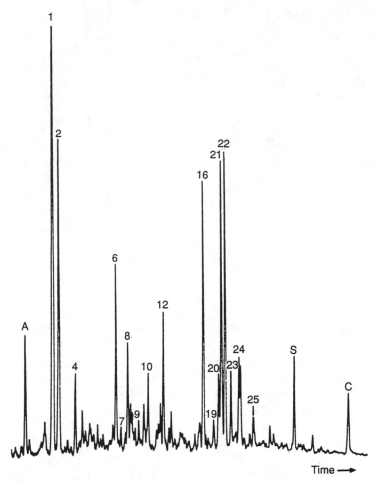

FIGURE 2.18 Gas chromatographic profiles of urinary steroid metabolites. Reproduced, with permission, from ref. 33. (a) Steroid profile from a urine sample from a 21-year-old man. Peaks 1 to 8 represent C_{19} metabolites of steroids of testicular and adrenal origin. The remaining peaks represent C_{21} metabolites, largely of adrenal origin, except for A, S, and C, which are the added internal standards. (b) Steroid profile of a urine sample from a 10-year-old girl. The major metabolites, 16, 21, and 22 are derived from cortisol. Comparison with (a) shows the relative unimportance of C_{19} steroids here. (c) Steroid profile from a urine sample from a 5-year-old girl with an adrenal tumor. The relatively small peaks for the internal standards (A, S, and C) emphasize the massive amounts of steroid, including C_{19} steroids which are being excreted.

Steroids: 1, androsterone; 2, etiocholanolone; 4, DHEA; 6, 11β-hydroxyandrosterone; 7, 11β-hydroxyetiocholanolone; 8, 16α-hydroxy-DHEA; 9, pregnanediol; 10, 5β-pregnan-3α,17α,20α-triol; 12, 5-androstene-3B,16α,17B-triol; 16, tetrahydrocortisone (THE); 19, tetrahydrocorticosterone; 20, allotetrahydrocorticosterone; 21, tetrahydrocortisol (THF); 22, allo-tetrahydrocortisol; 23, α-cortolone (20α reduced THE); 24, β-cortolone (20β reduced THE); 25, α-cortol (20α reduced THF).

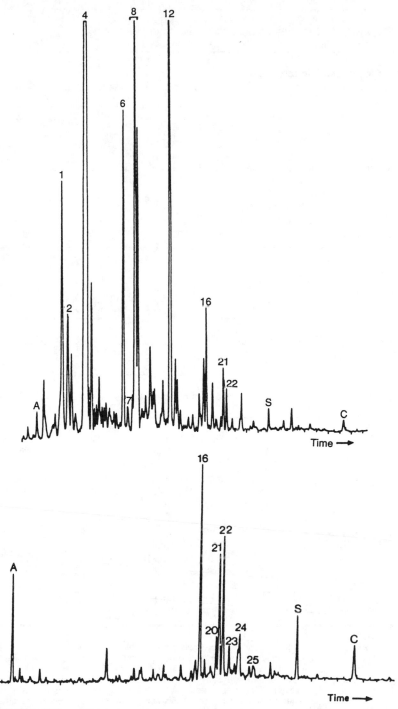

FIGURES 2.18B and C

TABLE 2.1 Kinetics of adrenal steroids in plasma

Total Plasma Concentration (c) nmol/l	Cortisol	Corticosterone	Aldosterone
AM	828	86	
PM	82	8	
Mean	415	35	0.4
Mean extra adrenal pool (nmol)	6215	290	20
Turnover time (time constant) (min)	130	43	43
Virtual volume of distribution (at plasma concentration) (l)	15	9	47
Metabolic clearance rate (l/min)	0.1	0.2	1.1
24-hr secretion rate (μmol/day)	60	10	0.7

Data from ref. 65.

TABLE 2.2 Examples of plasma concentrations of adrenal steroids in normal human subjects

Steroid	Concentration	Reference
Aldosterone	100–400 pmol/l	64
Androstenedione	adults 2–13 nmol/l	64
	children < 2 nmol/l	
Cortisol	adults, 07.00–09.00 hr: 280–720 nmol/l 21.00–24.00 hr 60–340 nmol/l	
Corticosterone	2.3–23 nmol/l	71
18-OH-DOC	156 ± 20 pmol/l	72
18-OH-Corticosterone	635 ± 60 pmol/1	73
DHEAS	adults 2–11 μmol/l	64
	children < 2 μmol/l	
DOC	202 ± 70 pmol/l	74
11-deoxycortisol	1.51 ± 0.83 nmol/l	74

It should be noted that these values may show considerable variation for a variety of reasons. First, all adrenal steroids may show a marked diurnal rhythm (see Chap. 3), as exemplified here by the cortisol data. Second, according to some studies, marked differences may also be seen at different stages of the ovarian cycle in women [74]. Third, environmental, nutritional, age, and health factors can also affect the data. Although the values for steroids given here are usually thought directly to reflect adrenal secretion, this is not always the case. Some androstenedione and DHEAS may be secreted by the gonads, for example, and extra-adrenal sites of metabolism, (for example 21-hydroxylation, [75], can conceivably contribute to overall steroid levels.

TABLE 2.3 Typical concentrations of adrenocortical hormones in circulating plasma in the rat.

Steroid	Concentration in plasma in normal animals (nmol/l)
Corticosterone	115 ± 27
18-OH-DOC	52 ± 7
DOC	0.39 ± 0.07
Aldosterone	0.25 ± 0.05
18-OH-B	0.75 ± 0.03
Progesterone	1.12 ± 0.08

In this species, in which 17α-hydroxylase activity is low in the adrenal, corticosterone is the major secretory product, but 18-hydroxydeoxycorticosterone (18-OH-DOC) is also secreted in substantial amounts. The high level of 18-hydroxylase activity which this reflects, mostly in the inner adrenocortical zones (see Chap. 9), is a unique feature of adrenocortical function in the rat, which has not been fully interpreted in functional terms. In contrast, 18-OH-B, a glomerulosa product, is present in concentrations similar to those seen in the human species (cf. Table 2.2). In a small species like the rat, it is often difficult to sample circulating plasma without stressing the animal, and thus stimulating the hypothalamo-pituitary adrenal system. For this reason, as well as for the others listed for Table 2.2, some published values may vary considerably from those given here. The animals used in these studies were male (Wistar) rats, and the progesterone in plasma is, therefore, probably of entirely adrenocortical origin. Data from refs. 76, 77.

made, the data may be slightly inaccurate, the differences between the glucocorticoids, which bind to transcortin, and aldosterone, which does not, are clear. The 24-hour secretion rate for aldosterone is thus approximately 1 percent of that of cortisol. However, the circulating concentration is no more than 0.1 percent. The metabolic clearance rate of aldosterone, approximately ten times that of cortisol, consolidates this data.

Further data on circulating plasma concentrations in humans of steroids of predominantly adrenal origin are shown in Table 2.2. Some species differences occur, and it is only in humans and the primates that DHEAS is such a major product. Data on circulating steroids in the rat are also given (Table 2.3) for further comparison. Here, the secretion of large amounts of 18-OH-DOC is a unique characteristic of this species.

References

1. Dorfman, R.I., and Ungar, F. 1968. *Metabolism of Steroid Hormones*. New York: Academic Press.

2. Kime, D.E., and Norymberski, J.K. 1976. Structure and nomenclature of steroids. In: I. Chester Jones and I.W. Henderson, eds. *General, Comparative and Clinical Endocrinology of the Adrenal Cortex*. vol. I, New York and London: Academic Press, pp. 1–24.

3. Kellie, A.E. 1984. Structure and nomenclature. In: H.L.J. Makin, ed. *Biochemistry of Steroid Hormones*. 2nd ed. Oxford: Blackwell, pp. 1–19.

4. Goad, L.J. 1984. Cholesterol biosynthesis and metabolism. In: H.L.J. Makin, ed. *Biochemistry of Steroid Hormones*. 2nd. ed. Oxford: Blackwell, pp. 20–70.

5. Kovanen, P.T., Faust, J.R., Brown, M.S., and Goldstein, J.L. 1979. Low-density lipoprotein receptors in bovine adrenal cortex. I. Receptor mediated uptake of low-density lipoprotein and utilization of its cholesterol for steroid synthesis in cultured adrenocortical cells. *Endocrinology*. 104:599–609.

6. Carr, B.R., Porter, J.C., MacDonald, P.C., and Simpson, E.R. 1980. Metabolism of low-density lipoprotein by human fetal adrenal tissue. *Endocrinology*, 107:1034–40.

7. Ohashi, M., Carr, B.R., and Simpson, E.R. 1981. Binding of high-density lipoprotein to human fetal adrenal membrane preparations. *Endocrinology*, 109:783–89.

8. Gwynne, J.T., Mahaffee, D., Brewer, H.B., and Ney, R.L. 1976. Adrenal cholesterol uptake from plasma lipoproteins: Regulation by corticotrophin. *Proc. Nat. Acad. Sci., U.S.A.* 73:4329–33.

9. Gwynne, J.T., and Hess, B. 1980. The role of high-density lipoproteins in rat adrenal cholesterol metabolism and steroidogenesis. *J. Biol. Chem.* 255:10875–83.

10. Kovanen, P.T., Schneider, W.J., Hillman, G.M., Goldstein, J.L., and Brown, M.S. 1979. Separate mechanisms for the uptake of high- and low-density lipoproteins by mouse adrenal gland *in vivo*. *J. Biol. Chem.*, 254:5498–505.

11. Brown, M.S., Kovanen, P.T., and Goldstein, J.L. 1979. Receptor-mediated uptake of lipoprotein-cholesterol and its utilization for steroid synthesis in the adrenal cortex. *Recent Progr. Hormone Res.* 35:215–49.

12. Lloyd, B.J. 1972. Rates of adrenal cholesterol formation by the hamster, sheep, guinea pig, and rat from labeled pyruvate *in vitro*. *Gen. Comp. Endocr.* 19:428–31.

13. Lehoux, J.-G. 1980. Phylogeny of sterol biosynthesizing systems. In: G. Pethes and V.L. Frenyo, eds. *Advances in Physiological Sciences*. vol. 20. *Advances in Animal and Comparative Endocrinology*. Oxford: Pergamon Press, pp. 337–44.

14. Sandor, T., Fazekas, A.G., and Robinson, B.H. 1976. The biosynthesis of corticosteroids throughout the vertebrates. In: I. Chester Jones and I.W. Henderson, eds. *General, Comparative and Clinical Endocrinology of the Adrenal Cortex*. vol. I. New York and London: Academic Press. pp. 24–142.

15. Lehoux, J.-G., Lefebvre, E., de Medicis, E., Bastin, M., Belisle, S., and Bellabarba, D. 1987. Effect of ACTH on cholesterol and steroid synthesis in adrenocortical tissues. *J. Steroid Biochem.* 27:1151–60.

16. Boyd, G.S., McNamara, B., Suckling, K.E., and Tocher, D.R. 1983. Cholesterol metabolism in the adrenal cortex. *J. Steroid Biochem.* 19:1017–27.

17. Jamal, Z., Suffolk, R.A., Bord, G.S., and Suckling, K.E. 1985. Metabolism of cholesteryl ester in monolayers of bovine adrenal cortical cells. *Biochim. Biophys. Acta*. 834:230–37.

18. Gower, D.B. 1984. Biosynthesis of the corticosteroids. In: H.L.J. Makin, ed. *Biochemistry of Steroid Hormones*, 2nd ed. Oxford: Blackwell, pp. 117–69.

19. Gower, D.B. 1984. The role of cytochrome P450 in steroidogenesis and properties of some of the steroid-transforming enzymes. In: H.L.J. Makin, ed. *Biochemistry of Steroid Hormones*, 2nd ed. Oxford: Blackwell, pp. 230–92.

20. Jefcoate, C.R. 1986. Cytochrome P450 enzymes in sterol biosynthesis and

metabolism. In: P.R. Ortiz de Montellano, ed. *Cytochrome P450 Structure, Mechanism and Biochemistry*. New York and London: Plenum Press, pp. 387–428.

21. Nebert, D.W., Nelson, D.R., and Feyersen, R. 1989. Evolution of the cytochrome P450 genes. *Xenobiotica*. 19:1149–60.

22. Williams, R.T. 1973. *Detoxication Mechanisms*. London: Chapman and Hall.

23. Thomas, P.E., Korzeniowski, D., Ryan, D., and Levin, W. 1979. Preparation of monospecific antibodies against two forms of rat liver cytochrome P450 and quantitation of these antigens in microsomes. *Arch. biochem. physiol.* 192:524–32.

24. Black, S.D., and Coon, M.J. 1986. Comparative structures of P450 cytochromes. In: P.R. Ortiz de Montellano, ed. *Cytochrome P450 Structure, Mechanism, and Biochemistry*. New York and London: Plenum Press, pp. 161–216.

25. Dawson, J.H. 1988. Probing structure-function relations in heme-containing oxygenases and peroxidases. *Science*. 240:433–39.

26. McMurry, T.J., and Groves, J.T. 1986. Metalloporphyrin models for cytochrome P450. In: P.R. Ortiz de Montellano, ed. *Cytochrome P450 Structure, Mechanism, and Biochemistry*. New York and London: Plenum Press, pp. 1–28.

27. Ortiz de Montellano, P.R. 1986. Oxygen activation and transfer. In: P.R. Ortiz de Montellano, ed. *Cytochrome P450 Structure, Mechanism and Biochemistry*. New York and London: Plenum Press, pp. 217–71.

28. Kimura, T., Nakamura, S., Huang, J.J., Chu, J.W., Wang, II.P., and Tsernoglou, D. 1973. Electron transport system for adrenocortical mitochondrial steroid hydroxylation reactions: The mechanism of the hydroxylation reactions and properties of the flavoprotein-iron-sulfur protein complex. *Ann. N.Y. Acad. Sci.* 212:94–106.

29. Waterman, M.R., and Simpson, E.R. 1985. Regulation of the biosynthesis of cytochromes P450 involved in steroid hormone synthesis. *Mol. Cell. Endocr.* 39:81–89.

30. Wada, A., Okamoto, M., Nonaka, Y., and Yamano, T. 1984. Aldosterone biosynthesis by a reconstituted cytochrome $P450_{11\beta}$ system. *Biochim. Biophys. Res. Comm.* 119:365–71.

31. Lauber, M., Sugano, S., Ohnishi, T., Okamoto, M., and Müller, J. 1987. Aldosterone biosynthesis and cytochrome $P450_{11\beta}$: Evidence for two different forms of the enzyme in rats. *J. Steroid Biochem*. 26:693–98.

32. Imai, M., Shimada, H., Okada, Y., Matsushima-Hibiya, Y., Ogishima, T., and Ishimura, Y. 1990. Molecular cloning of a cDNA encoding aldosterone synthase cytochrome P450 in rat adrenal cortex. *FEBS. lett.* 263:299–302.

33. Kominami, S., Hara, H., Ogishima, T., and Takemori, S. 1984. Interaction between cytochrome P450 ($P450_{C21}$) and NADPH-cytochrome P450 reductase from adrenocortical microsomes in a reconstituted system. *J. Biol. Chem.* 259:2991–99.

34. Takemori, S., and Kominami, S. 1984. The role of cytochromes P450 in adrenal steroidogenesis. *Trends in Biochem. Sci.* 9:393–96.

35. Nakajin, S., Shively, J.E., Yuan, P.M., and Hall, P.F. 1981. Microsomal cytochrome P450 from neonatal pig testis: two enzymatic activities 17α-hydroxylase and $C_{17.20}$-lyase) associated with one protein. *Biochemistry*. 20:4037–42.

36. Nakajin, S., Shinoda, M., Haniu, M., Shively, J.E., and Hall, P.F. 1984. C_{21}

steroid side-chain cleavage enzyme from porcine adrenal microsomes. *J. Biol. Chem.* 259:3971–76.

37. Shinzawa, K., Kominami, S., and Takemori, S. 1985. Studies on cytochrome P450 (P450$_{17\alpha,lyase}$) from guinea pig adrenal microsomes. Dual function of a single enzyme and effect of cytochrome b$_5$. *Biochim. Biophys. Acta.* 833, 151–60.

38. Haynes, R.C., and Berthet, L. 1957. Studies on the mechanism of action of the adrenocorticotropic hormone. *J. Biol. Chem.* 225:115–24.

39. Simpson, E.R., and Frenkel, R. 1969. Substrate induced efflux of anions from bovine adrenal cortex mitochondria and its relationship to steroidogenesis. *Biochim. Biophys. Res. Comm.* 35:765–70.

40. Cammer, W., and Estabrook, R.W. 1967. Respiratory activity of adrenal cortex mitochondria during steroid hydroxylation. *Arch. Biochem. Biophys.* 122:721–34.

41. Peron, F.G., and McCarthy, J.L. 1968. Corticosteroidogenesis in the rat adrenal gland. In: K.W. McKerns, ed. *Functions of the Adrenal Cortex.* vol. 2, Amsterdam: North Holland Publishing Co., pp. 261–337.

42. Kowal, J., Forchielli, E., and Dorfman, R.I. 1964. The Δ^5-3β-hydroxysteroid dehydrogenases of corpus luteum and adrenal. I. Properties, substrate specificity and cofactor requirements. *Steroids.* 3:531–49.

43. Kowal, J., Forchielli, E., and Dorfman, R.I. 1964. The Δ^5-3β-hydroxysteroid dehydrogenases of corpus luteum and adrenal. II. Interaction of C$_{19}$ and C$_{21}$ substrates. *Steroids.* 4:77–100.

44. Geynet, P., de Paillerets, C., and Alfsen, A. 1976. Lipid requirement of membrane-bound 3-oxosteroid Δ^4-Δ^5 isomerase. Studies on beef adrenocortical microsomes. *Eur. J. Biochem.* 71:607–12.

45. Alfsen, A., Baulieu, E.E., and Claquin, M.J. 1965. Isolation of an adrenal 3-oxo-5-cholestene isomerase. *Biochim. Biophys. Acta.* 20:251–55.

46. Neville, A.M., Orr, J.C., and Engel, L.L. 1969. The Δ^5-3β-hydroxysteroid dehydrogenase of bovine adrenal microsomes. *J. Endocr.* 43:599–608.

47. Yates, J., and Deshpande, N. 1975. Evidence for the existence of a single 3β-hydroxysteroid dehydrogenase/$\Delta^{5,4}$-3-oxosteroid isomerase complex in the human adrenal gland. *J. Endocr.* 64:195–96.

48. Geynet, P. 1977. Heterogeneity of membrane bound Δ^5-3-oxosteroid isomerase. Studies on bovine adrenocortical microsomes. *Biochim. Biophys. Acta.* 486:369–77.

49. Ford, H.C., and Engel, L.L. 1974. Purification and properties of the Δ^5-3β-hydroxysteroid dehydrogenase-isomerase systems of sheep adrenal cortical microsomes. *J. Biol. Chem.* 249:1363–68.

50. Ishii-Ohba, H., Saiki, N., Inano, H., and Tamaoki, B.-I. 1986. Purification and characterization of rat adrenal 3β-hydroxysteroid dehydrogenase with steroid 5-ene-4-ene-isomerase. *J. Steroid Biochem.* 24:753–60.

51. Monder, C., and Shackleton, C.H.L. 1984. 11β-hydroxysteroid dehydrogenase: Fact or fancy? *Steroids.* 44:373–427.

52. Lakshmi, V., and Monder, C. 1985. Evidence for independent 11-oxidase and 11- reductase activities of 11β-hydroxysteroid dehydrogenase: Enzyme latency, phase transitions, and lipid requirements. *Endocrinology.* 116:552–60.

53. Stewart, P.M., Wallace, A.M., Valentino, R., Burt, D., Shackleton, C.H.L., and Edwards, C.R.W. 1987. Mineralocorticoid activity of liquorice: 11-beta-hydroxysteroid dehydrogenase deficiency comes of age. *Lancet.* II:821–23.

54. Vinson, G.P., and Whitehouse, B.J. 1970. Comparative aspects of adrenocortical

function. In: M.H. Briggs, ed. *Advances in Steroid Biochemistry and Pharmacology*. vol. I. New York and London: Academic Press, pp. 163–342.

55. Williamson, D.G. 1979. In: R. Hobkirk, ed. *Steroid Biochemistry*. Boca Raton, Florida: CRC Press, p. 83.

56. Ulick, S. 1976. Diagnosis and nomenclature of the disorders of the terminal portion of the aldosterone biosynthetic pathway. *J. Clin. Endocr.* 43:92–96.

57. Aupetit, B., Accarie, C., Emeric, N., Vonarx, V., and Legrand, J.C. 1983. The final steps of aldosterone biosynthesis requires reducing power. It is not a dehydrogenation. *Biochim. Biophys. Acta.* 752:73–78.

58. Vahouny, G.V., Chanderbhan, R., Stewart, R., Tombes, R., Keyeyune-Nyombi, E., Fiskum, G., and Scallen, T.J. 1985. Phospholipids, sterol carrier protein$_2$ and adrenal steroidogenesis. *Biochim. Biophys. Acta.* 834:324–30.

59. Chanderbhan, R.F., Kharroubi, A.T., Noland, B.J., Scallen, T.J., and Vahouny, G.V. 1986. Sterol carrier protein$_2$: Further evidence for its role in adrenal steroidogenesis. *Endocrine Res.* 12:351–70.

60. Vinson, G.P., and Whitehouse, B.J. 1973. The biosynthesis and secretion of aldosterone by the rat adrenal zona glomerulosa, and the significance of the compartmental arrangement of steroids. *Acta Endocr.* 72:746–52.

61. Laird, S.M., Vinson, G.P., and Whitehouse, B.J. 1988. Steroid sequestration within the rat adrenal zona glomerulosa. *J. Endocr.* 117:191–96.

62. Kime, D.E., Vinson, G.P., Major, P.W., and Kilpatrick, R. 1980. Adrenal-gonad relationships. In: I. Chester Jones and I.W. Henderson, eds. *General, Comparative and Clinical Endocrinology of the Adrenal Cortex.* vol. III, New York and London: Academic Press, pp. 183–264.

63. Chaudhun, G., Verheugen, C., Pardridge, W.M., and Judd, H.L. 1987. Selective availability of protein bound estrogen and estrogen conjugates to the rat kidney. *J. Endocr. Invest.* 10:283–90.

64. Cook, B., and Beastall, G.H. 1987. Measurement of steroid hormone concentrations in blood, urine and tissues. In: B. Green and R.E. Leake, eds. *Steroid Hormones a Practical Approach.* Oxford and Washington: IRL Press, pp. 1–65.

65. Yates, F.E., Marsh, D.J., and Maran, J.W. 1980. The adrenal cortex. In: V.B. Mountcastle, ed. *Medical Physiology.* St. Louis: C.V. Mosby, pp. 1558–60.

66. Pardridge, W.M. 1987. Plasma protein-mediated transport of steroid and thyroid hormones. *Am. J. Physiol.* 252:E157–E164.

67. Avvakumov, G.V., and Strel'chyonok, O.A. 1988. Evidence for the involvement of the transcortin carbohydrate moiety in the glycoprotein interaction with the plasma membrane of human placental syncytiotrophoblast. *Biochim. Biophys. Acta.* 938:1–6.

68. Hryb, D.J., Khan, M.S., Romas, N.A., and Rosner, W.A. 1990. The control of the interaction of sex hormone-binding globulin with its receptor by steroid hormones. *J. Biol. Chem.* 265:6048–54.

69. Mendel, C.M. 1989. The free hormone hypothesis: A physiologically based mathematical model. *Endocr. Rev.* 10:232–74.

70. Ekins, R. 1990. Measurement of free hormones in blood. *Endocr. Rev.* 11:5–46.

71. Mason, P.A., and Fraser, R. 1975. Estimation of aldosterone, 11-deoxycorticosterone, corticosterone, 18-hydroxy-11-deoxycorticosterone, corticosterone, cortisol and 11-deoxycortisol in human plasma by gas-liquid chromatography with electron capture detection. *J. Endocr.* 64:277–88.

References

72. Williams, G.H., Braley, L.M., and Underwood, R.H. 1976. The regulation of plasma 18-hydroxy 11-deoxycorticosterone in man. *J. Clin. Invest.* 58:221–29.

73. Kater, C.E., Biglieri, E.G., Rost, C.R., Schambelan, M., Hiral, J., Chang, B.C.F., and Brust, N. 1985. The constant plasma 18-hydroxycorticosterone to aldosterone ratio: An expression of the efficacy of corticosterone methyloxidase type II activity in disorders with variable aldosterone production. *J. Clin. Endocr.* 60:225–28.

74. Schoneshofer, M., and Wagner, G.G. 1977. Sex differences in corticosteroids in man. *J. Clin. Endocr.* 45:814–17.

75. Winkel, C.A., Parker, C.R., Simpson, E.R., and MacDonald, P.C. 1980. Production rate of deoxycorticosterone in women during the follicular and luteal phases of the ovarian cycle: The role of extraadrenal 21-hydroxylation of circulating progesterone in deoxycorticosterone production. *J. Clin. Endocr.* 51:1354–58.

76. Chabert, P., Guelpha-Decorzant, C., Riondel, A.M., and Valloton, M.B. 1984. Effect of spironolactone on electrolytes, renin, ACTH and corticosteroids in the rat. *J. Steroid Biochem.* 20:1253–59.

77. Imaizumi, N., Yamamoto, I., Kamei, M., Yoshida, I., Miyauchi, E., Kigoshi, T., Hosojima, H., Uchida, K., and Morimoto, S. 1987. High performance liquid chromatographic separation of corticosteroids in plasma of rats. *Hormone Research.* 27:53–60.

3

Physiological and Cellular Aspects of the Control of Adrenocortical Hormone Secretion

In the mammals, in which aldosterone and cortisol (or corticosterone) have entirely distinct physiological effects (see Chap. 4), the separate control of their secretion is of paramount importance. In part, this has been achieved through cellular differentiation and specialization (see Chap. 9), in that glucocorticoid (and androgen) secretion is a function of the inner zones, the zonae fasciculata and reticularis, while mineralocorticoid (aldosterone) secretion is the primary function of the zona glomerulosa.

3.1 Control of zona fasciculata/reticularis function

3.1.1 Glucocorticoid secretion

The control of glucocorticoid secretion by the zona fasciculata is effected almost exclusively by the activity of the hypothalamo–pituitary axis. Following hypophysectomy, glucocorticoid levels in the adrenal vein blood fall to around 5 percent of previous basal values within 2 hours of the operation (Fig. 3.1). This can be prevented by the administration of corticotropin (ACTH), a peptide hormone secreted by the anterior pituitary [2,3].

Structure of ACTH Corticotropin is derived from a large precursor molecule, pro-opiomelanocortin, which is synthesized in the corticotroph cells of the anterior pituitary, and also gives rise to the melanotrophins,

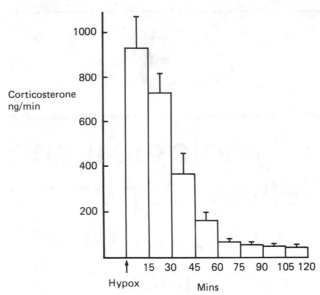

FIGURE 3.1 The secretion rate of corticosterone into the rat adrenal vein declines rapidly following hypophysectomy. Data taken from ref. 1.

lipotrophins, and endorphins (Fig. 3.2). ACTH is a single chain peptide consisting of 39 amino acid residues (Fig. 3.3), although the full biological activity resides within the first 24 residues. It is thought that the amino acid sequence lys-lys-arg-arg that occurs at positions 15–18 is the important region for the binding of ACTH to its receptor. In mammals there is almost complete structural homology of the first 24 amino acid residues of ACTH, but there is considerable interspecies variation in residues 25 to 39 [5]. This "tail" of the molecule is not important in the action of the peptide on the adrenal gland, but is thought to confer some degree of protection against proteolytic breakdown [6], although other suggestions have been made.

Secretion of ACTH Corticotropin is secreted by the anterior pituitary gland in an episodic manner, and with a distinct circadian rhythm. In humans, ACTH levels peak between 6–9 AM, falling during the day to reach a low point in the evening (Fig. 3.4). In the rat, a nocturnal animal, this rhythm is reversed, with a peak of ACTH secretion occurring in the late afternoon. The major physiological stimulus to ACTH secretion is usually termed "stress," and includes emotional stress as well as the physiological stresses of exercise, hypoglycemia, hypoxia, hypotension, and traumas such as hemorrhage, electric shock, and surgical procedures. In each case, the ACTH increment is directly proportional to the strength of the stimulus [7].

Several factors act on the anterior pituitary gland to produce an appropriate level of ACTH secretion. These factors include: corticotropin releasing factor (CRF 41) which is a peptide hormone secreted by the hypothalamus into

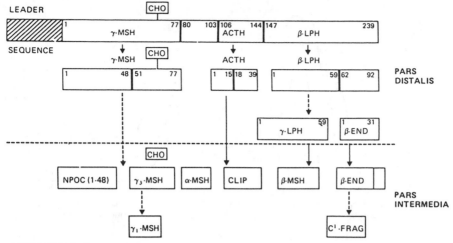

FIGURE 3.2 Bovine pro-opiomelanocortin. This precursor protein, synthesized in the anterior pituitary, gives rise to ACTH and a range of other biologically active peptides. In some species, part of the peptide processing takes place in the intermediate lobe of the pituitary gland. MSH = melanocyte stimulating hormone, ACTH = corticotropin, LPH = lipotropin, END = endorphin, NPOC = N-terminal region of pro-opiomelanocortin, CLIP = corticotropinlike intermediate lobe peptide.

the hypothalamo-pituitary portal system; arginine vasopressin; and certain neurotransmitters, notably epinephrine, in the portal blood [7]. Additionally, there is a negative feedback component, where corticosteroids act on both the hypothalamus and anterior pituitary to inhibit CRF 41 and ACTH secretion. Clinically, ACTH secretion is stimulated by controlled hypoglycemia, usually induced by insulin. Measurement of plasma cortisol levels following this procedure forms the basis of a clinical test for hypopituitarism. For recent reviews on the control of ACTH secretion see ref. 7.

FIGURE 3.3 Primary structure of human (and bovine) corticotropin and the melanocyte stimulating hormones produced from pro-opiomelanocortin (adapted from ref. 4).

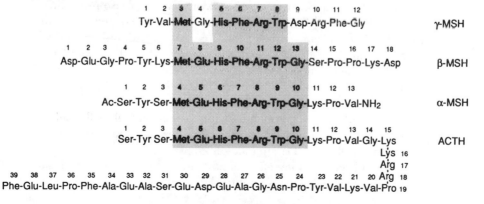

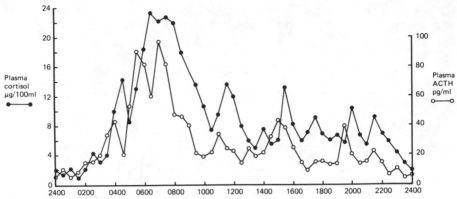

FIGURE 3.4 Circadian rhythm of plasma concentrations of ACTH and cortisol in humans. There is a characteristic peak in the morning, with an evening nadir.

Actions of ACTH The adrenal cortex is the principal target organ for ACTH. In all mammalian species, and most other species, ACTH is the major physiological stimulus to corticosteroid secretion (see Chap. 9). The effects of this peptide have been demonstrated in many species both in vivo and in vitro.

Acutely, the actions of ACTH on the adrenal are twofold, causing an increase in the secretion rate of adrenal steroids (which is preceded by a reduction in the adrenal content of ascorbic acid) and also causing vascular changes which result in an increase in the rate of blood flow through the gland.

Time course of ACTH action

One of the most rapid effects of ACTH administration is to decrease the adrenal content of ascorbic acid. The adrenal glands contain a high concentration of ascorbic acid; the highest of any tissue in the body. The exact role of this substance in the adrenal gland is unknown, but it has been proposed to act as an inhibitor of steroidogenesis in the absence of ACTH [8]. There is a temporal relationship between the release of ascorbic acid and steroids into the adrenal vein, with ascorbic acid release preceding the rise in steroidogenesis [9]. An increased ascorbic acid concentration in rat adrenal vein plasma occurs within 1 minute of ACTH administration, while corticosteroid output increases up to 4 minutes later [9]. A similar time course of steroidogenic response has been reported by other authors using perfused rat and canine adrenal preparations, with a delay of 2 to 5 minutes between ACTH administration and increased steroid secretion [10–12]. In continuous superfusion experiments with adrenocortical cells, an even shorter delay has been reported, with corticosteroid output rising within 1 minute of the onset of ACTH stimulation [13,14].

While steroid secretion increases almost immediately, a maximal rate of output is only achieved several minutes later, often peaking after removal

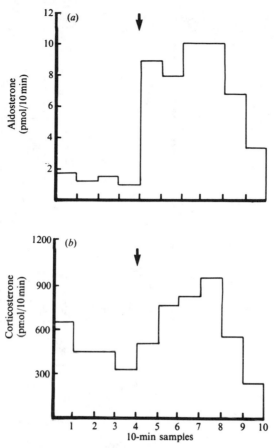

FIGURE 3.5 Time course of the (a) aldosterone and (b) corticosterone responses to a bolus dose of ACTH (300 fmol). Secretion rates of both steroids increase within 10 minutes after administration of ACTH, but a maximum rate of corticosterone secretion is only seen about 30 minutes later. Data were obtained using an in situ isolated perfused adrenal preparation. (Figure reproduced with permission from ref. 16.)

of ACTH (Fig. 3.5) [12,15,16]. Both the amplitude and the time course of the steroidogenic response depend on the dose of ACTH administered [10].

Vascular effects of ACTH

Corticotropin has profound vascular effects, causing a marked increase in the rate of blood flow through the adrenal gland of several species, including calf [17], rat [18–20], sheep [21], rabbit [22], and dog [23–26]. The adrenal gland is a highly vascular organ (see Chap. 1). In the rat, while the adrenal glands comprise only about 0.02 percent of the body weight, they receive around 0.14 percent of the cardiac output, even under basal conditions, and following stimulation with ACTH, the rate of blood flow may increase 100 percent (Fig. 3.6) [20,27].

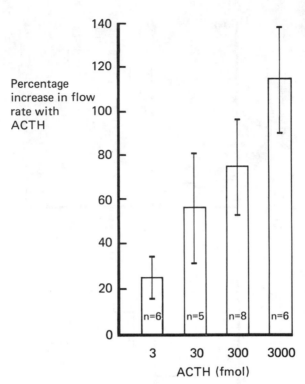

FIGURE 3.6 ACTH causes a dose-related increase in the rate of perfusion medium (or blood) flow through the intact perfused rat adrenal gland in situ. Numbers of experiments are indicated in bars. A significant increase in flow rate was obtained with 3 fmol ACTH ($p < 0.05$). (Figure reproduced with permission from ref. 20.)

The mechanism of ACTH-induced flow rate changes in the adrenal is not fully elucidated, although it is clear that the effect is of vasodilation rather than the vasoconstriction originally suggested [28,29]. It is unlikely that corticosteroids mediate the flow rate increases as the vascular response to ACTH is seen even in the presence of inhibitors of steroidogenesis [19]. It has been suggested that prostaglandins may regulate adrenal blood flow, but the evidence is somewhat contradictory [26,30].

Recent evidence suggests that, in the rat, the vascular effects of ACTH are mediated by mast cells in the connective tissue capsule of the adrenal. ACTH has been shown to release the mast cell products, serotonin and histamine, which both have profound vascular effects in the adrenal [31,32]. Furthermore, administration of disodium cromoglycate, which prevents mast cell degranulation, abolishes the vascular response to ACTH (Fig. 3.7) [32].

An increase in the rate of blood flow through the gland obviously has implications for the supply of oxygen to the adrenal and for the efficient removal of secretory products, but it appears that steroid secretion may be stimulated by increased vascular flow through the gland, even in the absence

Control of Adrenocortical Hormone Secretion   Chapter 3

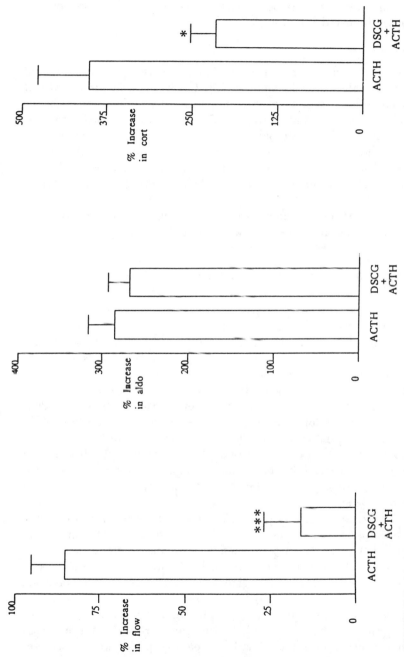

FIGURE 3.7 Administration of disodium cromoglycate, a mast cell stabilizer, to the intact perfused rat adrenal virtually abolishes the flow increment in response to ACTH (*** $p < 0.001$), and significantly attenuates the corticosterone response (* $p < 0.05$), while the aldosterone reponse is unaffected. These results strongly suggest a role for adrenal mast cells in the vascular response to ACTH, and further suggest that this vascular response is an essential component of the steroidogenic response to stimulation. (Figure reproduced with permission from ref. 32.)

of any known adrenocortical stimulant. In studies using the intact perfused rat adrenal, this effect is quite specific to zona fasciculata/reticularis function, and appears to be independent of such factors as oxygen supply, removal of end-product inhibition, and so on [33].

The vascular effect of ACTH is an important aspect of the action of this peptide: if the flow rate increase is blocked, the corticosterone response is significantly attenuated [32] (Fig. 3.7).

Long-term effects

Chronic ACTH administration results in an increase in both weight and blood content of the adrenal gland [34–36]. The increase in adrenal weight may be seen within 1 to 2 days after the onset of treatment, and is due to both cell proliferation and hypertrophy [37] (see Chap. 1). Corticotropin is required to maintain the steroidogenic capacity of the zona fasciculata/reticularis: following hypophysectomy the cytochrome-P450 content of adrenocortical mitochondria and microsomes falls dramatically [38]. Administration of ACTH on the other hand induces synthesis of the steroid biosynthetic enzymes [39,40].

The effects of ACTH on the morphology, vasculature and ultrastructure of the adrenal cortex are considered in detail in Chap. 1.

3.1.2 Control of adrenal androgen secretion

Introduction   On a quantitative basis, androgens are the major group of steroids secreted by the human adrenal, and they are secreted in varying amounts by the adrenal glands of many other species (see Chap. 9). Of the androgens, dehydroepiandrosterone sulphate (DHEAS) is the most abundant, although the range of androgens secreted includes androstenedione, 11β-hydroxyandrostenedione, testosterone, and dehydroepiandrosterone (DHEA) (see Chap. 2). The relative secretion rates of these steroids is shown in Table 2.2. While the total androgenic activity of the hormones secreted by the adrenal cortex is small by comparison with the testes, the adrenal androgens are converted peripherally to more potent androgens and also to oestrogens. In women, the adrenal is the major source of androgens, and following peripheral conversion in postmenopausal women, the only source of oestrogens.

Adrenal androgen secretion rises throughout childhood, from the age of about 5 years, paralleling the increase in adrenal size during childhood, and peaks at puberty (Fig. 3.8), a phenomenon termed adrenarche by Albright [42]. Secretion of these hormones remains high until 40 to 50 years of age, and this is followed by a gradual decline in secretion rate, which is termed adrenopause.

In the rat, guinea pig, and human, both the zona fasciculata and the zona reticularis produce androgens in addition to glucocorticoids [43], although the zona reticularis of humans and guinea pig preferentially secretes androgens, at least under basal conditions [44,45]. The sulphation of steroids such as dehydroepiandrosterone is specific to the zona reticularis, however [46,47].

Peptide stimulation of adrenal androgen secretion Similar to the other adrenal steroids, the secretion rate of adrenal androgens is increased by the administration of ACTH in vivo [48,49], although cells from the zona reticularis are relatively unresponsive to ACTH stimulation in vitro [44,50]. There are however, situations in which the secretion of androgens is dissociated from ACTH and cortisol. During adrenarche, for example, when adrenal androgen secretion rates are rising, cortisol secretion is unchanged [51]. This dissociation between androgens and glucocorticoids has been widely cited as evidence for the existence of a specific hormone which stimulates adrenal androgen secretion [52]. Several candidates have been put forward, the two most notable being prolactin and a novel hormone, termed the adrenal androgen stimulating hormone (AASH) [53]. Prolactin has little intrinsic steroid stimulating activity on its own, but can act to potentiate the effects of ACTH on adrenal androgen secretion by human adrenal cells in monolayer culture [54,55]. Nussdorfer's group has also shown that prolactin causes hypertrophy of the zona reticularis in gonadecto-mized rats, with a concomitant rise in plasma androgen levels [56]. Studies on hyperprolactinaemic patients have proved inconclusive, however, as adrenal androgen secretion does not consistently correlate with high-prolactin levels [54,57–59].

A postulated adrenal androgen stimulating hormone has been isolated from bovine pituitaries and characterized as a glycoprotein with apparent molecular weight around 60,000. This pituitary factor has been reported selectively to stimulate androgen secretion by canine adrenal cells in vitro, in the absence of any change in cortisol production [53]. This remains unconfirmed, however.

Intra-adrenal control of androgen secretion As it has proved so difficult to identify a specific adrenal androgen stimulating hormone, Anderson [60] has suggested that this elusive factor simply does not exist. He proposed instead that intra-adrenal steroid concentrations play an important role in controlling the maturation of the zona reticularis and determining the nature of the secreted steroid. According to this hypothesis, the normal adrenal growth and increased biosynthetic capacity seen during childhood, together with the centripetal nature of the adrenal blood flow (Chap. 1), result in the inner adrenocortical cells being exposed to increasing concentrations of cortisol, which cause morphological and functional changes in the innermost cells, comprising the zona reticularis, and favor the production of androgens (for a review see ref. 61). This theory is supported by reports that adrenocortical enzyme activities can be substantially altered in the presence of various steroids [62], but it does make certain assumptions about steroid movement within the adrenal, which have not been validated.

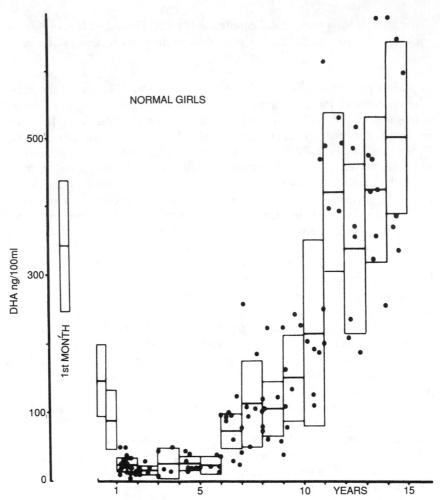

FIGURE 3.8A Adrenal androgen (dehydroepiandrosterone) secretion increases throughout childhood, and peaks at puberty. This peak is termed "adrenarche". (Figure reproduced with permission from ref. 41.) Data for boys is shown in Fig 3.8B

3.2 The control of zona glomerulosa function and aldosterone secretion

3.2.1 Introduction: Aldosterone and electrolyte balance

Aldosterone is the most potent mineralocorticoid produced by the mammalian adrenal cortex. In mammals it is secreted exclusively by the zona glomerulosa cells of the adrenal cortex, and exerts its major action on the distal convoluted tubule of the kidney nephron to promote the resorption of sodium from the urine (see Chap. 4). Physiologically, the most important stimulus to aldosterone secretion is altered electrolyte and fluid status, and

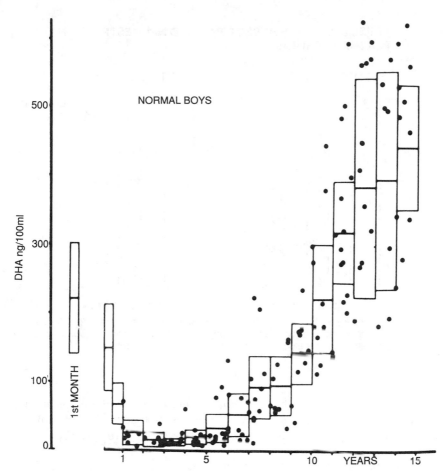

FIGURE 3.8B See Fig 3.8A for legend.

experimentally dietary sodium restriction or extracellular fluid volume reduction (by controlled hemorrhage or use of diuretics) is commonly used to stimulate aldosterone secretion (for a review see ref. 63). Restriction of dietary sodium intake causes hypertrophy of the zona glomerulosa, with an increased capacity for aldosterone secretion, resulting in a rise in the circulating concentration of aldosterone without affecting circulating corticosterone or cortisol concentrations (Fig. 3.9).

The mechanism of control of aldosterone secretion is not, however, simply caused by a direct effect of reduced plasma sodium on the zona glomerulosa cells: There is instead a complex interaction between several factors, both stimulatory and inhibitory, to achieve a balanced hormone secretion.

The renin-angiotensin system Arguably, the most important effect of reduced sodium balance is the activation of the renin-angiotensin system. Sodium depletion causes parallel rises in plasma renin activity, angiotensin

EFFECT OF HYPOPHYSECTOMY AND Na⁺ RESTRICTION ON PLASMA STEROIDS

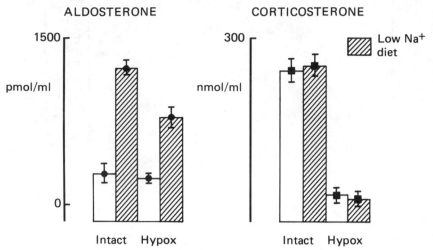

FIGURE 3.9 Dietary sodium restriction in the rat causes an increase in circulating concentrations of aldosterone without affecting corticosterone levels. Hypophysectomy significantly attenuates the aldosterone response to sodium depletion.

II and aldosterone in both human subjects and experimental animals, and the increase in aldosterone secretion can be totally blocked by an inhibitor of angiotensin II formation such as captopril (see Fig. 3.10) [64,65].

Renin and the production of the angiotensins

Renin is a proteolytic enzyme secreted by the "juxtaglomerular" cells which are located in the afferent arteriole of the kidney. There are several factors known to stimulate the release of this enzyme; a decreased renal perfusion pressure, decreased sodium delivery to the distal convoluted tubule, and sympathetic stimulation of the juxtaglomerular cells [66,67]. Renin is secreted directly into the bloodstream and has a half life of about 15 minutes in plasma, being degraded mainly in the liver.

Renin substrate, more commonly called angiotensinogen, is found in the α_2 globulin fraction of plasma proteins. It is secreted by the liver into the circulatory system, where it is cleaved by renin to form a decapeptide, angiotensin I, which has relatively little biological activity (Fig. 3.11). A converting enzyme, found predominantly in the lungs, acts on angiotensin I, cleaving a leucine and a histidine residue, to form an octapeptide, angiotensin II. The structure of angiotensin II is illustrated in Figure 3.10, together with the mechanism of formation of this peptide from angiotensinogen. Angiotensinogen may also be directly converted to angiotensin II by the action of a serine protease, tonin, which is found in the

Angiotensinogen

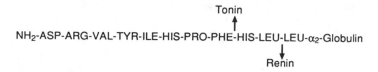

Angiotensin I

Angiotensin II

NH$_2$-ASP-ARG-VAL-TYR-ILE-HIS-PRO-PHE
↑
Aminopeptidase A

Angiotensin III

ARG-VAL-TYR-ILE-HIS-PRO-PHE

FIGURE 3.10 Primary structure and formation of the angiotensins. Circulating angiotensinogen, secreted by the liver, is the substrate for renin, an enzyme secreted by the kidney, and converted to angiotensin I. A "converting enzyme" found predominantly in the lung cleaves two residues from angiotensin I to yield angiotensin II. Captopril inhibits the activity of this enzyme. Angiotensin II may be further processed by aminopeptidase A to give angiotensin III.

venous effluent of the submaxillary gland [69]. Tonin appears to be active in the rat in vivo, but its physiological significance is unclear [70].

The half life of angiotensin II is very short, around 2 minutes in plasma, and it is rapidly broken down by aminopeptidases located in various tissues, particularly red blood cells, but also in target tissues such as the adrenal [71]. One of these aminopeptidases, called aminopeptidase A, cleaves the N-terminal aspartate residue from angiotensin II to form angiotensin III (Fig. 3.10).

Intra-adrenal production of angiotensin II

The kidney is not the exclusive source of renin, and the presence of reninlike activity has been demonstrated in a variety of tissues, including the adrenal gland of several mammalian species [72–75]. The physiological role of intra-adrenal renin has not been established, but it has been shown that the concentration of this enzyme in the zona glomerulosa of the rat varies with the electrolyte status of the animal [76] and, moreover, increases

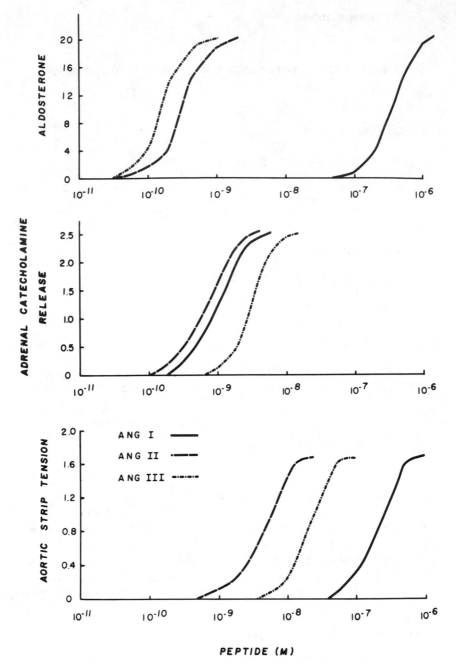

FIGURE 3.11 Relative potencies of angiotensins I, II, and III, in stimulating aldosterone secretion (in ng/10^5 cells/hr) by rabbit zona glomerulosa cell suspensions (top panel), catecholamine release (in μg) from the retrogradely perfused rabbit adrenal (center panel) and response (in g of developed tension) of rabbit aortic strip (bottom panel). (Figure reproduced with permission from ref. 68.)

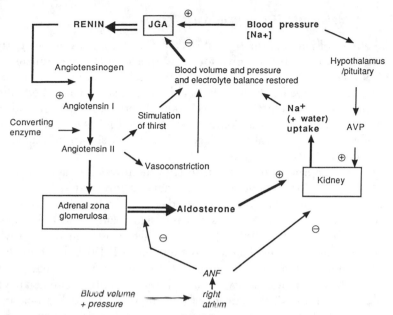

FIGURE 3.12 The role of the renin-angiotensin-aldosterone system in the regulation of blood pressure and electrolyte balance. Key: ANF = atrial natriuretic factor, JGA = juxta-glomerular apparatus, AVP = arginine vasopressin, + = stimulatory pathway, − = inhibitory pathway.

about 16-fold in response to nephrectomy [77]. The production of this enzyme activity at sites other than the kidney may partly account for the observation that aldosterone secretion may still be stimulated by dietary sodium depletion in nephrectomized rats [78]. Similar to renin, angiotensin converting enzyme activity has been found within the adrenal gland, and in the rat is located in the capillary endothelial cells of the adrenal cortex [79].

Extra-adrenocortical effects of angiotensin II

Angiotensin II is a powerful vasoconstrictor, being around 40 times more potent than noradrenaline. A single intravenous dose of angiotensin II causes a rise in mean arterial pressure within 10 seconds [80]. Angiotensin II also acts on the adrenal medulla to stimulate the release of catecholamines, enhancing its hypertensive actions [68] (Fig. 3.11). This peptide has a direct action on the kidney where it may regulate renal blood flow [81], and also acts to inhibit the secretion of renin [82]. Circulating angiotensin II also stimulates the subfornical organ to stimulate thirst [83]. Thus, angiotensin II initiates a complex series of physiological events which work together to restore plasma volume, pressure, and sodium concentration (Fig. 3.12). However, probably the most important aspect of the action of angiotensin II is its effect on the adrenal cortex.

Actions of angiotensin II on the adrenal cortex

The ability of angiotensin II to stimulate aldosterone secretion, both in vivo and in vitro, has been shown for many species including humans [84], rat [85,86], dog [87,88], sheep [89], and rabbit [22]. An intravenous infusion of angiotensin II specifically stimulates aldosterone secretion in both sheep and humans, without affecting the secretion of cortisol (Fig. 3.13) and thus appears to be a specific zona glomerulosa stimulant [89,90]. A direct effect of angiotensin II on adrenal steroidogenesis was first shown by Kaplan and Bartter [91].

While the importance of angiotensin II in the physiological regulation of aldosterone secretion is generally acknowledged for most species, it was several years before this was confirmed in the rat. Administration of angiotensin II to anesthetized rats did not significantly affect the plasma aldosterone concentration [92–94], and no effect was seen in vitro when adrenal segments were incubated with angiotensin II [91,95], although the chronic administration of angiotensin II to rats in vivo enhanced the aldosterone biosynthetic capacity of adrenals subsequently incubated in vitro [95,96]. It was not until the advent of enzyme-dispersed cell preparations that a direct effect of angiotensin II on the rat adrenal could be demonstrated [97]. This was later confirmed in vivo using physiological concentrations of the peptide, when it was also shown that the effect of angiotensin II was severely attenuated by anesthesia [85].

The use of enzyme-dispersed preparations of rat adrenocortical cells has enormously facilitated the study of adrenocortical function, and the use of purified cell populations has revealed a direct action of angiotensin II on the zona glomerulosa [86]. Pure angiotensin II does not stimulate steroid secretion by cells prepared from the inner zones of the rat adrenal [98]. There are also some species differences in responsiveness of zona fasciculata cells; cortisol production is stimulated by angiotensin II in bovine and human adrenals [91,99,100], but the zona fasciculata generally is less responsive to this peptide than the zona glomerulosa.

Angiotensin II receptors

Angiotensin II exerts its effects on the cells of the zona glomerulosa via an interaction with specific receptors located in the plasma membrane [71,101]. Two classes of angiotensin II receptor have been identified in the plasma membrane of adrenocortical cells: a high-affinity receptor, $Kd10^{-9}$ mols/l and a lower affinity receptor, $Kd10^{-8}$ mols/l. There is, usually, a predominance of the higher affinity receptor, and the low-affinity sites are thought to be of minor importance [102]. Although there are specific receptors for angiotensin II in the inner adrenocortical zones, they are much less numerous than in the zona glomerulosa [100,103]. The ratio of angiotensin II receptor content of zona glomerulosa to inner zones varies between different species (Fig. 3.14) [104].

In addition to activating the renin-angiotensin system, changes in electro-

Dose Response Relationships Between Angiotensin II and Corticosteroids in Normal Sodium Replete Man

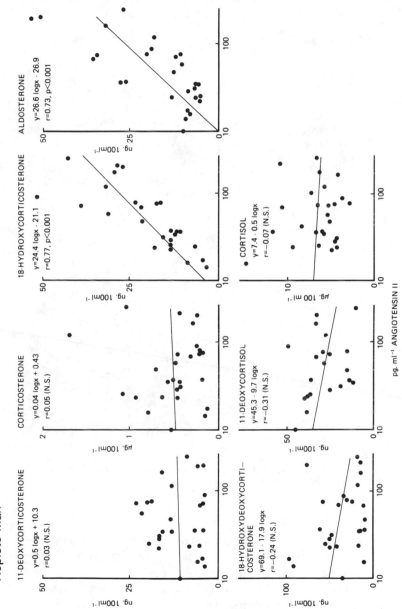

FIGURE 3.13 Angiotensin II infusion into normal human specifically stimulates zona glomerulosa function, resulting in an increase in the plasma concentrations of the zona glomerulosa products, 18-hydroxycorticosterone and aldosterone. Plasma concentrations of the other corticosteroids are not affected. (Figure reproduced with permission from ref. 90.)

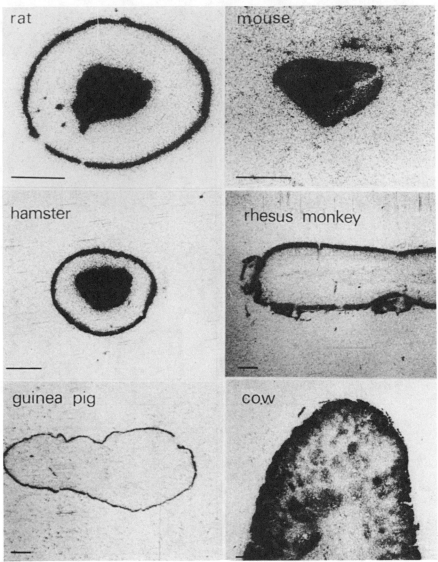

FIGURE 3.14 Autoradiograms of angiotensin II receptor sites in adrenals of different species. The highest concentrations are seen in the zona glomerulosa and the medullary chromaffin tissue of the rat and hamster, but there is species variation. (Figure reproduced with permission from ref. 104.)

lyte balance also modify the adrenocortical response to circulating angiotensin II, increasing both the concentration and affinity of angiotensin II receptors. These changes are reflected in the increased sensitivity of the adrenal cortex to angiotensin II stimulation following dietary sodium depletion, with a shift to the left of the dose-response curve (Fig. 3.15) [105]. The amplitude of

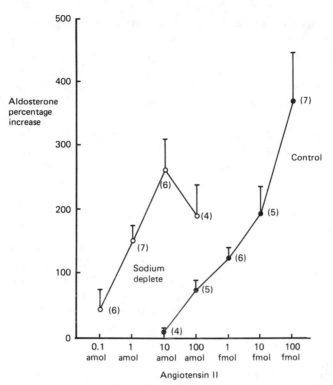

FIGURE 3.15 Dietary sodium depletion causes an increase in sensitivity of the zona glomerulosa to angiotensin II stimulation, shown by a shift to the left of the dose response curve. These data were obtained using the isolated perfused rat adrenal in situ. Numbers of determinations for each point are shown in parentheses. Stimulation was significant at a dose of 100 amol in the control animals ($p < 0.05$), and at 1 amol in the sodium deplete group ($p < 0.01$). While the percent increase in aldosterone secretion in response to angiotensin II was lower in the sodium deplete group, this group had a much higher basal rate of aldosterone secretion than the controls, and the maximum rate of aldosterone secretion was also higher in the sodium deplete group. (Figure reproduced with permission from ref. 105.)

the aldosterone response to angiotensin II is also greatly increased by sodium depletion, reflecting the increased biosynthetic capacity of the zona glomerulosa [106–108].

These changes may be mediated by angiotensin II itself as angiotensin II has been reported to up-regulate its own receptors. For example, an infusion of angiotensin II into hypophysectomized rats, sufficient to cause a threefold increase in plasma angiotensin II concentration, caused a twofold increase in receptor number over the 6-day infusion period [109]. Factors which cause a decrease in angiotensin II receptor concentration include chronic ACTH administration and dietary sodium loading [110,111].

Angiotensin III

In most species angiotensin II appears to be the most physiologically important stimulus to aldosterone secretion, although another peptide in the angiotensin family, angiotensin III, may be important in some species. This peptide is a potent stimulus to aldosterone secretion in several species, including the dog [112], rat [113], cat [114], sheep [115], and humans [116]. In most cases, the two angiotensins appear to be equipotent in their aldosterone-stimulating activities (Fig. 3.11), although angiotensin III has much less pressor activity than angiotensin II in both sheep and humans [68,115,117].

The ratio of circulating angiotensin II to angiotensin III shows interspecies variation; in the rat about 58 percent of the angiotensin immunoreactivity is angiotensin III, while this figure is only about 12 percent in humans and dogs [118,119]. During dietary sodium deficiency, however, while both angiotensin II and III are elevated, in the dog there is a relatively greater increase in angiotensin III to around 25 percent of the total angiotensin immunoreactivity [119].

It has been proposed that angiotensin II is inactive and needs to be converted to angiotensin III at the site of the receptor [120,121] but this is no longer thought to be the case, although there is evidence for aminopeptidase A activity in the adrenal gland of the rat [79]. Thus all the mechanisms exist within the adrenal for the production of both angiotensin II and angiotensin III, and it is, therefore, possible that the adrenal gland may, to a degree, itself regulate aldosterone secretion.

The role of the pituitary gland   Although the renin-angiotensin system is probably the most important mechanism for the control of aldosterone secretion, it is by no means exclusive, and the pituitary gland appears to have a significant role in the control of zona glomerulosa function, although this role is not as clear as for the control of zona fasciculata function. Thus, while hypophysectomy causes a 95 percent decrease in basal glucocorticoid secretion it has a more variable effect on basal aldosterone output [63]. Although basal aldosterone levels may be unchanged by hypophysectomy or hypopituitarism, there is evidence that a pituitary factor is necessary to maintain the adrenal responsiveness to various stimulators of aldosterone secretion, such as infusion of exogenous angiotensin II [122]. It is clear that an intact pituitary gland is also required for a full response to changes in electrolyte balance: the aldosterone response to sodium depletion is significantly impaired both in hypophysectomized rats and in human subjects with pituitary insufficiency [123]. Unlike the renin-angiotensin system, there is no evidence for increased secretion of any pituitary peptide in response to sodium depletion, and many studies have investigated the aldosterone-stimulating properties of various pituitary peptides, and also looked for changes in adrenocortical sensitivity in response to sodium depletion.

Corticotropin

While ACTH is undoubtedly the major factor controlling inner zone function, its role in the physiological control of aldosterone secretion is unclear. Acute infusion of ACTH in vivo causes a rise in circulating aldosterone in most species (see Chap. 9). In in vitro studies ACTH was also shown to cause an increase in the rate of aldosterone secretion by adrenal capsular tissue [91,124]. It was not clear at first whether this represented a direct action of ACTH on the zona glomerulosa as early studies in vitro used adrenocortical slices, contaminated with inner zone cells, rather than pure zona glomerulosa preparations. It was, therefore, thought possible that ACTH was stimulating the production of aldosterone precursors from the zona fasciculata cells present, rather than exerting a direct effect on the zona glomerulosa. However, the use of purified zona glomerulosa cells confirmed a direct action of ACTH on aldosterone synthesis [86,125]. In preparations of collagenase-dispersed rat zona glomerulosa cells, ACTH is the most effective stimulus to aldosterone secretion, giving a maximal response about three times greater than the response to any other zona glomerulosa stimulant [97,98,126].

As acute administration of ACTH clearly stimulates aldosterone secretion, this peptide was an obvious candidate for the pituitary factor required in sodium depletion. There have been reports of an increased adrenocortical sensitivity to ACTH in sodium deplete animals, with a shift to the left of the dose-response curve [105,127], although in other preparations the response is unchanged [128,129].

Chronic administration of ACTH to sodium deplete animals, however, causes a decrease in the production of aldosterone [130], and an infusion of ACTH is not able to restore the adrenal responsiveness in hypophysectomized rats [131]. Thus it appears most unlikely that ACTH is the pituitary factor required to maintain adrenocortical responsiveness to sodium depletion.

Other pro-opiomelanocortin derived peptides

This group of peptides, which includes the melanocyte-stimulating hormones (MSH peptides), β-endorphin, and CLIP, is synthesized in the pituitary gland from the same precursor molecule as ACTH (Fig. 3.2). The structures of some of the MSH peptides are illustrated in Fig. 3.3. In most vertebrates these peptides are produced by the intermediate lobe of the pituitary, but in adult humans, where the intermediate lobe is absent, the anterior pituitary is the site of α-MSH and desacetyl α-MSH production [132].

It has been shown that low concentrations of α-MSH stimulate aldosterone secretion by rat zona glomerulosa cells incubated in vitro, but have no effect on the zona fasciculata [133,134]. An infusion of α-MSH into intact rats has been reported to cause an acute increase in the circulating concentration of aldosterone, and chronically, to stimulate growth and steroidogenic capacity of the zona glomerulosa [135,136]. A possible role for the MSH peptides in the control of adrenocortical function in humans, is suggested by the recent finding that disorders of the pituitary-adrenal axis are associated with greatly altered pituitary concentrations of α-MSH and desacetyl α-MSH [137]. Further evidence is provided by the observation that desacetyl α-MSH, at

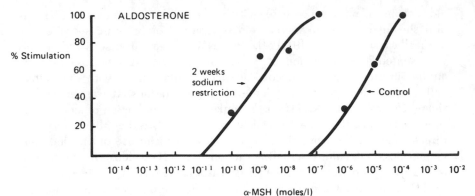

FIGURE 3.16 Dietary sodium depletion causes a marked increase in adrenocortical sensitivity to α-MSH, such that the threshold concentration falls within the normal circulating concentration of this peptide. Data were obtained using a collagenase-dispersed preparation of zona glomerulosa cells incubated in vitro. (Figure reproduced with permission from ref. 129.)

concentrations close to plasma levels, stimulate aldosterone secretion by human adrenocortical cells in vitro [138].

Although concentrations of α-MSH higher than those found in circulating blood are needed to stimulate adrenals from sodium replete animals, a period of dietary sodium depletion causes a marked increase in adrenocortical sensitivity to α-MSH such that the threshold concentration required for stimulation is within the physiological concentration range of the peptide (Fig. 3.16) [105,129,139]. A persuasive argument for a role for α-MSH in the adrenocortical response to sodium depletion comes from the data presented by Shenker and co-workers who demonstrated that an infusion of α-MSH into hypophysectomized rats selectively restored the aldosterone response to sodium depletion [131].

There have also been reports of a specific zona glomerulosa stimulation by β-MSH which, like α-MSH is enhanced by prior sodium depletion [128,140]. Another peptide, pro-γ-MSH potentiates the adrenal response to ACTH but is ineffective alone [134,141,142]. It has been suggested that β-lipotropin may stimulate aldosterone in the rat [143], but this effect may be caused by contamination of the preparation with ACTH [134,144]. Experiments with β-lipotropin and human adrenal cells have been inconclusive [145,146]. β-endorphin has been shown to stimulate aldosterone secretion in hypophysectomized, nephrectomized dogs [147] although this peptide has no effect on human adrenal cells [146], and experiments with rats have been inconclusive [148,149].

Direct effects of sodium on aldosterone secretion   Although the major effects of altered sodium balance on zona glomerulosa function are indirect, by activation of the renin-angiotensin system, and so on, (Fig. 3.12), there is evidence for a direct action of sodium on aldosterone secretion. In many reports, it is not possible to distinguish the effects of changes in extracellular sodium concentration from the effects of altered osmolarity, which are very similar [150]. An acute effect of sodium on aldosterone secretion was shown in rats by using peritoneal dialysis with glucose, to reduce the plasma sodium concentration while maintaining a constant osmolarity. A decrease in plasma sodium from 150 to 130 mmol/l is sufficient to stimulate aldosterone secretion [151]. Increases in plasma sodium concentration, induced by infusion of sodium chloride, cause a reduction in the rate of aldosterone secretion, but do not reverse the effects of chronic sodium depletion [152].

Role of potassium Sodium depletion is not the only electrolyte imbalance to stimulate aldosterone secretion; the zona glomerulosa is also sensitive to changes in potassium concentration. In many respects these two electrolytes are inversely related: sodium depletion is often associated with raised plasma potassium levels, and experimentally, dietary potassium loading may be used to stimulate aldosterone secretion [153,154]. In humans, an increased potassium intake, from 40 to 200 mmol/day for 8 days, caused a five-fold increase in plasma aldosterone concentration, while plasma potassium concentration was only slightly raised, from 4.0 to 4.2 mmol/l [153].

A direct effect of potassium ions on the adrenal cortex was first shown in vitro, where increasing concentrations of potassium ions, from 5.6 to 8.6 mmol/l had a stimulatory effect on aldosterone secretion [124]. Later it was shown that an infusion of potassium chloride or potassium sulphate selectively stimulates aldosterone secretion in hypophysectomized, nephrectomized dogs [155], normal humans [156], and other species [16,89]. In isolated rat zona glomerulosa cells, increasing the potassium ion concentration of the incubation medium from 3.6 mmol/l (the normal plasma potassium concentration in the rat) to 8.4 mmol/l causes a dose-dependent increase in aldosterone secretion, with a response threshold about 4.0 mmol/l [157–159]. At higher potassium concentrations aldosterone secretion declines [160].

The most physiologically important effect of potassium ions on the adrenal cortex is probably the increased sensitivity to angiotensin II caused by small increases in plasma potassium concentration [111]. During experimental potassium loading, however, the plasma potassium concentration of experimental animals may exceed the stimulation threshold for aldosterone secretion measured in vitro [154], and in these cases it is likely that potassium has a direct action on the zona glomerulosa.

Atrial natriuretic peptides Most of the mechanisms for the control of aldosterone secretion in relation to electrolyte balance involve the stimulation of zona glomerulosa function. There is, however, one mechanism which is activated by high-plasma sodium levels, and raised blood volume, and which acts to inhibit aldosterone secretion, thus permitting natriuresis and diuresis.

A number of peptides with potent diuretic, natriuretic and vasodilatory effects have recently been isolated from atrial extracts [161]. These are termed "atrial natriuretic peptides" (ANPs). ANPs inhibit vasopressin release

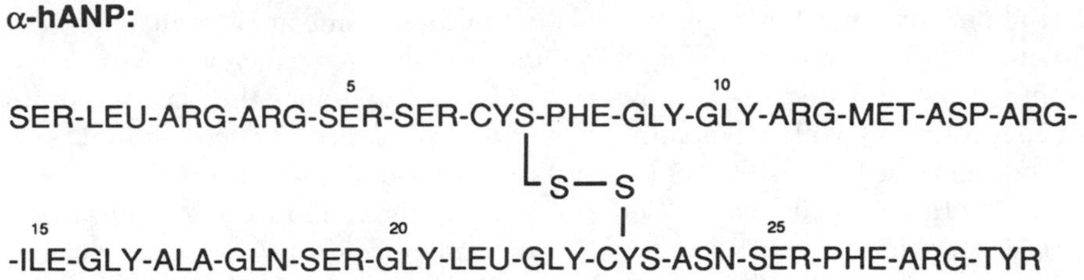

α-hANP:

$$
\begin{array}{c}
5 10 \\
\text{SER-LEU-ARG-ARG-SER-SER-CYS-PHE-GLY-GLY-ARG-MET-ASP-ARG-} \\
\llcorner_{S-S} \\
| \\
15 20 25 \\
\text{-ILE-GLY-ALA-GLN-SER-GLY-LEU-GLY-CYS-ASN-SER-PHE-ARG-TYR}
\end{array}
$$

FIGURE 3.17 Primary structure of alpha-human atrial natriuretic peptide (ANP). This peptide is one of a family of related peptides secreted principally by the right cardiac atrium, which are collectively termed "atrial natriuretic factor" (ANF). (Modified from ref. 162.)

from the pituitary, and act directly on the kidney to inhibit renin release and promote salt and water loss. ANPs also inhibit aldosterone secretion.

Structure and secretion of ANP

Atrial natriuretic peptides are synthesized and stored in mammalian atrial muscle cells. In the rat, the major storage form of ANP is a prohormone of 126 amino acids, termed atriopeptigen, which is stored linked to a leader sequence comprising a further 25 residues. Proteolytic cleavage of atriopeptigen yields a range of low-molecular weight peptides with natriuretic activity. The human peptide, termed alpha human atrial natriuretic peptide, has 28 amino acid residues (Fig. 3.17), while at least four peptides have been isolated from rat atria. One of these peptides, termed cardionatrin, has 28 residues and corresponds to the human peptide. This is now thought to be the predominant form in the circulation. For reviews see refs. [163] and [164].

Atrial natriuretic peptides are released from storage granules in mammalian atria in response to a variety of stimuli, including high plasma sodium concentration, atrial tachycardia, water immersion, and atrial stretching caused by volume expansion. (For a review, see ref. 165.)

Actions of ANP on the zona glomerulosa

Infusion of ANP in dogs and humans causes a decrease in the plasma levels of aldosterone [166,167]. While in part this almost certainly reflects a decrease in plasma renin activity, infusion of ANP in vivo also inhibits the aldosterone response to a simultaneous infusion of angiotensin II [168,169]. Experiments using isolated zona glomerulosa cells in vitro have confirmed a direct action of ANP on aldosterone secretion [170]. This effect is specific for aldosterone secretion as an infusion of ANP in dogs and humans has no effect on cortisol or corticosterone secretion [166,167], and ANP does not affect corticosterone secretion by isolated zona fasciculata cells incubated in vitro [168]. In rat zona glomerulosa cells, both basal and stimulated aldosterone secretion rates may be attenuated by ANP [168,171], although in bovine adrenal cells basal steroid secretion is unaffected [170].

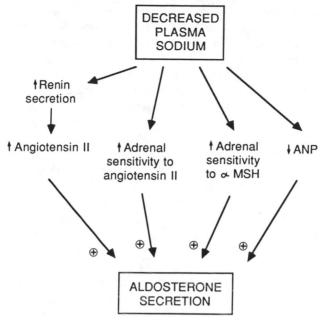

FIGURE 3.18 Summary of the mechanisms for the control of aldosterone secretion in response to changes in sodium balance. α-MSH = α-melanocyte stimulating hormone, ANP = atrial natriuretic peptide.

The mechanism of ANP action is unclear, although it is believed to act through specific receptors and blocks the conversion of cholesterol to pregnenolone [172,173]. ANP does not appear to affect the binding of other zona glomerulosa stimulants [174].

The mechanisms for the control of aldosterone secretion in response to changes in sodium balance are summarized in Figure 3.18.

3.2.2 Other factors which influence aldosterone secretion

In addition to the factors which exert a physiological control over aldosterone secretion in response to altered electrolyte status, there are many other factors which have been reported to influence aldosterone secretion. These factors include inhibitors as well as stimulators of zona glomerulosa function. Several of these substances are known to be neurotransmitters or neuromodulators in other systems, thus posing questions about the functional significance of the innervation of the adrenal cortex (Chap. 1). Of these agents, most attention has been given to serotonin and the catecholamines, which will, therefore, be considered in some detail. A brief summary of the other factors is given, although in general their physiological relevance remains unclear.

Serotonin   The aldosterone-stimulating effects of this monoamine are well documented in the literature [97,175–177]. In vitro low concentrations of serotonin have a specific stimulatory action on the zona glomerulosa, causing a rise in aldosterone secretion [178]. In an intact gland preparation, however, the effects of serotonin are not confined to the zona glomerulosa, and serotonin stimulates inner zone function in the dog adrenal [179] and also in the rat adrenal [32]. Serotonin is thought to act through specific receptors in the zona glomerulosa, which have been characterized as serotonin type 2 receptors [180].

The physiological significance of this action of serotonin has been questioned, however, as it is not possible to reproduce these effects in vivo, probably because of the rapid uptake of serotonin by platelets in blood [181]. It has been reported that infusion rates of serotonin sufficient to stimulate aldosterone secretion are lethal to the animal [3]. It therefore seemed unlikely that serotonin could have any physiological role in the control of adrenocortical function unless it was released locally. Recently it has been suggested that serotonin may be locally released, from the mast cells which have been identified in the connective tissue capsule surrounding the adrenal gland (Fig. 3.19), and may in addition have a role in the control of blood flow through the adrenal gland [32].

Studies in the frog have shown that serotonin can stimulate adrenal steroidogenesis, and moreover, is found in the chromaffin tissue intermingled with the steroidogenic tissue, so that a paracrine action of serotonin in the frog adrenal is quite likely [182].

Catecholamines   There are several reports of catecholamines stimulating adrenocortical steroid secretion, in addition to the many reports of the inhibitory effects of dopamine. It has been found that both adrenaline and noradrenaline stimulate the release of aldosterone from bovine zona glomerulosa cells maintained in culture for 3 days. Use of β blockers has shown this to be a beta-adrenergic effect [183]. The effects of catecholamines were not seen when freshly prepared bovine adrenocortical cell suspensions were used [183,184], although a response to catecholamines is seen in freshly dispersed rat zona glomerulosa cell suspensions [185].

There is a small amount of catecholamine which can be extracted from the adrenal cortex, comprising between 2 and 4 percent of the total adrenal content of catecholamines in the rat [186,187]. As the ratio of adrenaline to noradrenaline is the same in the cortical and medullary extracts, it has been proposed that the cortical catecholamines may derive from the medulla [187], and it has, therefore, also been proposed that the adrenal medulla may exert some control over adrenocortical function.

In species such as the frog, where the cortical and medullary cells are adjacent, it is possible to envisage a role for medullary products in the control of adrenocortical function. In higher vertebrates such as rat, cow, and humans, where the arrangement of the adrenal gland is such that the medulla is enclosed by cortical tissue and with the rapid centripetal blood

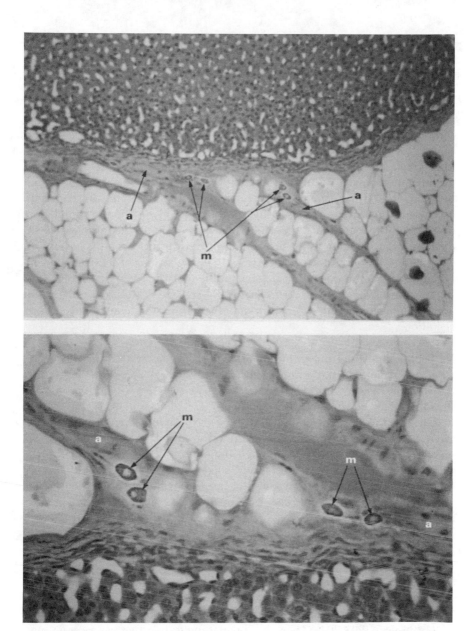

FIGURE 3.19 Toluidine blue staining of 3 μm sections through a perfusion-fixed rat adrenal gland clearly reveals the presence of mast cells (m) in close proximity to adrenal arterioles (a) as they penetrate the connective tissue capsule of the adrenal gland. (Micrograph reproduced with permission from ref. 32.)

3.2 The control of zona glomerulosa function and aldosterone secretion **91** flow carrying blood from the cortex to the medulla (see Chap. 1), it is difficult to imagine a situation whereby products of the medulla could be carried backwards, against the blood flow, to affect the most outer part of the cortex, the zona glomerulosa. A recent report has put forward an alternative proposal, suggesting that part of the adrenocortical innervation may arise in the medulla, and in this way the adrenal medulla may affect cortical function [188].

Acetylcholine   Cholinergic stimulation of inner zone function is well recognized in the bovine adrenal [189,190]. More recently, acetylcholine has been shown to stimulate aldosterone secretion by perifused frog adrenocortical cells, acting through a muscarinic receptor [191]. Cholinergic stimulation of aldosterone secretion by bovine zona glomerulosa cells also has been demonstrated [192], and specific cholinergic muscarinic receptors have been identified in the bovine adrenal cortex [193]. The actions of acetylcholine on adrenocortical steroidogenesis are compatible with the reports of cholinergic innervation of the adrenal cortex in several species [194,195], and suggest a possible role for adrenocortical innervation in the control of steroid secretion.

Vasoactive intestinal polypeptide (VIP)   VIP has been identified in nerve terminals supplying the adrenal capsule and zona glomerulosa [196] (Fig. 3.20, see also Chap. 1). VIP stimulates aldosterone secretion when infused into the intact rat, apparently potentiating the effects of ACTH, as VIP is inactive in the absence of ACTH [197]. Experiments in vitro have not been conclusive: one group demonstrated a stimulatory effect of VIP on aldosterone secretion [188], although other groups have failed to show a direct effect of this peptide on rat adrenocortical cells in vitro [198].

In amphibia, VIP has been found in the chromaffin tissue of the adrenal gland, and stimulates steroid secretion by perifused interrenal tissue in vitro [199].

Vasopressin and oxytocin   Both these peptides have been immunocytochemically located in the adrenal cortex of several species, including the Brattleboro rat, which is unable to produce vasopressin in its pituitary [200–202]. The acute administration of vasopressin stimulates aldosterone secretion in the intact perfused rat adrenal preparation, but is much less potent in vitro [203,204]. There has been a suggestion that vasopressin may act synergistically with ACTH [205]. Vasopressin appears to act through specific receptors on the adrenocortical cells, which are linked to the phosphatidylinositol second messenger system (see Sec. 3.3) [203,206]. Oxytocin stimulates aldosterone secretion by the intact perfused rat adrenal (Fig. 3.21), but has no effect on dispersed cell preparations [204].

Aldosterone stimulating factor   In 1975, a factor was isolated from normal human urine which was found to cause hypertension in rats [207]. This

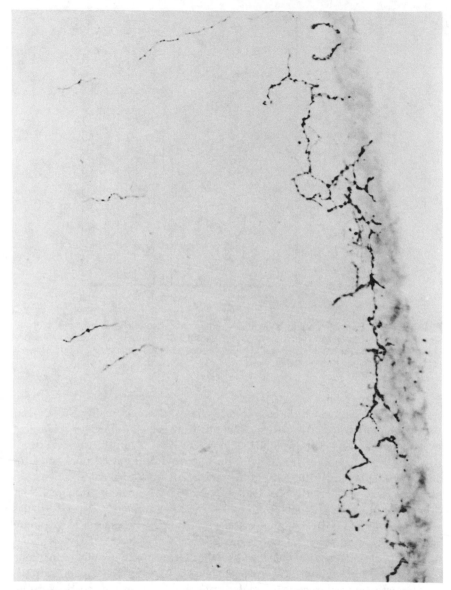

FIGURE 3.20 Vasoactive intestinal polypeptide (VIP) immunoreactivity in the rat adrenal gland. VIP-staining fibers are particularly concentrated in the zona glomerulosa. Final magnification ×200. (Photograph kindly supplied by M.A. Holzwarth.)

factor, identified as a glycoprotein of 26,000 molecular weight, stimulates aldosterone secretion and has been termed "aldosterone stimulating factor." In vitro, ASF is equipotent with angiotensin II [208]. These studies await confirmation by other groups. A review of the findings on ASF was recently published [209].

3.2 The control of zona glomerulosa function and aldosterone secretion **93**

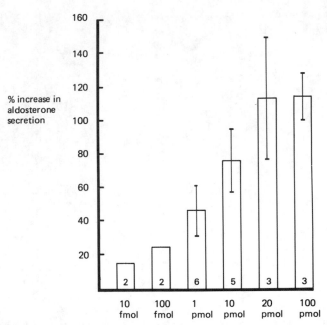

FIGURE 3.21 Oxytocin causes a dose-dependent increase in the rate of aldosterone secretion by the isolated, perfused rat adrenal gland. Numbers of experiments are shown in bars. Stimulation was significant at a dose level of 1 pmol ($p < 0.01$). (Figure reproduced with permission from ref. 204.)

Somatostatin   Somatostatin is a peptide of 14 amino acid residues, secreted by the hypothalamus where its main function is to inhibit the release of growth hormone by the pituitary, but it also acts to inhibit the release of several other hormones, notably gut hormones. In the rat adrenal, somatostatin acts specifically to inhibit the angiotensin II-stimulated release of aldosterone [210], and this observation has also been made in humans [211]. Somatostatin exerts this effect at concentrations around 10^{-10} mol/l. It has no effect on basal steroidogenesis or on ACTH stimulated aldosterone secretion, even at high concentrations.

When administered to intact rats, somatostatin causes atrophy of the zona glomerulosa and significantly lowered plasma aldosterone concentrations, without affecting either zona fasciculata morphology or plasma corticosterone levels [212]. There are specific somatostatin receptors in the zona glomerulosa of several species, including the rat, cow, and guinea pig (Fig. 3.22) [104,213,214]. However, somatostatin is without effect on steroid secretion by the interrenal gland of the frog [215].

The discovery of high levels of somatostatinlike immunoreactivity in the connective tissue capsule surrounding the adrenal gland led Aguilera and co-workers to suggest that this peptide may be released from local stores in the capsule, and they proposed that the action of somatostatin may be a local mechanism for the fine control of aldosterone secretion [210]. More

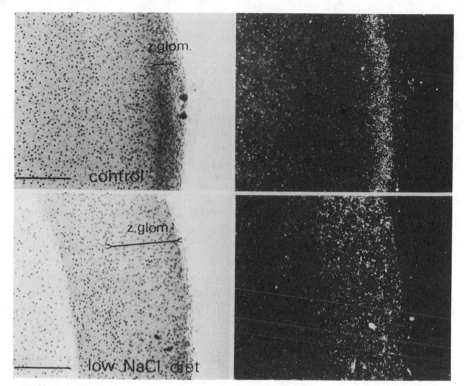

FIGURE 3.22 Somatostatin receptors in the rat adrenal zona glomerulosa. Dietary sodium restriction causes a marked increase in the width of the zona glomerulosa (left), with a concomitant increase in the number of somatostatin receptor sites (right). (Figure reproduced with permission from ref. 104.)

recent studies have failed to support this theory, as only low levels of somatostatinlike immunoreactivity could be found [216].

Dopamine It is thought that aldosterone secretion may be under tonic inhibition by dopamine. This theory was proposed by Carey et al., [217], following the earlier observation that administration of a dopamine agonist, bromocriptine, inhibits the stimulation of aldosterone secretion by frusemide, a diuretic [218]. It has also been shown that administration of metoclopramide, a dopamine antagonist, causes a rise in plasma aldosterone in humans [219]. In the rat there are reports that metoclopramide stimulates aldosterone secretion [220], but this was by no means a universal finding [221–223]. While most studies have been performed using either agonists or antagonists of dopamine action, there are some reports of a direct inhibitory action of high concentrations of dopamine on aldosterone secretion, and specific receptors for dopamine have been characterized in the zona glomerulosa [224].

It is known that urinary dopamine excretion is proportional to dietary sodium intake, and it has been suggested that the rise in aldosterone secretion during sodium depletion may be accounted for by removal of dopamine inhibition [225]. The adrenocortical content of dopamine does not change, however, during changes in dietary sodium intake, and so this mechanism seems unlikely [186]. Recent reports indicate that dopamine increases the rate of renal clearance of angiotensin II, and may act to inhibit aldosterone secretion in vivo by decreasing the effective biological half-life of its major stimulant angiotensin II rather than having a direct effect on the adrenal [186,226]. An alternative view is that dopamine may exert its primary action on the pituitary, modulating circulating concentrations of α-MSH [227,228].

Prostaglandins   Several studies have reported an effect of prostaglandins on aldosterone secretion. In most in vitro adrenocortical preparations, the E series prostaglandins have been shown to stimulate zona glomerulosa function [229–231] while the F series are reported to be either without effect [229,231], or to inhibit aldosterone secretion [232]. Studies in vivo are complicated by the observation that prostaglandins may influence renin secretion [233], but there is evidence that prostaglandin E_2 may stimulate aldosterone secretion in the absence of a rise in plasma renin activity [234]. There are also a few reports on the effects of A, B, H, and I series prostaglandins, but the results are difficult to interpret because of the variability of the response between different preparations (for a review see ref. 63).

Circadian rhythm of aldosterone secretion   Aldosterone, like cortisol, is secreted episodically with a clear rhythm over a 24-hour period. Aldosterone levels in humans peak in the morning between 6 and 9 AM, and decrease during the day, reaching a nadir in the evening. The mechanism of this diurnal rhythm is unclear; it was originally thought that the close correlation of cortisol and aldosterone diurnal rhythms provided evidence for a physiological role for ACTH in the control of aldosterone secretion, but the rhythm persists in the absence of ACTH, for example, in patients with ACTH suppressed by dexamethasone treatment [235,236]. The early morning peak of aldosterone is not correlated with plasma potassium or sodium levels, but may be related to changes in renin secretion, as the rhythm is abolished in anephric subjects. If the circadian rhythm of renin secretion is disrupted, however, or if angiotensin II production is inhibited by the administration of captopril, a converting enzyme inhibitor, the rhythm of aldosterone secretion persists [237]. The administration of propranolol, a beta adrenergic antagonist, blocks the circadian rhythm of aldosterone secretion in both humans and rats, suggesting the existence of a neural component to the rhythm [238,239].

Postural changes in aldosterone secretion   It is well recognized that aldosterone secretion is stimulated by upright posture, and this observation forms the basis of clinical tests of aldosterone secretion. There is a close correlation between plasma renin activity and plasma aldosterone concentration in response to postural changes and it seems likely that the renin-angiotensin system is the principal mediator of the aldosterone response to upright posture [240,241].

There are also postural changes in the metabolic clearance rate of aldosterone, but these changes are small by comparison with the change in secretion rate of aldosterone [242].

3.3 Intracellular control of steroidogenesis

3.3.1 Introduction

Similar to other peptide hormones, those which act on the adrenal cortex, including ACTH and angiotensin II, do not need to enter a target cell in order to exert their effects. Instead they bind to specific receptors in the plasma membrane, thus initiating a series of intracellular events which result in increased steroidogenesis and, eventually, growth (see Chap. 1). The events which take place between the binding of the peptide hormone to its receptor and the stimulated activity are initiated by the generation of intracellular signals. These signals have been termed "second messengers," where the "first messenger" is the hormone acting on the plasma membrane receptor (Fig. 3.23). There are several criteria which must be fulfilled by putative second messengers. A second messenger is generated by the action of the stimulus, which is usually a peptide hormone, but may also be a neurotransmitter for instance, at the level of the plasma membrane receptor. Also, the second messenger, when administered to the endocrine tissue, elicits the same response as administration of the "first messenger," and inhibition of the generation of the second messenger impairs the stimulatory effect of the hormone. The second messengers in general act by activating a protein kinase, which phosphorylates cytosolic proteins and thus increases steroidogenic activity (Fig. 3.23). There is additionally, a class of membrane receptors linked to tyrosine kinase, which acts directly to phosphorylate proteins without the requirement for generation of second messengers (Fig. 3.23) [243,244]. The receptors for epidermal growth factor, which have been identified in the adrenal cortex, are an example of this [245].

Several different second messenger systems have been identified in endocrine tissues, including cyclic AMP, calcium ions, products of phospho-inositide metabolism, and arachidonic acid metabolites [246–250]. Where a gland is the target tissue for more than one hormone, it is not uncommon for several of these second messenger systems to occur, each linked to the action of a different stimulant. The adrenal cortex is no exception and several second-messenger systems have been identified in adrenocortical cells, linked with varying degrees of specificity to the different factors which affect adrenocortical function. Thus, while the stimulants of adrenal steroidogenesis

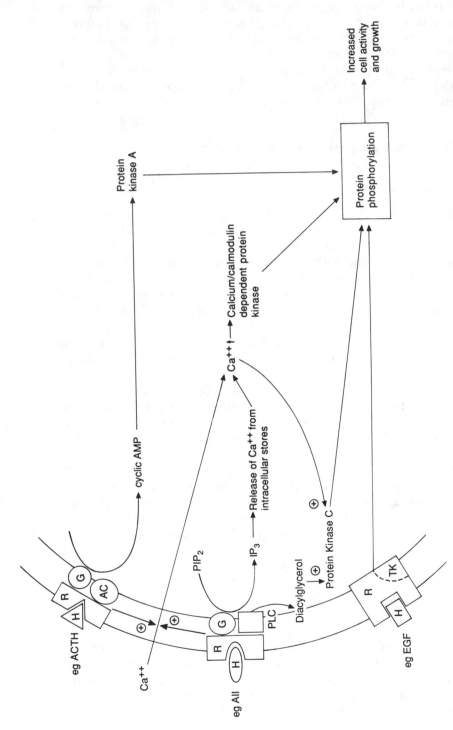

FIGURE 3.23 The role of second messengers in the mechanism of response of a typical endocrine cell to hormonal stimulation. Each hormone type interacts with a specific receptor and activates an intracellular sequence resulting in protein phosphorylation. AC = adenylate cyclase, PLC = phospholipase C, TK = tyrosine kinase, PIP_2 = phosphatidylinositol bisphosphate, IP_3 = inositol 1,4,5-trisphosphate, A-kinase = cAMP-dependent protein kinase R = receptor, G = GTP binding protein, H = hormone, AII = angiotensin II, EGF = epidermal growth factor. There may be additional systems, not illustrated here, involving arachidonic acid metabolism.

each act through their own second messenger systems, there is a degree of overlap, such that integrated adrenocortical function is achieved.

3.3.2 ACTH and cyclic-AMP

Cyclic-AMP was the first agent to be positively identified as an intracellular second messenger. In the adrenal, Haynes [251], showed that ACTH administration caused an increase in intracellular levels of cyclic-AMP, and later showed that the addition of cyclic-AMP to adrenal cells mimicked the steroidogenic effects of adding ACTH itself [252]. This finding was confirmed by Grahame-Smith and associates [253], who also reported that the production of cyclic-AMP in response to ACTH was dose-dependent, and preceded the increase in steroidogenesis.

ACTH receptors The early studies of Schimmer and co-workers [254], using ACTH linked to a large molecule, established that ACTH does not need to enter adrenocortical cells in order to exert its steroidogenic effect, and suggest that ACTH acts through a receptor located in the plasma membrane. Since this time there have been several attempts to characterize the ACTH receptors, using ACTH analogues and [^{125}I]-labeled ACTH, with different preparations of adrenocortical tissue. Until recently, studies were restricted by practical constraints, as it proved difficult to obtain a preparation of [^{125}I]-labeled ACTH with a high specific-activity that retained its biological activity. This problem was eventually overcome using a labeled analogue of ACTH, Phe^2, Nle^4-ACTH. This analogue had the advantages of full biological activity combined with a high specific-activity. Using this peptide Ramachandran and co-workers investigated ACTH receptors on both rat zona fasciculata cells and human adrenocortical cells maintained in primary culture [255,256]. Very similar results were obtained for both species; in each case a single class of ACTH receptor was described, with a Kd of around 1.5 nM and a binding capacity of 3500 to 4000 sites per cell [255,256].

Using fresh rat zona fasciculata cells, however, Gallo-Payet & Escher [257] demonstrated two classes of ACTH receptor, in agreement with the early studies in this field. They described a high-affinity receptor (Kd 1.1×10^{-11} mols/l) and a low-affinity site (Kd 2.9×10^{-9} mols/l), with a much greater binding capacity than that described by Ramachandran: 7200 high-affinity sites per cell and 6.3×10^5 low-affinity sites. The differences between these observations may be explained by the findings of Gallo-Payet & Escher [257], that cells maintained in primary culture for 1 day show a decrease in binding capacity compared with fresh cells. In view of the circulating concentration of ACTH in the rat, of between about 3 pmol/l to 100 pmol/l [258,259], the Kd value of 1.5 nmol/l described by Ramachandran appears to be rather high.

ACTH receptors in the rat zona glomerulosa have also been described [257]. These receptors also show two distinct binding affinities comparable to those reported by this group for the zona fasciculata, but perhaps

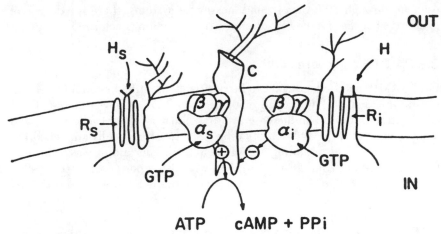

FIGURE 3.24 The receptor (R), G protein and adenylate cyclase (C) assembly. The receptors are highly hydrophobic, with seven membrane-spanning domains. The β-subunits of G_s and G_i are embedded in the membrane bilayer. The α subunits are bound to the β-subunits but are themselves hydrophilic. The α subunit interacts with R_s and the catalyst C, probably through unique sequences in the cytoplasmic domains. (Figure reproduced with permission from ref. 263.)

surprisingly the zona glomerulosa appeared to have more sites of each affinity per cell than the zona fasciculata. Despite the very high number of ACTH receptors on adrenal cells, the occupancy of only a small proportion of these receptors is apparently sufficient to elicit a maximal steroidogenic response [260].

The ACTH receptor from Y1 cells, the murine adrenocortical tumor cell line, has been purified using immunoaffinity chromatography, and found to comprise four subunits with ACTH binding activity residing in the largest subunit (83 kDa) [261]. The purified receptor also was shown to possess two binding affinities, comparable to those reported by Gallo-Payet & Escher [257].

The generation of cyclic-AMP The ACTH receptor is linked to a plasma membrane adenylate cyclase, and causes an increase in intracellular cyclic-AMP by stimulating adenylate cyclase activity [262]. The system comprises at least three distinct components: the hormone receptor, a catalytic subunit, and a guanine nucleotide-binding regulatory protein (Fig. 3.24). The guanine nucleotide-binding regulatory protein exists in two forms, termed G_s and G_i (sometimes also called N_s and N_i). The G_s form mediates the action of stimulants of the adenylate cyclase system, while the G_i form mediates the action of inhibitors [264,265].

Activation of the catalytic subunit requires the binding of an agonist to the hormone receptor which results in a conformational change in the receptor, permitting a noncovalent interaction between the receptor and the catalytic subunit [266]. Activation of the receptor also causes the guanine nucleotide regulatory protein to change from binding GDP to binding magnesium-complexed GTP (Fig. 3.24). The binding of Mg-GTP is necessary for the catalytic subunit to convert ATP to cyclic-AMP [267]. Intracellular levels of cyclic-AMP are controlled both by the action of adenylate cyclase, and by a specific phosphodiesterase which inactivates cyclic-AMP by hydrolysis to AMP.

There is evidence for a dose-dependent inhibitory effect of angiotensin II on both basal and ACTH-stimulated adenylate cyclase activity in rat and bovine zona glomerulosa cells, but the physiological relevance of this effect is not known [268,269]. It is believed that angiotensin II acts through a G_i protein to inhibit cyclic-AMP formation [270].

Action of cyclic-AMP   Cyclic-AMP interacts with a specific binding protein located mainly in the cytosol of adrenocortical cells [271,272]. This binding protein is closely associated with a catalytic protein, and together these two subunits comprise a cyclic-AMP-dependent protein kinase. The binding of cyclic-AMP to its receptor protein causes a physical dissociation of the two subunits, thus activating the protein kinase [273,274]. The activated protein kinase phosphorylates several cytosolic and microsomal proteins, including the enzyme cholesterol ester hydrolase [275]. Phosphorylation activates this enzyme, which causes an increase in the pool of free, metabolically active cholesterol in the adrenal cell [39].

It currently appears that all the intracellular effects of ACTH which have been investigated, including the increased rate of steroid biosynthesis, effects on protein synthesis, RNA transcription, enzyme induction, and so on, can all be mimicked by cyclic-AMP, suggesting that this nucleotide has the capacity to mediate all the intracellular effects of ACTH stimulation, which will be considered in detail later.

Other mediators for ACTH   Some evidence suggests that cyclic-AMP is not the exclusive second messenger for ACTH stimulation: the concentrations of ACTH required for half-maximal stimulation of cyclic-AMP production and steroidogenesis are very different, and at low concentrations of ACTH there may be an increased rate of steroidogenesis in the absence of any measurable rise in intracellular cyclic-AMP [276] (Fig. 3.25). One explanation for the dissociation between cAMP and steroidogenic responses is based on the observation that only a small proportion of the available receptors need to be occupied in order to elicit a maximal steroidogenic response, implying that only a fraction of the cAMP generated by high concentrations of ACTH is needed for maximal steroidogenesis [260]. This hypothesis cannot account for the observation that a higher concentration of ACTH is required for stimulation of cAMP than steroidogenesis [276], and so an alternative explanation is required.

The observed discrepancies may be explained by the proposed existence of two distinct classes of ACTH receptor (previously mentioned). It has been

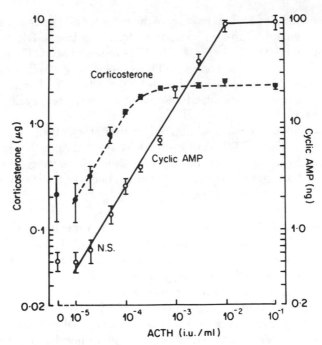

FIGURE 3.25 Effect of ACTH on corticosterone and cyclic AMP production by isolated adrenal cells. The dissociation between steroid and cyclic AMP production seen at the lowest concentration of ACTH suggests that another second messenger may be involved in the mechanism of action of ACTH. (Figure reproduced with permission from ref. 276.)

suggested that low concentrations of ACTH stimulate steroidogenesis by interacting with a high-affinity receptor, which is not linked to adenylate cyclase and is thus independent of cyclic-AMP, while high concentrations of ACTH stimulate cyclic-AMP by interacting with a low-affinity receptor [277]. This theory requires that the effects of low concentrations of ACTH are mediated by a second messenger other than cyclic-AMP. Both the calcium messenger system and arachidonic acid metabolites have been implicated in the actions of ACTH, and proposed as alternatives to cAMP in mediating the intracellular effects of ACTH (following).

Other effectors and cyclic-AMP   Serotonin is also believed to exert its steroidogenic actions through cyclic-AMP [278], and there is evidence that the generation of cyclic-AMP is one component of the intracellular response of potassium ion stimulation of zona glomerulosa cells [279–281]. Although potassium ions cause an increase in adenylate cyclase activity, the aldosterone response to potassium is more closely dependent on calcium than is the response to ACTH, which is mediated primarily through cyclic-AMP [282,283]. It has been proposed that the function of the potassium-induced elevation of cyclic-AMP is to serve as a "positive sensitivity modulator" of the calcium message [281].

3.3.3 The calcium/phosphatidylinositol system

Calcium It is well recognized that extracellular calcium is an important factor in the steroidogenic responses to all of the principal zona glomerulosa stimulants, as the removal of calcium or the addition of calcium channel blockers, markedly attenuates the aldosterone response [282,284–286]. The calcium requirements of the different adrenocortical stimulants are very different however; angiotensin II and potassium stimulation being more closely dependent on calcium than ACTH stimulation. While the mechanism of angiotensin II action requires the hydrolysis of phospholipids, and the subsequent release of calcium from intracellular stores (following), the mechanisms of action of ACTH and potassium are different.

Role of calcium in the response to ACTH

In the case of ACTH stimulation, extracellular calcium is required to facilitate the hormone-receptor interaction and to maintain receptor occupancy [287,288]. An influx of extracellular calcium is required for the generation of cyclic-AMP in response to ACTH stimulation [289], resulting in a rapid, but transient, increase in the concentration of cytosolic free calcium [290]. The effects of ACTH on cytosolic free calcium do not appear to involve the release of calcium from intracellular stores [290]. The role of calcium in the action of cyclic-AMP is still undetermined; removal of extracellular calcium attenuates but does not abolish the steroidogenic response to exogenous cyclic-AMP [282,288].

It has also been demonstrated that low concentrations of ACTH, which stimulate steroidogenesis, but which are subthreshold for cyclic-AMP stimulation, cause an increased calcium uptake. It was therefore proposed that calcium may be an alternative second messenger for ACTH [291], and this idea is supported by the finding that verapamil, a calcium channel blocker, inhibits the steroidogenic effects of physiological concentrations of ACTH in humans, but has no effect on the response to either dibutyryl cyclic-AMP or a high dose of ACTH [292].

Angiotensin II and calcium

It has been known for several years that the zona glomerulosa response to angiotensin II is highly dependent on calcium. Stimulation with angiotensin II causes both an influx of calcium ions into the adrenal zona glomerulosa cell and an efflux of ^{45}Ca from preloaded cells [286,293–295]. These two gross effects are now known to reflect two distinct pathways of calcium movement within the cell, which both occur in response to angiotensin II stimulation. The rapid formation of inositol-1,4,5-trisphosphate causes an immediate increase in the concentration of cytosolic free calcium, liberated

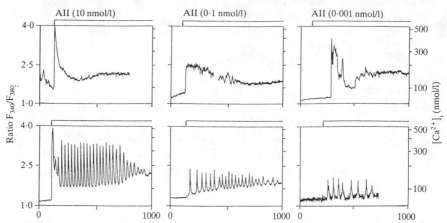

FIGURE 3.26 The two types of cytosolic free Ca^{2+} ([Ca^{2+}]$_i$) response to various angiotensin II (AII) concentrations. Individual bovine adrenal glomerulosa cells were loaded with the Ca^{2+} indicator fura-2 and microperfused with AII at 10, 0·1 or 0·001 nmol/l as indicated. The time of addition of AII is shown in each panel by the arrow and the duration of the AII microperfusion by the horizontal bar at the top of each panel. In all panels, extracellular [Ca^{2+}] was 1·2 mmol/l. The responses without oscillations in [Ca^{2+}]$_i$ are shown in the upper panels and oscillatory [Ca^{2+}]$_i$ responses are shown in the lower panels. F_{340}/F_{380} is the ratio of fluorescence emitted at 340 and 380 nm excitation. Reproduced with permission from [298].

from intracellular stores (see section on p. 108). This effect is only transient, however, and an influx of extracellular calcium is needed for a sustained steroidogenic response to angiotensin II [296].

Zona glomerulosa tissue, incubated in calcium-free medium shows an increase in cytosolic free calcium in response to angiotensin II, apparently reaching a peak in 30 seconds, and declining to basal values within 5 minutes [297], although a dose-related delay in the calcium response has also been reported, ranging from 2.6 ± 0.3 sec at 5×10^{-8} mol/l to 181 ± 27 seconds at 5×10^{-12} mol/l. After the delay there is an abrupt increase in the concentration of cytosolic free calcium, followed by a series of oscillations which last up to 13 minutes [298]. In the absence of extracellular calcium, the oscillatory phase is abolished, but the initial increase in cytosolic calcium persists, and is therefore presumed to result from liberation of calcium from intracellular stores [298]. The initial rise in calcium precedes the increase in aldosterone secretion, and there is a strong correlation between the amplitude of the transient calcium rise, and the magnitude of the aldosterone response [297]. The mean basal level of cytosolic calcium is around 80 nmol/l, rising to between 150 to 240 nmol/l on stimulation [299]. The sequence of intracellular calcium changes in zona glomerulosa cells in response to angiotensin II is illustrated in Figure 3.26.

Following removal of angiotensin II stimulation, the rate of calcium influx falls immediately, although the total cell calcium content continues to rise.

The intracellular calcium is first taken up into a compartment that is not sensitive to inositol-1,4,5-trisphosphate, and the sensitive pool is only slowly restored, thus a second exposure of the cells to angiotensin II does not cause release of intracellular calcium unless there is an interval longer than 20 minutes between the two periods of exposure [300,301].

Role of calcium in the response to potassium ions

Stimulation with potassium ions has no effect on the rate of calcium efflux from adrenocortical cells, but causes a marked increase in calcium influx [294,302]. Recent evidence has shown that adrenal zona glomerulosa cells behave like particularly sensitive excitable neuroendocrine cells with a highly negative resting potential which is significantly altered by changes in extracellular potassium ion concentration [303,304]. Potassium-induced depolarization activates voltage-dependent calcium channels, permitting an influx of calcium ions which causes a net rise in cytosolic free calcium, that can be totally blocked by a calcium channel blocker such as nitrendipine, nifedipine, or verapamil [296,304–307]. The rise in cytosolic free calcium occurs within a few seconds after extracellular potassium ion concentration is increased, and is sustained for several minutes [308]. Subsequent addition of nifedipine causes an immediate reversal of the effect of potassium, and a rapid drop in the concentration of cytosolic free calcium [305]. Potassium ion stimulation does not apparently cause mobilization of calcium from intracellular pools, as dantrolene, an inhibitor of intracellular calcium mobilization, has no effect on potassium stimulation of zona glomerulosa cells [302].

Mechanism of action of calcium ions

There are several routes by which intracellular events may be influenced by the raised levels of cytosolic free calcium which occur in response to stimulation. Many of the actions of calcium are mediated by calmodulin which is a specific calcium-binding protein without intrinsic enzymatic activity [246,309]. Calmodulin has been identified in the adrenal cortex [310], and undergoes a conformational change when bound to calcium ions which activates a specific protein kinase which phosphorylates proteins, notably in the mitochondria [311,312]. It has been shown that inhibitors of calmodulin abolish the aldosterone secretory response to angiotensin II, and significantly attenuate the response to cyclic-AMP [313], suggesting a role for calmodulin in the second messenger actions of calcium in the adrenal cortex.

Calcium ions also exert a direct effect on some enzymes, independently of calmodulin; the activation of protein kinase C by diacylglycerol is calcium-dependent, and this is thought to be a major role of raised cytosolic free calcium in adrenocortical cells. It has also been suggested that cytosolic calcium may facilitate the binding of protein kinase-C to diacylglycerol in the plasma membrane [246]. A range of other enzymes are also directly activated by calcium ions; some evidence suggests that these may include cholesterol side-chain cleavage and 11β-hydroxylase activities [314,315].

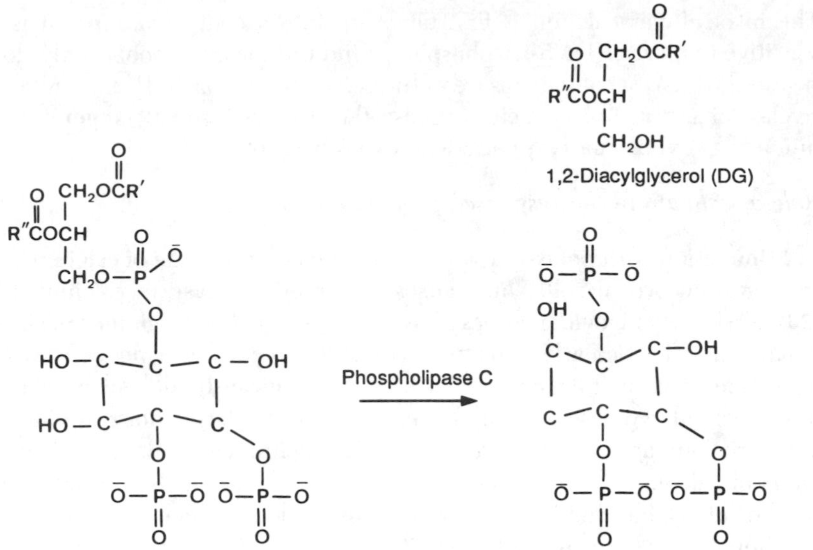

FIGURE 3.27 The action of phospholipase C on phosphatidyl inositol 4,5-bisphosphate yields diacylglycerol and inositol 1,4,5-trisphosphate.

Phosphatidylinositol metabolism For many years it has been recognized that a wide range of hormones and neurotransmitters exert a receptor-mediated stimulation on phosphatidylinositol turnover in their target cells (for a review see refs. 248,250). Michell [316] proposed that the hydrolysis of phosphatidylinositol may be related to the opening of cell surface calcium channels, thus providing a possible link between receptor activation and elevated calcium levels. More recent evidence has shown that it is primarily polyphosphinositides, rather than phosphatidylinositol, which are hydrolyzed in response to receptor activation [317]. The hydrolysis of these polyphosphoinositides yields inositol trisphosphate and diacylglycerol, which act as second messengers in a wide range of cell types, including the zona glomerulosa cells of the adrenal cortex [249].

Formation of inositol trisphosphate and diacylglycerol

In place of adenylate cyclase, the angiotensin II receptors are believed to be linked to a phospholipase, usually termed phospholipase C, but also known as phosphoinositidase or phosphoinositide phosphodiesterase. This enzyme catalyzes the hydrolysis of phosphatidylinositides to inositol trisphosphate and diacylglycerol (Fig. 3.27), which both function as second messengers in adrenal glomerulosa cells [318]. It seems likely that, in common with other cell types, the receptor-mediated activation of phospholipase C occurs via a guanine nucleotide regulatory protein, termed G_p, which is

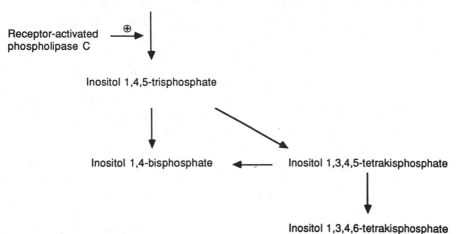

Phosphatidylinositol 4,5-bisphosphate

Receptor-activated phospholipase C ⊕

Inositol 1,4,5-trisphosphate

Inositol 1,4-bisphosphate ← Inositol 1,3,4,5-tetrakisphosphate

Inositol 1,3,4,6-tetrakisphosphate

FIGURE 3.28 Proposed pathways of inositol phosphate metabolism in the adrenal cortex. It is possible that other pathways, not illustrated here, may also exist. (From refs. 322 and 323.)

distinct from the inhibitory protein (G_i) mediating the effects of angiotensin II on adenylate cyclase [270,319–322].

There are several possible pathways of phosphoinositide metabolism in zona glomerulosa cells (see Fig. 3.28), but the most important event in the intracellular signalling mechanism is the hydrolysis of phosphatidylinositol 4,5-bisphosphate to yield inositol 1,4,5-trisphosphate and diacylglycerol [318]. These two agents act as second messengers in adrenocortical cells, and other cell types.

Formation of inositol-1,4,5-trisphosphate in zona glomerulosa cells occurs rapidly after the onset of angiotensin II stimulation, reaching a maximum within 5 to 10 seconds [322,324]. Both the rate of increase of inositol-1,4,5-trisphosphate formation and the magnitude of the response depend on the concentration of angiotensin II [322]. In adrenal zona glomerulosa cells, the effect on phosphoinositide metabolism is relatively specific to angiotensin II stimulation: the other major aldosterone secretagogues do not exert such a pronounced effect on inositol-1,4,5-trisphosphate turnover [324–326]. There are other inositol phosphates produced in zona glomerulosa cells in response to angiotensin II stimulation: these are illustrated in Figure 3.28. It has been shown that inositol-1,4,5-trisphosphate is inactivated by successive dephosphorylations giving inositol-1,4-bisphosphate, then inositol-4-phosphate and finally inositol [327]. An alternative route of metabolism has been demonstrated in several cell types, including the adrenal zona glomerulosa, where the first event is a further phosphorylation which yields inositol-1,3,4,5-tetraphosphate, then dephosphorylation to give inositol-1,3,4-trisphosphate which may itself be further phosphorylated or dephosphorylated [323,328–331].

Action of 1,2-diacylglycerol

Diacylglycerol is thought to act on a calcium-activated, phospholipid-dependent protein kinase enzyme, termed protein kinase C (PK-C), which has been identified in the adrenal cortex [332]. The exact mechanism of PK-C activation in response to diacylglycerol in the adrenal cortex has not been directly investigated and it is generally assumed that diacylglycerol in adrenocortical cells acts in a manner similar to other cells, where hormonal stimulation causes an increase in the diacylglycerol content of the plasma membrane resulting in increased binding of PK-C to the inner surface of the plasma membrane [249,250,333]. The activated PK-C is thought to phosphorylate several cytosolic proteins, and substrates for this enzyme have recently been identified in the rat adrenal zona glomerulosa [312].

In bovine adrenocortical cells, as demonstrated, angiotensin II causes a subcellular redistribution of PK-C, analogous to that seen in other cell types in response to diacylglycerol and its agonists, with a reduction in cytosolic PK-C activity and an increase in PK-C activity in the particulate fraction [334,335]. The phorbol ester, 12-O-tetradecanoyl phorbol-13-acetate (TPA) is commonly used to activate PK-C, but attempts to correlate steroid production with PK-C activation in rat adrenal cells have been largely unsuccessful [336]. A slight stimulatory effect of TPA has been reported in bovine adrenal cells, and it has been suggested that PK-C may have a role in the sustained phase of the aldosterone secretory response to angiotensin II, which is associated with an influx of extracellular calcium [318]. A more recent report has demonstrated a significant effect of TPA on aldosterone secretion by intact adrenal capsular tissue, but not by collagenase-dispersed zona glomerulosa cells, suggesting that the pathway of PK-C activation is disrupted by the cell-dispersal process [337].

Action of inositol-1,4,5-trisphosphate

The major action of inositol-1,4,5-trisphosphate is believed to be to elevate cytosolic free calcium levels by causing the release of calcium from intracellular stores [338]. There are intracellular binding sites in the zona glomerulosa, which are specific for inositol-1,4,5-trisphosphate, and have very low affinity for other inositol phosphates [339,340]. These receptors have been identified in both microsomal preparations and plasma membrane preparations from the adrenal cortex [339,340]. The inositol phosphate-sensitive store of calcium is known to be extra-mitochondrial in most tissues, including the adrenal zona glomerulosa [341], and is most likely to be the endoplasmic reticulum [342].

3.3.4 Arachidonic acid metabolites

The phospholipids in the plasma membrane of adrenocortical cells are rich in arachidonic acid, and the observation that stimulation of the adrenal cortex may result in the liberation of arachidonic acid and its metabolites has led to

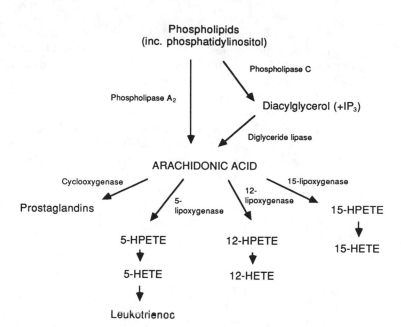

FIGURE 3.29 Pathways of arachidonic acid release and metabolism. IP_3 = inositol 1,4,5-trisphosphate, HPETE = hydroperoxyeicosatetraeinoic acid, HETE = hydroxyeicosatetraeinoic acid.

the hypothesis that this may constitute a second messenger system in the adrenal. This hypothesis is supported by reports suggesting that prostaglandins stimulate adrenal steroid secretion (see Sec. 3.2.2; subsec. Prostaglandins), and by the evidence for a second messenger role for arachidonic acid and its metabolites in other endocrine systems (e.g., ref. 343). Arachidonic acid metabolism follows one of two major pathways: the lipoxygenase pathway which yields leukotrienes and HPETES, or the cyclooxygenase pathway which yields prostaglandins (Fig. 3.29) [344]. The variety of specific inhibitors of each of these pathways which is now available has made it possible to investigate these arachidonic acid metabolites individually.

Arachidonic acid production There are two pathways in the adrenal through which arachidonic acid may be released. First by the action of phospholipase A_2 on phospholipids, and second by the action of diglyceride lipase on the diacylglycerol formed by the phospholipase C-catalyzed breakdown of phosphatidylinositol (Fig. 3.29) [345]. In the adrenal cortex, angiotensin II is known to exert its effects through the activation of phospholipase C and the formation of diacylglycerol, but little attention has been paid to this potential route for arachidonic acid release, and most studies have looked at phospholipase A_2 activity in response to stimulation of the adrenal cortex. Studies on cat adrenocortical cells have shown a calcium-dependent activation of phospholipase A_2 by ACTH, together with an increased turnover of arachidonic acid in membrane phospholipids [346,347]. In bovine zona glomerulosa cells, however, Kojima and co-workers [348] found that arachidonic acid is released from phosphatidylinositol (presumably via diacylglycerol production), but not from other phospholipids, suggesting that phospholipase A_2 is not activated by angiotensin II [348].

Cyclooxygenase products Bergstrom [349] first suggested that prostaglandins may be produced as a consequence of ACTH action on the adrenal gland, and several groups have subsequently investigated prostaglandin production by adrenocortical cells in response to stimulation. There is a great deal of variation in the findings reported; while most reports are in agreement that PGE_2 production is increased in response to ACTH, the response of PGA and the F series is uncertain [350–352]. More recently, Swartz and Williams [353] failed to measure any significant change in prostaglandin production by rat adrenals in response to ACTH, angiotensin II, or potassium stimulation, while Campbell and Gomez-Sanchez [354] found that angiotensin II stimulated the production of a range of prostaglandins by rat zona glomerulosa cells.

Reports of experiments using indomethacin, a cyclooxygenase inhibitor, have also been inconclusive [348,353,355–359].

Lipoxygenase products There have been far fewer studies on the role of lipoxygenase products than on cyclooxygenase products, but the subject is no clearer. A general lipoxygenase inhibitor (BW755c) causes inhibition of the steroidogenic response to angiotensin II and ACTH [348,360,361]. It has not proved possible, however, to identify a lipoxygenase product which reverses this inhibition, and the effect of BW755c may therefore be nonspecific.

In view of the total lack of consensus on the subject, it seems unlikely that either cyclooxygenase or lipoxygenase products have a significant second messenger role in adrenocortical steroidogenesis.

3.4 Mechanism of increased steroidogenesis

3.4.1 Introduction

In 1954, Hechter and Stone suggested that the rate-limiting step in adrenal steroidogenesis is the conversion of cholesterol to pregnenolone, and furthermore, that this is the site of action of ACTH [362]. Since this original proposal, it has been demonstrated that, in both the zona glomerulosa and the zonae fasciculata/reticularis, all the stimulants of steroidogenesis which have been investigated share the common feature of acting to increase the rate of the cholesterol side-chain cleavage reaction in the mitochondria [363]. In addition to this action, specific zona glomerulosa stimulants also exert a "late pathway" effect, stimulating the synthesis of aldosterone from its immediate precursors (see Chap. 2). The mechanism by which the rate of cholesterol side-chain cleavage is increased has been studied in some detail with respect to ACTH stimulation. While the mode of action of the other stimulants has not been closely studied, a similarity to ACTH may be inferred from the finding that all the second messenger systems of the stimuli investigated appear to ultimately involve the activation of a protein kinase (see Sec. 3.3), whether this is A-kinase for cyclic-AMP, C-kinase for diacylglycerol or calcium/calmodulin dependent kinase (Fig. 3.23). An interesting possibility is that A-kinase may phosphorylate, and thus activate, proteins controlling cholesterol availability and transport, while C-kinase and calcium/calmodulin dependent kinase may additionally activate "late pathway" mechanisms in the zona glomerulosa [337]. The immediate intracellular consequences of the activation of these kinases is unclear, and the specific proteins which are phosphorylated have yet to be identified. Most of the studies on the mechanism of increased steroidogenesis have focussed on the acute actions of ACTH on cholesterol metabolism and the mechanism of increased mitochondrial pregnenolone formation.

3.4.2 Actions of ACTH on cholesterol metabolism

The major sources of cholesterol for adrenal steroidogenesis have been described in Chap. 2. These are, briefly, lipid droplets within the adrenocortical cells, receptor-mediated uptake of plasma lipoproteins and de novo synthesis of cholesterol from acetate (Fig. 3.30). The relative contribution of each of these sources of cholesterol varies according to species and according to the state of stimulation of the gland. Under conditions of acute ACTH stimulation, the intracellular store of esterified cholesterol in the lipid droplets is probably the most important source of cholesterol to supplement the cholesterol in the mitochondrial membranes [275,364]. This esterified cholesterol is metabolically inactive and requires enzymic hydrolysis by cholesterol ester hydrolase to yield free cholesterol for the side-chain cleavage reaction (see Chap. 2). This enzyme, found in the cytosol of adrenocortical cells, is known to be activated by phosphorylation by cyclic-AMP dependent protein kinase in response to ACTH stimulation [275,365]. The action of this enzyme on adrenal lipid droplets causes an increase in the free (nonesterified) cholesterol content of the lipid droplets [275]. This free cholesterol is then transported from the lipid droplets to the mitochondria for steroid biosynthesis, causing the depletion of lipid droplet volume seen following ACTH stimulation (see Chap. 1).

Adrenocortical cells take up lipoprotein-cholesterol from the plasma by a receptor-mediated mechanism [366,367]. In the rat, around 80 percent of circulating cholesterol is associated with high density lipoproteins (HDL), while in the human around 70 percent of plasma cholesterol is found in low density lipoprotein (LDL). Rat adrenal cells have a specific, ACTH-dependent receptor-mediated mechanism for HDL uptake, and ACTH stimulates binding, internalization and degradation of plasma lipoproteins

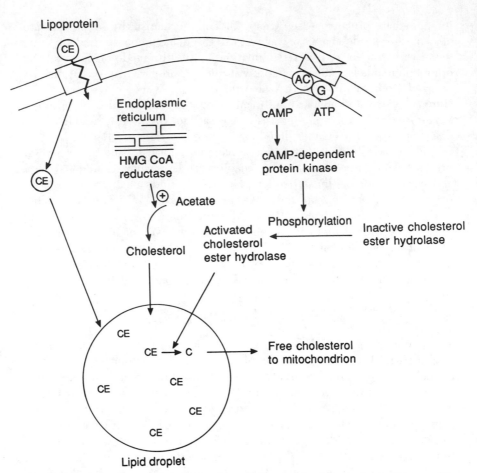

FIGURE 3.30 Pathways of cholesterol metabolism in adrenocortical cells. It is possible that the actions of cAMP may include the stimulation of lipoprotein uptake and HMG CoA reductase activity. Lipoprotein uptake may also be stimulated by a fall in the cholesterol ester content of the lipid droplets. CE = cholesterol ester, C = free cholesterol.

[368,369]. It is not clear, however, whether the acute action of ACTH on lipoprotein uptake is a direct effect, or secondary to depletion of intracellular cholesterol stores.

The de novo synthesis of cholesterol in adrenocortical cells may only be important when the intracellular stores have been depleted, and there is no cholesterol available from plasma lipoproteins [366]. This suggestion is supported by the finding that the rate of ^{14}C acetate conversion to cholesterol was very low in control animals, but greatly enhanced in lipoprotein deficient animals [370]. The activity of HMG CoA reductase is the rate-limiting step in the conversion of acetate to cholesterol (see Chap. 2). The activity of this enzyme is increased in response to ACTH in human fetal adrenal cells in culture, but this increase is blocked by an inhibitor of $P450_{scc}$, which prevents the depletion of intracellular cholesterol, suggesting that the action of ACTH on HMG CoA reductase may be an indirect effect, secondary to intracellular cholesterol depletion [40,371]. This may be an important regulatory mechanism in species which do not rely on the uptake of cholesterol from plasma.

3.4.3 The role of protein synthesis in the response to ACTH

As early as 1963 it was shown that puromycin, an inhibitor of protein synthesis, abolished the steroidogenic response to ACTH when added to an in vitro incubation of rat adrenal quarters or bovine adrenal slices [372]. This initial finding has been confirmed and extended, using other protein synthesis inhibitors, including cycloheximide and chloramphenicol, both in vivo and in vitro [373,374], leading to the conclusion that protein synthesis is an obligatory step in the adrenocortical response to ACTH stimulation. These early studies also showed that the acute actions of ACTH on steroidogenesis are independent of RNA synthesis. Actinomycin-D, a specific inhibitor of mRNA synthesis, when given in concentrations sufficient to totally inhibit ^{14}C-uridine incorporation into rat adrenals, had no effect on the steroidogenic response to ACTH [374]. This finding is consistent with the rapid onset of the steroidogenic response to ACTH, which supports the hypothesis that ACTH modulates protein synthesis from extant mRNA, rather than by inducing the synthesis of RNA itself (for a review see ref. 375).

The initial experiments in vitro suggested that protein synthesis was only required for an "initiation step," as addition of cycloheximide 30 minutes after addition of ACTH did not alter the steroidogenic response, while simultaneous addition of ACTH and the inhibitor totally abolished the response [372]. Subsequent experiments in vivo did not support these findings, however, as administration of cycloheximide at the time when the steroidogenic response to ACTH was maximal, caused inhibition of steroidogenesis with $t_{\frac{1}{2}}$ of 7 to 10 minutes, leading the authors to conclude that ACTH caused the synthesis of a rapidly induced stimulatory protein which was obligatory for the response to ACTH [374]. This factor has been termed the "labile protein," and has been the subject of many studies attempting to isolate and purify it.

3.4.4 Site of action of the labile protein

The rate-limiting step for the stimulation of adrenal steroidogenesis by ACTH is the conversion of cholesterol to pregnenolone by the enzyme complex termed cytochrome-$P450_{scc}$ which is located on the inner mitochondrial membrane (see Chap. 2). This step in the biosynthetic pathway is also the cycloheximide sensitive reaction [376]. There are essentially three ways in which ACTH may regulate the rate of this reaction. First by increasing the total intracellular pool of metabolically free cholesterol, second by increasing the activity of the cytochrome-P450$_{scc}$ complex, and third by altering the conditions in the mitochondrion to favor the side-chain cleavage reaction, by increasing the delivery of substrate to the enzyme complex.

It is clear that the acute administration of ACTH does not affect the concentration of cytochrome-P450$_{scc}$ in the mitochondrion, neither does it alter the V$_{max}$ of the reaction [377,378]. While ACTH has profound effects on the intracellular concentration of free cholesterol, by stimulating the activity of cholesterol ester hydrolase (see above), this effect is not sensitive to cycloheximide [379], and therefore not the step controlled by the labile protein.

It was thought that cycloheximide may act to block the transport of cholesterol to the mitochondria, from the lipid droplets and cytosol, but it has been shown that mitochondrial cholesterol content increases in the presence of cycloheximide [380]. Clearly then, a different mechanism, which is independent of protein synthesis, is operating to transport cholesterol to the mitochondrion. Some experimental evidence supports the theory that microtubules and microfilaments are involved in this process, as adrenal steroidogenesis is inhibited by inhibitors of microtubule and microfilament formation, such as cytochalasin B [39,381,382].

The finding that mitochondrial cholesterol levels may be high, although steroidogenesis is inhibited in the presence of cycloheximide, suggests that high-mitochondrial cholesterol is not in itself sufficient to drive steroidogenesis, and that protein synthesis is required to facilitate the association between substrate and enzyme, possibly by enhancing cholesterol movement within the mitochondrion. Evidence to support this suggestion has been presented by several groups [383–385] who have shown that cycloheximide blocks the transport of cholesterol from the outer to the inner mitochondrial membrane, causing an accumulation of cholesterol in the outer membrane, where it is not available to the cytochrome P450$_{scc}$. In the presence of aminoglutethimide, which blocks the active site of cytochrome P450$_{scc}$, there is an accumulation of cholesterol in the inner mitochondrial membrane. Thus it appears that the rate-limiting cycloheximide-sensitive step in the ACTH stimulation of adrenal steroidogenesis is the movement of cholesterol from the outer to the inner mitochondrial membrane, and this accumulation of cholesterol in the inner mitochondrial membrane is sufficient to drive steroidogenesis [386,387].

Since the initial finding that ACTH-stimulated steroidogenesis is inhibited by cycloheximide there have been many attempts to identify the labile protein involved, and it was in the course of these investigations that much of the process described has been elucidated. The earliest approach to identifying the cycloheximide factor was to take the well-characterized sterol carrier protein (SCP$_2$) from the liver, and look at its effects on pregnenolone synthesis by reconstituted subcellular fractions of the adrenal cortex. In this way it was shown that hepatic SCP$_2$ can enhance the formation of pregnenolone from cholesterol by an acetone powder of adrenocortical mitochondria, and some evidence was presented for the existence of a protein similar to SCP_2 in the adrenal [388]. After this study in the early 1970s, little progress was made in this area for several years, until there was a sudden resurgence of interest in the 1980s. Within a few years several quite distinct factors, mostly proteins, had been put forward as the "cycloheximide-sensitive labile factor" mediating the response to ACTH.

Vahouny and co-workers continued the earlier studies of Kan et al., on sterol carrier protein, and demonstrated a stimulatory effect of the hepatic SCP_2 on cholesterol transport from lipid droplets to mitochondria, after both cellular components had been isolated by centrifugation from adrenal homogenates and incubated in vitro [389]. Later both this group and others showed that hepatic SCP_2 enhanced cholesterol transport from the outer to the inner mitochondrial membrane, reversing the inhibitory effect of cycloheximide [383,390]. Their proposal that an adrenal protein similar to hepatic SCP_2 may mediate the response to ACTH was considerably strengthened when such a protein was identified [390,391]. More recently it has been demonstrated that ACTH treatment of rat adrenocortical cells in monolayer culture induces the formation of an adrenal sterol carrier protein with a molecular weight of 11,300, apparently synthesized via a higher molecular weight precursor [392]. While this adrenal SCP_2 appears to satisfy most of the criteria for the labile protein, the time course for its induction, as measured in monolayer culture, with a half-life of around 30 hours in the presence of ACTH, is probably not compatible with its proposed role as the "labile protein."

Orme-Johnson and co-workers identified two proteins quite distinct from SCP_2. Protein p is found only in unstimulated adrenocortical cells, while protein i (mol wt 28,000) is produced in response to ACTH, with a time course and dose-dependency closely paralleling the steroidogenic response [393]. Administration of cycloheximide blocks the formation of protein i. Protein i does not appear to be produced by a posttranslational modification of protein p, but the two proteins may be related by cotranslational modification, probably a phosphorylation [394]. This finding is particularly interesting in view of the observation that the activation of protein kinases occurs in response to adrenocortical stimulation.

Another peptide (mol wt 2200) has been identified in rat adrenals by Pedersen and Brownie [395]. This peptide, much smaller than either SCP_2 or protein i, is induced by ACTH and the increase is blocked by cycloheximide. The time-course and dose-dependency of the response was not measured. Furthermore, addition of this peptide to adrenal mitochondria enhanced the rate of pregnenolone formation. This peptide, like the others, seems to fulfill many of the criteria required for the labile protein.

Evidence suggesting that the labile factor is a phospholipid rather than a protein has been presented by Solano and co-workers [396], who propose a mechanism whereby cycloheximide inhibits the ACTH-independent formation of a phospholipid exchange protein (mol wt 28,000). This exchange protein solubilizes an ACTH-dependent phospholipid which enhances the rate of cholesterol side-chain cleavage.

Most recently another peptide (mol wt 8200) has been identified in bovine zona fasciculata cells by Yanagibashi and others [397]. The ACTH-dependency and cycloheximide sensitivity of this peptide were not investigated, but like SCP_2 it had a marked effect on cholesterol transport within the mitochondria and significantly enhanced the rate of pregnenolone formation.

Since the early 1980s, a range of "labile factors" has been described, with widely differing properties and molecular weights (see ref. 398). It seems clear from these reports that, far from the initial concept of a single "labile factor," it now appears that ACTH may induce a range of different factors in adrenocortical cells, which can act together, possibly through a variety of mechanisms, to increase the rate of steroid biosynthesis in the adrenal cortex.

3.4.5 Long-term effects of ACTH on steroid biosynthesis

Following removal of endogenous ACTH by hypophysectomy, there is a marked decrease in the adrenocortical content of the cytochrome-P450s involved in steroid biosynthesis [399]. Chronic ACTH treatment causes increased biosynthesis of the steroidogenic enzymes, by increasing the expression of the genes for all of the cytochrome-P450 species involved in steroid biosynthesis as well as the iron-sulphur protein, adrenodoxin [400]. This effect of ACTH is mimicked by cyclic-AMP and is blocked by cycloheximide, suggesting the involvement of a regulatory protein [400]. ACTH treatment of bovine adrenocortical cells in culture also brings about increased synthesis of LDL receptors and SCP_2, thus it appears that in this tissue ACTH induces an increase in the enzymes involved in all the stages of increasing cholesterol supply and metabolism in the adrenal cortex, from the mechanisms for increasing intracellular cholesterol, to the enzymes of the steroid biosynthetic pathway. This effect of ACTH is specific to these steroidogenic enzymes as ACTH does not appear to affect the enzymes of gluconeogenesis and glycolysis in adrenocortical cells [40].

References

1. Porter, J.C., and Klaiber, M.S. 1964. Relationship of input of ACTH to secretion of corticosterone in rats. *Am. J. Physiol.* 207:789–92.

2. Li, C.H. 1962. Synthesis and biological properties of ACTH peptides. *Rec. Prog. Horm. Res.* 18:1–40.

3. Liddle, G.W., Island, D., and Meador, C.K. 1962. Normal and abnormal regulation of corticotropin secretion in man. *Rec. Prog. Horm. Res.* 18:125–66.

4. Sawyer, T.K., Hruby, V.J., Hadley, M.E., and Engel, M.H. 1983. α-Melanocyte stimulating hormone: Chemical nature and mechanism of action. *Amer. Zool.* 23:529–40.

5. Baker, B.I. 1979. The evolution of ACTH, α-MSH and LPH structure, function and development. In: E.J.W. Barrington, ed. *Hormones and Evolution* vol. II Academic Press: New York, pp. 643–722.

6. Baumann, G., and Felber, J.P. 1976. Prolonged corticotropic action of synthetic human ACTH in man. *J. Clin. Endocr. Metab.* 42:160–63.

7. Jones, M.T., and Gillam, B. 1988. Factors involved in the regulation of adrenocorticotropic hormone/β-lipotropic hormone. *Physiol. Rev.* 68:743–818.

8. Kitabchi, A.E. 1967. Ascorbic acid in steroidogenesis. *Nature.* 215:1385–86.

9. Lipscomb, H.S., and Nelson, D.H. 1960. Dynamic changes in ascorbic acid and corticosteroids in adrenal vein after ACTH. *Endocrinology.* 66:144–46.

10. Papaikonomou, E. 1977. Rat adrenocortical dynamics. *J. Physiol.* 265:119–31.

11. Urquhart, J., and Li, C.H. 1968. The dynamics of adrenocortical secretion. *Am. J. Physiol.* 214:73–85.

12. Porter, J.C., and Klaiber, M.S. 1965. Corticosterone secretion in rats as a function of ACTH input and adrenal blood flow. *Am. J. Physiol.* 209:811–14.

13. Lowry, P.J., and McMartin, C. 1974. Measurement of the dynamics of stimulation and inhibition of steroidogenesis in isolated rat adrenal cells by using column perfusion. *Biochem. J.* 142:287–94.

14. Schulster, D., and Jenner, C. 1975. A counter streaming centrifugation technique for the superfusion of adrenocortical cell suspensions stimulated by ACTH. *J. Steroid. Biochem.* 6:389–94.

15. Keller-Wood, M.E., Shinsako, J., and Dallman, M.F. 1983. Integral as well as proportional adrenal responses to ACTH. *Am. J. Physiol.* 245:R53–R59.

16. Hinson, J.P., Vinson, G.P., Whitehouse, B.J., and Price, G. 1985. Control of zona glomerulosa function in the isolated perfused rat adrenal gland in situ. *J. Endocr.* 104:387–95.

17. Balfour, W.E. 1953. Changes in the hormone output of the adrenal cortex of the young calf. *J. Physiol.* 122:59P.

18. Holzbauer, M., and Vogt, M. 1957. Functional changes produced in the adrenal cortex of the rat by administration or by release of corticotropin. *J. Physiol.* 138:449–59.

19. Maier, R., and Staehelin, M. 1968. Adrenal responses to corticotropin in the presence of an inhibitor of protein synthesis. *Acta Endocr.* 58:619–29.

20. Hinson, J.P., Vinson, G.P., Whitehouse, B.J., and Price, G. 1986. Effects of stimulation on steroid output and perfusion medium flow rate in the isolated perfused rat adrenal gland in situ. *J. Endocr.* 109:279–85.

21. Wright, R.D. 1963. Blood flow through the adrenal gland. *Endocrinology.* 72:418–28.

22. Braverman, B., and Davis, J.O. 1973. Adrenal steroid secretion in the rabbit; sodium depletion, angiotensin II and ACTH. *Am. J. Physiol.* 225:1306–10.

23. Hartman, F.A., Brownell, K.A., and Liu, T.Y. 1955. Blood flow through the dog adrenal. *Am. J. Physiol.* 180:375–77.

24. Nichols, J., and Richardson, A.W. 1960. Effect of DDD on steroid response and blood flow through the adrenal after ACTH. *Proc. Soc. Exp. Biol. Med.* 104:539–42.

References   **117**

25. Stark, E., Varga, B., Ac,s Z., and Papp, M. 1965. Adrenal blood flow response to adrenocorticotropic hormone and other stimuli in the dog. *Pflug. Arch. Physiol.* 385:296–301.

26. Gerber, J.G., and Nies, A.S. 1979. The failure of indomethacin to alter ACTH-induced hyperaemia or steroidogenesis in the anaesthetised dog. *Br. J. Pharm.* 67:217–20.

27. Sapirstein, L.A., and Goldman, H.A. 1959. Adrenal blood flow in the albino rat. *Am. J. Physiol.* 196:159–62.

28. Harrison, R.G., and Hoey, M.J. 1960. *The Adrenal Circulation.* Oxford: Blackwell.

29. Vinson, G.P., Pudney, J.A., and Whitehouse, B.J. 1985. The mammalian adrenal circulation and the relationship between adrenal blood flow and steroidogenesis. *J. Endocr.* 105:285–94.

30. Varga, B., Stark, E., and Folly, G. 1979. Inhibition of the stimulatory effect of ACTH on adrenal and ovarian blood flow by indomethacin in the dog. *Act. Physiol. Acad. Sci. Hung.* 54:123–28.

31. Irman-Florjanc, T., and Erjavec, F. 1984. The effect of adrenocorticotropin on histamine and 5-hydroxytryptamine secretion from rat mast cells. *Agents and Actions.* 14:454–57.

32. Hinson, J.P., Vinson, G.P., Pudney, J., and Whitehouse, B.J. 1989. Adrenal mast cells modulate vascular and secretory responses in the intact adrenal gland of the rat. *J. Endocr.* 121:253–60.

33. Hinson, J.P., Vinson, G.P., and Whitehouse, B.J. 1986. The relationship between perfusion medium flow rate and steroid secretion in the isolated perfused rat adrenal gland in situ. *J. Endocr.* 111:391–96.

34. Maier, F., and Staehelin, M. 1968. Adrenal hyperaemia caused by corticotropin. *Acta Endocr.* 58:613–18.

35. Nussdorfer, G.G., Mazzocchi, G., and Rebonato, L. 1971. Long-term trophic effect of ACTH on rat adrenocortical cells. *Z. Zellforsch.* 115:30–45.

36. Pudney, J., Price, G.M., Whitehouse, B.J., and Vinson, G.P. 1984. Effects of chronic ACTH stimulation on the morphology of the rat adrenal cortex. *Anat. Rec.* 210:603–15.

37. Vazir, H., Whitehouse, B.J., Vinson, G.P., and McCredie, E. 1981. Effects of prolonged ACTH treatment on adrenal steroidogenesis and blood pressure in rats. *Acta Endocr.* 97:533–42.

38. Purvis, J.L., Canick, J.A., Mason, J.I., Estabrook, R.W., and McCarthy, J.L. 1973. Lifetime of adrenal cytochrome P450 as influenced by ACTH. *Ann. N.Y. Acad. Sci.* 212:319–42.

39. Simpson, E.R., and Waterman, M.R. 1983. Regulation by ACTH of steroid hormone biosynthesis in the adrenal cortex. *Can. J. Biochem. Cell Biol.* 61:692–707.

40. Simpson, E.R., and Waterman, M.R. 1988. Regulation of the synthesis of steroidogenic enzymes in adrenal cortical cells by ACTH. *Ann. Rev. Physiol.* 50:427–40.

41. de Peretti, E., and Forest, M.G. 1976. Unconjugated dehydroepiandrosterone plasma levels in normal subjects from birth to adolescence in human: The use of a sensitive radioimmunoassay. *J. Clin. Endocrinol. Metab.* 43:982–91.

42. Albright, F. 1947. Osteoporosis. *Ann. Int. Med.* 27:861–82.

43. Bell, J.B.G., Bhatt, K., Hyatt, P.J., Tait, J.F., and Tait, S.A.S. 1980. Properties

of adrenal zona reticularis cells. In A.R. Genazzani, J.H.H. Thijssen, and P.J. Siiteri, eds. *Adrenal Androgens.* New York: Raven Press, pp. 1–6.

44. Davison, B., Large, D.M., Anderson, D.C., and Robertson, W.R. 1983. Basal steroid production by the zona reticularis of the guinea pig adrenal cortex. *J. Steroid Biochem.* 18:285–90.

45. O'Hare, M.J., Nice, E.C., and Neville, A.M. 1980. Regulation of androgen secretion and sulpho-conjugation in adult human adrenal cortex: Studies with primary monolayer cultures. In: A.R. Genazzani, J.H.H. Thijssen, and P.J. Siiteri, eds. *Adrenal Androgens.* New York: Raven Press, pp. 7–25.

46. Cameron, E.H.D., Jones, T., Jones, D., Anderson, A.B.M., and Griffiths, K. 1969. Further studies on the relationship between C_{19} and C_{21} steroid synthesis in the human adrenal gland. *J. Endocr.* 45:215–30.

47. Jones, T., and Griffiths, K. 1968. Ultramicrochemical studies on the site of formation of dehydroepiandrosterone sulphate in the adrenal cortex of the guinea pig. *J. Endocr.* 42:559–65.

48. Rosenfeld, R., Hellman, L., Roffwarg, H., Weitzman, E., Fukushima, D., and Gallagher, T. 1971. Dehydroepiandrosterone is secreted episodically and synchronously with cortisol by normal men. *J. Clin. Endocrinol. Metab.* 33:87–92.

49. Irvine, W., Toft, A., Wilson, K., Fraser, R., Wilson, A., Young, J., Hunter, W., Ismail, A., and Burger, P. 1974. The effect of synthetic corticotropin on adrenocortical, anterior pituitary and testicular function. *J. Clin. Endocrinol. Metab.* 39:522–29.

50. Bell, J.B.G., Gould, R.P., Hyatt, P.J., Tait, J.F., and Tait, S.A.S. 1979. Properties of rat adrenal zona reticularis cells: Production and stimulation of certain steroids. *J. Endocr.* 83:435–47.

51. Forest, M.G. 1978. Age-related response to plasma testosterone, Δ^4 androstene dione and cortisol to adrenocorticotropin in infants, children and adults. *J. Clin. Endocrinol. Metab.* 47:931–37.

52. Odell, W., and Parker, L. 1980. Control of adrenal androgen secretion. In: A.R. Genazzani, J.H.H. Thijssen, and P.K. Siiteri, eds. *Adrenal Androgens.* New York: Raven Press, pp. 27–42..

53. Parker, L.N., Lifrak, E.T., and Odell, W.D. 1983. A 60,000 molecular weight human pituitary glycopeptide stimulates adrenal androgen secretion. *Endocrinology.* 113:2092.

54. Serio, M., Forti, G., Giusti, G., Bassi, F., Giannotti, P., Calabresi, E., Mantero, F., Armato, U., Fiorelli, G., and Pinchera, A. 1980. In vivo and in vitro effects of prolactin on adrenal androgen secretion. In: A.R. Genazzani, J.H.H. Thijssen, and P.K. Siiteri, eds. *Adrenal Androgens.* New York: Raven Press, pp. 71–81.

55. Higuchi, K., Nawata, H., Maki, T., Higashizima, M., Kato, K-I., and Ibayashi, H. 1984. Prolactin has a direct effect on adrenal androgen secretion. *J. Clin. Endocrinol. Metab.* 59:714–18.

56. Robba, C., Rebuffat, P., Mazzocchi, G., and Nussdorfer, G.G. 1985. Opposed effects of chronic prolactin administration on the zona fasciculata and zona reticularis of the rat adrenal cortex: an ultrastructural stereological study. *J. Submicrosc. Cytol.* 17:255–61.

57. Parker, L., Change, S., and Odell, W. 1978. Adrenal androgens in patients with chronic marked elevations of prolactin. *Clin. Endocrinol.* 8:1.

58. Varma, M.M., Huseman, C.A., Johanson, A.J., and Blizzard, R.M. 1977.

Effect of prolactin on adrenocortical and gonadal function in normal men. *J. Clin. Endocrinol. Metab.* 44:760.

59. Carter, J., Tyson, J., Warne, G., McNeilly, A., Faiman, C., and Friesen, H. 1977. Adrenocortical function in hyperprolactinaemic women. *J. Clin. Endocrinol. Metab.* 45:973.

60. Anderson, D.C. 1980. The adrenal androgen stimulating hormone does not exist. *Lancet.* II:454–56.

61. Dewis, P., and Anderson, D.C. 1985. The adrenarche and adrenal hirsutism. In: D.C. Anderson, and J.S.D. Winter, eds. *Adrenal Cortex.* London: Butterworths, pp. 96–119.

62. Hornsby, P.J. 1987. Physiological and pathological effects of steroids on the function of the adrenal cortex. *J. Steroid Biochem.* 27:1161–71.

63. Muller, J. 1987. *"Regulation of Aldosterone Biosynthesis"* Monographs on Endocrinology 29. Berlin: Springer-Verlag.

64. Swartz, S.L., Williams, G.H., Hollenberg, N.K., Dluhy, R.G., and Moore, T.J. 1980. Primacy of the renin-angiotensin system in mediating the aldosterone response to sodium restriction. *J. Clin. Endocrinol. Metab.* 50:1071–74.

65. Aguilera, G., and Catt, K.J. 1978. Regulation of aldosterone secretion by the renin-angiotensin system during sodium restriction in rats. *Proc. Natl. Acad. Sci. USA.* 75:4057–61.

66. Torretti, J. 1982. Sympathetic control of renin release. *Ann. Rev. Pharmacol. Toxicol.* 22:167–92.

67. Gibbons, G.H., Dzau, V.J., Farhi, E.R., and Barger, A.C. 1984. Interaction of signals influencing renin release. *Ann. Rev. Physiol.* 46:291–308.

68. Peach, M.J. 1977. The renin-angiotensin system. *Physiol. Rev.* 57:313–70.

69. Boucher, R., Demassieux, S., Garcia, R., and Genest, J. 1977. Tonin-angiotensin II system: A review. *Circ. Res.* 41, suppl. 1:1–26.

70. Schiffrin, E.L., Garcia, R., Gutkowska, J., Lis, M., and Genest, J. 1981. Effects of tonin on the adrenal secretion in the rat. *Hypertension.* 3 (suppl. II):II.9–II.13.

71. Lin, S.Y., and Goodfriend, T.L. 1970. Angiotensin receptors. *Am. J. Physiol.* 218:1319–28.

72. Ryan, J.W. 1972. Distribution of a reninlike enzyme in the bovine adrenal gland. *Experientia.* 29:407–08.

73. Ryan, J.W. 1967. Reninlike enzyme in the adrenal gland. *Science.* 158:1589–90.

74. Ganten, D., Hutchinson, J.S., Schelling, P., Ganten, U., and Fischer, H. 1976. The iso-renin angiotensin systems in extra-renal tissue. *Clin. Exp. Pharmacol. Physiol.* 3:102–26.

75. Doi, Y., Atarashi, K., Franco-Saenz, R., and Mulrow, P. 1983. Adrenal renin: A possible regulator of aldosterone production. *Clin. Exp. Hypertens* (A) 5:1119–26.

76. Ganten, D., Ganten, U., Kubo, S., Granger, P., Nowaczynski, W., Boucher, R., and Genest, J. 1974. Influence of sodium, potassium, and pituitary hormones on iso-renin in rat adrenal glands. *Am. J. Physiol.* 227:224–29.

77. Doi, Y., Atarashi, K., Franco-Saenz, R., and Mulrow, P.J. 1984. Effects of changes in sodium or potassium balance, and nephrectomy, on adrenal renin and aldosterone concentrations. *Hypertension.* 6 (suppl. 1):I-124-I-129.

78. Palmore, W.P., Marieb, N.J., and Mulrow, P.J. 1969. Stimulation of aldoste-

rone secretion by sodium depletion in nephrectomised rats. *Endocrinology.* 84:1342–51.

79. Del Vecchio, P.J., Ryan, J.W., Chung, A., and Ryan, U.S. 1980. Capillaries of the adrenal cortex possess aminopeptidase A and angiotensin-converting-enzyme activities. *Biochem. J.* 186:605–12.

80. Douglas, W.W. 1985. Polypeptides—angiotensin, plasma kinins and others. In: 7th ed. eds. *Goodman & Gilman's Pharmacological Basis of Therapeutics.* A.G. Goodman, L.S. Goodman, T.W. Rall, and F. Murad, New York: Macmillan. pp. 639–59.

81. Wright, F.S., and Briggs, J.P. 1979. Feedback control of glomerular blood flow, pressure, and filtration rate. *Physiol. Rev.* 59:958–1006.

82. Davis, J.O., and Freeman, R.H. 1976. Mechanisms regulating renin release. *Physiol. Rev.* 56:1–56.

83. Simpson, J.B., and Routenberg, A. 1973. Subfornical organs: Site of drinking illicitation by angiotensin II. *Science.* 181:1172–74.

84. Laragh, J.H., Angers, M., Kelly, W.G., and Lieberman, S. 1960. Hypotensive agents and pressor substances. The effect of epinephrine, norepinephrine, angiotensin II and others on the secretory rate of aldosterone in man. *J. Am. Med. Ass.* 174:234–40.

85. Coleman, T.G., McCaa, R. E., and McCaa, C.S. 1974. Effect of angiotensin II on aldosterone secretion in the conscious rat. *J. Endocr.* 60:421–27.

86. Tait, J.F., Tait, S.A.S., Gould, R.P., and Mee, M.S.R. 1974. The properties of adrenal zona glomerulosa cells after purification by gravitational sedimentation. *Proc. R. Soc. Lond. B.* 185:375–407.

87. Ganong, W.F., Mulrow, P.J., Boryczka, A., Cera, G. 1962. Evidence for a direct effect of angiotensin II on adrenal cortex of the dog. *Proc. Soc. Exp. Biol. Med.* 109:381–84.

88. Urquhart, J., Davis, J.O., and Higgins, J.T. 1963. Effects of prolonged infusion of angiotensin II in normal dogs. *Am. J. Physiol.* 205:1241–46.

89. Blair-West, J.R., Coghlan, J.P., Denton, D.A., Goding, J.R., Munro, J.A., Peterson, R.E., and Wintour, M. 1962. Humoral stimulation of adrenal cortical secretion. *J. Clin. Invest.* 41:1606–27.

90. Mason, P.A., Fraser, R., Morton, J.J., Semple, P.F., and Wilson, A. 1976. The effect of angiotensin II infusion on plasma corticosteroid concentrations in normal man. *J. Steroid Biochem.* 7:859–61.

91. Kaplan, N.M., and Bartter, F.C. 1962. The effect of ACTH, renin, angiotensin II and various precursors on biosynthesis of aldosterone by adrenal slices. *J. Clin. Invest.* 41:715–24.

92. Singer, B., Losito, C., and Salmon, S. 1964. Some studies on the effect of angiotensin II on adrenocortical secretion in hypophysectomised rats with renal pedicle ligation. *Endocrinology.* 74:325–32.

93. Marieb, N.J., and Mulrow, P.J. 1965. Role of the renin-angiotensin system in the regulation of aldosterone secretion in the rat. *Endocrinology.* 76:657–64.

94. Cade, R., and Perenich, T. 1965. Secretion of aldosterone by rats. *Am. J. Physiol.* 208:1026–30.

95. Glaz, E., and Sugar, K. 1962. The effect of synthetic angiotensin II on synthesis of aldosterone by the adrenals. *J. Endocr.* 24:299–302.

96. Marx, A.J., Deane, H.W., Mowles, T.F., and Sheppard, H. 1963. Chronic administration of angiotensin in rats: Changes in blood pressure, renal and

adrenal histophysiology and aldosterone production. *Endocrinology.* 73:329–37.

97. Haning, R., Tait, S.A.S., and Tait, J.F. 1970. In vitro effects of ACTH, angiotensins, serotonin and potassium on steroid output and conversion of corticosterone to aldosterone by isolated adrenal cells. *Endocrinology.* 87:1147–67.

98. Tait, J.F., Tait, S.A.S., Bell, J.B.G., Hyatt, P.J., and Williams, B.C. 1980. Further studies on the stimulation of rat adrenal capsular cells: Four types of response. *J. Endocr.* 87:11–27.

99. McKenna, T.J., Island, D.P., Nicholson, W.E. and Liddle, G.W. 1978. Angiotensin stimulates cortisol biosynthesis in human adrenal cells in vitro. *Steroids.* 32:127–36.

100. Douglas, J.G., Brown, G.P., and White, C. 1984. Angiotensin II receptors of human and primate adrenal fasciculata and glomerulosa: Correlations of binding and steroidogenesis. *Metabolism.* 33:685–88.

101. Glossman, H., Baukal, A.J., and Catt, K.J. 1974. Properties of angiotensin II receptors in the bovine and rat adrenal cortex. *J. Biol. Chem.* 249:825–34.

102. Catt, K.J., and Aguilera, G. 1980. Angiotensin II receptors. In: D. Schulster and A. Levitzki eds. *Cellular Receptors for Hormones and Neurotransmitters.* New York: Wiley & Sons, pp. 233–51.

103. Douglas, J.G., Aguilera, G., Kondo, T., and Catt, K.J. 1978. Angiotensin II receptors and aldosterone production in rat adrenal glomerulosa cells. *Endocrinology.* 102:685–96.

104. Maurer, R., and Reubi, J.C. 1986. Distribution and coregulation of three peptide receptors in adrenals. *Eur. J. Pharmacol.* 125:241–47.

105. Hinson, J.P., Vinson, G.P., and Whitehouse, B.J. 1988. Effects of dietary sodium restriction on peptide stimulation of aldosterone secretion by the isolated perfused rat adrenal gland in situ: A report of exceptional sensitivity to angiotensin II amide. *J. Endocr.* 119:83–88.

106. Aguilera, G., Hauger, R.L., and Catt, K.J. 1978. Control of aldosterone secretion during sodium restriction: Adrenal receptor regulation and increased adrenal sensitivity to angiotensin II. *Proc. Nat. Acad. Sci. USA.* 75:975–79.

107. Blair-West, J.R., Coghlan, J.P., Denton, D.A., Goding, J.R., Wintour, M., and Wright, R.D. 1965. Effect of variations of plasma sodium concentration on the adrenal response to angiotensin II. *Circ. Res.* 17:386–93.

108. Kenyon, C.J., Mosley, W., Hargreaves, G., Balment, R.J., and Henderson, I.W. 1978. The effects of dietary sodium restriction and potassium supplementation and hypophysectomy on adrenocortical function in the rat. *J. Steroid Biochem.* 9:337–44.

109. Aguilera, G., Menard, R.H., and Catt, K.J. 1980. Regulatory actions of angiotensin II on receptors and steroidogenic enzymes in adrenal glomerulosa cells. *Endocrinology.* 107:55–60.

110. Aguilera, G., Fujita, K., and Catt, K.J. 1981. Mechanisms of inhibition of aldosterone secretion by adrenocorticotropin. *Endocrinology.* 108:522–28.

111. Douglas, J.G., and Catt, K.J. 1976. Regulation of angiotensin II receptors in the rat adrenal cortex by dietary electrolytes. *J. Clin. Invest.* 58:834–43.

112. Fredlund, P., Saltman, S., and Catt, K.J. 1975. Aldosterone production by isolated adrenal glomerulosa cells: Stimulation by physiological concentrations of angiotensin II. *Endocrinology.* 97:1577–86.

113. Aguilera, G., Capponi, A., Baukal, A., Fujita, K., Hauger, R., and Catt, K.J.

1979. Metabolism and biological activities of angiotensin II and des-asp[1]-angiotensin II in isolated adrenal glomerulosa cells. *Endocrinology.* 104:1279–85.

114. Bravo, E.L., Khosla, M.C., and Bumpus, F.M. 1976. The role of angiotensins in aldosterone production. *Circ. Res.* 38, supp. 2:II104–07.

115. Blair-West, J.R., Coghlan, J.P., Denton, D.A., Fei, D.T.W., Hardy, K.J., Scoggins, B.A., and Wright, R.D. 1980. A dose-response comparison of the actions of angiotensin II and angiotensin III in sheep. *J. Endocr.* 87:409–17.

116. Kono, T., Oseko, F., Shimpo, S., Nanno, M., and Endo, J. 1975. Biological activity of des-asp[1] angiotensin II (angiotensin III) in man. *J. Clin. Endocrinol. Metab.* 41:1174–77.

117. Carey, R.M., Vaughan, E.D., Peach, M.J., and Ayers, C.R. 1978. Activity of [des-aspartyl[1]]-angiotensin II and angiotensin II in man. *J. Clin. Invest.* 61:20–31.

118. Semple, P.F., and Morton, J.J. 1976. Angiotensin II and angiotensin III in rat blood. *Circ. Res.* 38, supp. 2:II122–26.

119. Semple, P.F., Nicholls, M.G., Tree M., and Fraser, R. 1978. (Des-asp[1]) angiotensin II in the dog: blood levels and effect on aldosterone. *Endocrinology.* 103:1476–82.

120. Goodfriend, T.L., and Peach, M.J. 1975. Angiotensin III: (des-aspartic acid[1])-angiotensin II: Evidence and speculation for its role as an important agonist in the renin-angiotensin system. *Circ. Res.* 36, 37, supp 1:I-38–I-48.

121. Davis, J.O., and Freeman, R.H. 1977. The other angiotensins. *Biochem. Pharmacol.* 26:93–97.

122. Palmore, W.P., and Mulrow, P.J. 1967. Control of aldosterone secretion by the pituitary gland. *Science.* 158:1482–84.

123. Palmore, W.P., Anderson, R., and Mulrow, P.J. 1970. Role of the pituitary in controlling aldosterone production in sodium depleted rats. *Endocrinology.* 86:728–34.

124. Giroud, C.J.P., Stachenko, J., and Venning, E.H. 1956. Secretion of aldosterone by rat adrenal glands in vitro. *Proc. Soc. Exp. Biol. Med.* 92:154–58.

125. Tait, S.A.S., Tait, J.F., Gould, R.P., Brown, B.L., and Albano, J.D.M. 1974. The preparation and use of purified and unpurified dispersed adrenal cells and a study of the relationship of their cyclic AMP and steroid output. *J. Steroid Biochem.* 5:775–87.

126. Tait, J.F., Tait, S.A.S., and Bell, J.B.G. 1980. Steroid hormone production by mammalian adrenocortical dispersed cells. Ed. P.N. Campbell and R.D. Marshall. Biochemical Society and Academic Press, London. Essays in biochemistry 16:99–175.

127. Gordon, R.D., Nicholls, M.G., Tree, M., Fraser, R, and Robertson, J.I.S. 1980. Influence of sodium balance on ACTH/adrenal corticosteroid dose response curves in the dog. *Am. J. Physiol.* 238:E543–51.

128. Yamakado, M., Franco-Saenz, R., and Mulrow, P.J. 1983. Effect of sodium deficiency on β-Melanocyte stimulating hormone stimulation of aldosterone in isolated rat adrenal cells. *Endocrinology.* 113:2168–72.

129. Vinson, G.P., Whitehouse, B.J., Dell, A., Bateman, A., and McAuley, M.E. 1983. α-MSH and zona glomerulosa function in the rat. *J. Steroid Biochem.* 19:537–44.

130. Muller, J. 1970. Decreased aldosterone production by rat adrenal tissue in

vitro due to treatment with 9α-fluorocortisol, dexamethasone, and adrenocorticotropin in vivo. *Acta Endocr.* 63:1–10.

131. Shenker, Y., Villareal, J.Z., Sider, R.S., and Grekin, R.J. 1985. α-melanocyte-stimulating hormone stimulation of aldosterone secretion in hypophysectomized rats. *Endocrinology.* 116:138–41.

132. Coates, P.J., Doniach, I., Hale, A.C., and Rees, L.H. 1986. The distribution of immunoreactive α-melanocyte stimulating hormone in the human pituitary gland. *J. Endocr.* 111:335–42.

133. Vinson, G.P., Whitehouse, B.J., Dell, A., Etienne, A., and Morris, H.R. 1980. Characterization of an adrenal zona glomerulosa-stimulating component of posterior pituitary extracts as α-MSH. *Nature.* 284:464–67.

134. Vinson, G.P., Whitehouse, B.J., Dell, A., Etienne, A.T., and Morris, H.R. 1981. Specific stimulation of steroidogenesis in rat adrenal zona glomerulosa cells by pituitary peptides. *Biochem. Biophys. Res. Comm.* 99:65–72.

135. Nussdorfer, G.G., Mazzocchi, G., and Malendowicz, L.K. 1986. Acute effects of α-MSH on the rat zona glomerulosa in vivo. *Biochem. Biophys. Res. Comm.* 141:1279–84.

136. Robba, C., Rebuffat, P., Mazzocchi, G., and Nussdorfer, G.G. 1986. Long-term trophic action of α-melanocyte stimulating hormone on the zona glomerulosa of the rat adrenal cortex. *Acta Endocr.* 112:404–08.

137. Coates, P.J., McNicol, A.M., Doniach, I., and Rees, L.H. 1988. Increased production of α-melanocyte stimulating hormone in the pituitary gland of patients with untreated Addison's disease. *Clin. Endocr.* 29:421–26.

138. Henville, K.L., Hinson, J.P., Vinson, G.P., and Laird, S.M. 1989. Actions of desacetyl-α-melanocyte stimulating hormone on human adrenocortical cells. *J. Endocr.* 121:579–83.

139. Vinson, G.P., Whitehouse, B.J., and Thody, A.J. 1981. α-MSH at physiological concentrations stimulates "late pathway" steroid products in adrenal zona glomerulosa cells from sodium restricted rats. *Peptides.* 2:141–44.

140. Matsuoka, H., Mulrow, P.J., Franco-Saenz, R., and Li, C.H. 1981. Stimulation of aldosterone production by β-melanotropin. *Nature.* 291:155–56.

141. Al-Dujaili, E.A.S., Hope, J., Estivariz, F.E., Lowry, P.J., and Edwards, C.R.W. 1981. Circulating human pituitary pro-γ-melanotropin enhances the adrenal response to ACTH. *Nature.* 291:156–59.

142. Pedersen, R.C., and Brownie, A.C. 1980. Adrenocortical response to corticotropin is potentiated by part of the amino terminal region of pro-corticotropin/endorphin. *Proc. Nat. Acad. Sci. USA.* 77:2239–43.

143. Matsuoka, H., Mulrow, P.J., and Li, C.H. 1980. β-Lipotropin: a new aldosterone stimulating factor. *Science.* 209:307–08.

144. Washburn, D.D., Kem, D.C., Orth, D.N., Nicholson, W.E., Chretien, M., and Mount, C.D. 1982. Effect of β-lipotropin on aldosterone production in the isolated rat adrenal preparation. *J. Clin. Endocrinol. Metab.* 54:613–18.

145. Pham Huu Trung, M.T., Bogyo, A., Leneuve, P., and Girard, F. 1986. Compared effects of ACTH, angiotensin II and POMC peptides on isolated human adrenal cells. *J. Steroid Biochem.* 24:345–48.

146. Eggens, U., Bahr, V., Oelkers, W., and Li, C.H. 1987. Effects of β-lipotropin, β-endorphin, γ$_3$-melanotropin and corticotropin on steroid production by isolated human adrenocortical cells. *J. Clin. Chem. Clin. Biochem.* 25:779–83.

147. Gullner, H.G., and Gill, J.R. 1983. Beta endorphin selectively stimulates

aldosterone secretion in hypophysectomized, nephrectomized dogs. *J. Clin. Invest.* 71:124–28.

148. Shanker, G., and Sharma, R.K. 1979. β-endorphin stimulates corticosterone synthesis in isolated rat adrenal cells. *Biochem. Biophys. Res. Comm.* 86:1–5.

149. Szalay, K.S., and Stark, E. 1981. Effect of beta-endorphin on the steroid production of isolated zona glomerulosa and zona fasciculata cells. *Life Sci.* 29:1355–61.

150. Schneider, E.G., Radke, K.J., Ulderich, D.A., and Taylor, R.E. 1985. Effect of osmolality on aldosterone secretion. *Endocrinology.* 116:1621–26.

151. Balla, T., Nagy, K., Tarjan, E., Renczes, G., and Spät, A. 1981. Effect of reduced extracellular sodium concentration on the function of adrenal zona glomerulosa: Studies in conscious rats. *J. Endocr.* 89:411–16.

152. Blair-West, J.R., Coghlan, J.P., Denton, D.A., Goding, J.R., Wintour, M., and Wright, R.D. 1966. The direct effect of increased sodium concentration in adrenal arterial blood on corticosteroid secretion in sodium deficient sheep. *Aus. J. Exp. Biol. Med. Sci.* 44:455–74.

153. Dluhy, R.G., Axelrod, L., Underwood, R.H., and Williams, G.H. 1972. Studies of the control of plasma aldosterone concentrations in normal man. II. Effect of dietary potassium and acute potassium infusion. *J. Clin. Invest.* 51:1950–57.

154. Boyd, J.E., Palmore, W.P., and Mulrow, P.J. 1971. Role of potassium in the control of aldosterone secretion in the rat. *Endocrinology.* 88:556–65.

155. Davis, J.O., Urquhart, J., and Higgins, J.T. 1963. Effects of alterations of plasma sodium and potassium concentration on aldosterone secretion. *J. Clin. Invest.* 42:587–609.

156. Himathongkam, T., Dluhy, R.G., and Williams, G.H. 1975. Potassium aldosterone-renin interrelationships. *J. Clin. Endocrinol. Metab.* 41:153–59.

157. Boyd, J.F., Mulrow, P.J., Palmore, W.P., and Silvio, P. 1973. Importance of potassium in the regulation of aldosterone production. *Circ. Res.* 32, supp. 1:39–45.

158. Braley, L.M., and Williams, G.H. 1977. Rat adrenal sensitivity to angiotensin II, (1–24)-ACTH and potassium: A comparative study. *Am. J. Physiol.* 233:E402–06.

159. Tait, J.F., and Tait, S.A.S. 1976. The effect of changes in potassium concentration on the maximal steroidogenic response of purified zona glomerulosa cells to angiotensin II. *J. Steroid Biochem.* 7:687–90.

160. Muller, J. 1965. Aldosterone stimulation in vitro II. Stimulation of aldosterone production by monovalent cations. *Acta Endocr.* 50:301–09.

161. De Bold, A.J., Borenstein, H.B., and Veress, A.T. 1981. A rapid and potent natriuretic response to intravenous injection of atrial myocardial extract in rats. *Life Sci.* 28:89–94.

162. Kangawa, K., and Matsuo, H. 1984. Purification and complete amino acid sequence of α-human atrial natriuretic polypeptide (α-ANP). *Biochem. Biophys. Res. Comm.* 118:131–39.

163. Schwartz, D., Geller, D.M., Manning, P.T., Siegel, N.R., Fik, K.F., Smith, C.E., and Needleman, P. 1985. Ser-Leu-Arg-Arg-Atriopeptin III: the major circulating form of atrial peptide. *Science.* 229:397–400.

164. Thibault, G., Lazure, C., Schiffrin, E.L., Gutkowska, J., Chartier, L., Garcia, R., Sediah, N.H., Chretien, M., Genest, J., and Cantin, M. 1985. Identification

of a biologically active circulating form of rat atrial natriuretic factor. *Biochem. Biophys. Res. Comm.* 130:981–86.

165. Atlas, S.A. 1986. Atrial natriuretic factor: A new hormone of cardiac origin. *Rec. Prog. Horm. Res.* 42:207–49.

166. Maack, T., Marion, D.N., Camargo, M.J.F., Kleinert, H.D., Laragh, J.H., Vaughan, E.D., and Atlas, S.A. 1984. Effects of auriculin (atrial natriuretic factor) on blood pressure, renal function and the renin-aldosterone system in dogs. *Am. J. Med.* 77:1069–75.

167. Richards, A.M., Nicholls, M.G., Espiner, E.A., Ikram, H., Yandle, T.G., Joyce, S.L., and Cullens, M.M. 1985. Effects of alpha-human atrial natriuretic peptide in essential hypertension. *Hypertension.* 7:812–17.

168. Atarashi, K., Mulrow, P.J., and Franco-Saenz, R. 1985. Effect of atrial peptides on aldosterone production. *J. Clin. Invest.* 76:1807–11.

169. Chartier, L., Schiffrin, E., Thibault, G., and Garcia, R. 1984. Atrial natriuretic factor inhibits the stimulation of aldosterone secretion by angiotensin II, ACTH and potassium in vitro and angiotensin II-induced steroidogenesis in vivo. *Endocrinology.* 115:2026–28.

170. Goodfriend, T.L., Elliott, M.E., and Atlas, S.A. 1984. Actions of synthetic atrial natriuretic factor on bovine adrenal glomerulosa. *Life Sci.* 35:1675–82.

171. Kudo, T., and Baird, A. 1984. Inhibition of aldosterone production in the adrenal glomerulosa by atrial natriuretic factor. *Nature.* 312:756–57.

172. Campbell, W.B., Currie, M.G., and Needleman, P. 1985. Inhibition of aldosterone biosynthesis by atriopeptins in rat adrenal cells. *Circ. Res.* 57:113–18.

173. Hirose, S., Akiyama, F., Shinjo, M., Ohno, H., and Murakami, K. 1985. Solubilization and molecular weight estimation of atrial natriuretic factor receptor from bovine adrenal cortex. *Biochem. Biophys. Res. Comm.* 130:574–79.

174. De Lean, A., Racz, K., Gutkowska, J., Nguyen, T-T., Cantin, M., and Genest, J. 1984. Specific receptor-mediated inhibition by synthetic atrial natriuretic factor of hormone-stimulated steroidogenesis in cultured bovine adrenal cells. *Endocrinology.* 115:1636–38.

175. Rosencrantz, H. 1959. A direct effect of 5-hydroxytryptamine on the adrenal cortex. *Endocrinology.* 64:355–62.

176. Jouan, P. 1967. Properties adrenoglomerulotrophique de la 5-hydroxytryptamine. *Pathol. Biol.* 15:1145–53.

177. Tait, S.A.S., Tait, J.F., and Bradley, J.E.S. 1972. The effect of serotonin and potassium on corticosterone and aldosterone production by isolated zona glomerulosa cells of the rat adrenal cortex. *Aust. J. Exp. Biol. Med.* 50:833–46.

178. Muller, J., and Ziegler, W.H. 1968. Stimulation of aldosterone biosynthesis in vitro by serotonin. *Acta Endocr.* 59:23–35.

179. Verdesca, A.S., Westerman, C.D., Crampton, R.S., Black, W.C., Nedeljkovic, R.I., and Hilton, J.G. 1961. Direct adrenocortical stimulatory effect of serotonin. *Am. J. Physiol.* 201:1065–67.

180. Matsuoka, H., Ishii, M., Goto, A., and Sugimoto, T. 1985. Role of serotonin type 2 receptors in regulation of aldosterone production. *Am. J. Physiol.* 249:E234–E238.

181. Zinner, M.J., Kasher, F., and Jaffe, B.M. 1983. The hemodynamic effects of intravenous infusions of serotonin in conscious dogs. *J. Surg. Res.* 34:171–78.

182. Delarue, C., Lefebre, H., Idres, S., Leboulenger, F., Homo-Delarche, G., Lihrman, I., Feuilloly, M., and Vaudry, H. 1988. Serotonin stimulates corticosteroid secretion by frog adrenocortical tissue in vitro. *J. Steroid Biochem.* 29:519–25.

183. De Lean, A., Racz, K., McNicoll, N., and Desrosiers, M-L. 1984. Direct β-adrenergic stimulation of aldosterone secretion in cultured bovine adrenal subcapsular cells. *Endocrinology.* 115:485–92.

184. Sequeira, S.J., and McKenna, T.J. 1985. Examination of the effects of epinephrine, norepinephrine, and dopamine on aldosterone production in bovine glomerulosa cells in vitro. *Endocrinology.* 117:1947–52.

185. Horiuchi, T., Tanaka, K., and Shimizu, N. 1987. Effect of catecholamine on aldosterone release in isolated rat glomerulosa cell suspensions. *Life Sci.* 40:2421–28.

186. Inglis, G.C., Kenyon, C.J., Hannah, J.A.M., Connell, J.M.C., and Ball, S.G. 1987. Does dopamine regulate aldosterone secretion in the rat? *Clin. Sci.* 73:93–97.

187. Pratt, J.H., Turner, D.A., Bowsher, R.R., and Henry, D.P. 1987. Dopamine in rat adrenal cortex. *Life Sci.* 40:811–16.

188. Holzwarth, M.A., Cunningham, L.A., and Kleitman, N. 1987. The role of adrenal nerves in the regulation of adrenocortical functions. *Ann. N.Y. Acad. Sci.* 512:449–64.

189. Rosenfeld, G. 1955. Stimulative effect of acetylcholine on the adrenocortical function of isolated perfused calf adrenals. *Am. J. Physiol.* 183:272–78.

190. Hadjian, A.J., Guidicelli, C., and Chambaz, E.M. 1982. Cholinergic muscarinic stimulation of steroidogenesis in bovine adrenal cortex fasciculata cell suspensions. *Biochim. Biophys. Acta.* 714:157–63.

191. Benyamina, M., Leboulenger, F., Lirhmann, I., Delarue, C., Feuilloley, M., and Vaudry, H. 1987. Acetylcholine stimulates steroidogenesis in isolated frog adrenal gland through muscarinic receptors: Evidence for a desensitization mechanism. *J. Endocrinology.* 113:339–48.

192. Kojima, I., Kojima, K., Shibata, H., and Ogata, E. 1986. Mechanism of cholinergic stimulation of aldosterone secretion in bovine adrenal glomerulosa cells. *Endocrinology.* 119:284–91.

193. Hadjian, A.J., Ventre, R., and Chambaz, E.M. 1981. Cholinergic muscarinic receptors in bovine adrenal cortex. *Biochem. Biophys. Res. Comm.* 98:892–900.

194. Unsicker, K. 1969. Zur Innervation der Nebennierenrinde vom Goldenhamster. *Zeitschr. zellforsch. mikrosc. anat.* 95:608–19.

195. Robinson, P.M., Perry, R.A., Hardy, K.J., Coghlan, J.P., and Scoggins, B.A. 1977. The innervation of the adrenal cortex in the sheep. *Ovis ovis. J. Anat.* 124:117–29.

196. Holzwarth, M.A. 1984. The distribution of vasoactive intestinal peptide in the rat adrenal cortex and medulla. *J. Autonom. Nerv. Sys.* 11:269–83.

197. Nussdorfer, G.G., and Mazzocchi, G. 1987. Vasoactive intestinal peptide (VIP) stimulates aldosterone secretion by rat adrenal glands in vivo. *J. Steroid Biochem.* 26:203–06.

198. Enyedi, P., Szabo, B., and Spät, A. 1983. Failure of vasoactive intestinal peptide to stimulate aldosterone production. *Acta Physiol. Hung.* 61:77–79.

199. Leboulenger, F., Leroux, P., Delarue, C., Tonon, M.C., Charnay, Y., Dubois, P.M., Coy, D.H., and Vaudry, H. 1983. Co-localization of vasoactive intestinal

peptide (VIP) and enkephalins in chromaffin cells of the adrenal gland of amphibia. Stimulation of corticosteroid production by VIP. *Life Sci.* 32:375–83.

200. Nicholson, H.D., Swann, R.W., Burford, G.D., Wathes, D.C., Porter, D.G., and Pickering, B.T. 1984. Identification of oxytocin and vasopressin in the testis and in adrenal tissue. *Reg. Peptides.* 8:141–46.

201. Nussey, S.S., Ang, V.T.Y., Jenkins, J.S., Chowdrey, H.S., and Bisset, G.W. 1984. Brattleboro adrenal contains vasopressin. *Nature.* 310:64–66.

202. Hawthorn, J., Nussey, S.S., Henderson, J.R., and Jenkins, J.S. 1987. Immuno-histochemical localization of oxytocin and vasopressin in the adrenal glands of rat, cow, hamster, and guinea pig. *Cell Tiss. Res.* 250:1–6.

203. Woodcock, E.A., McLeod, J.K., and Johnston, C.I. 1986. Vasopressin stimulates phosphatidiylinositol turnover and aldosterone synthesis in rat glomerulosa cells: Comparison with angiotensin II. *Endocrinology.* 118:2432–36.

204. Hinson, J.P., Vinson, G.P., Porter, I.D., and Whitehouse, B.J. 1987. Oxytocin and arginine vasopressin stimulate steroid secretion by the isolated perfused rat adrenal gland. *Neuropeptides.* 10:1–7.

205. Payet, N., and Lehoux, J-G. 1982. Aldosterone and corticosterone stimulation by ACTH in isolated rat adrenal glomerulosa cells: interaction with vasopressin. *J. Physiol. Paris.* 78:317–21.

206. Balla, T., Enyedi, P., Spät, A., and Antoni, P.A. 1985. Pressor-type vasopressin receptors in the adrenal cortex: Properties of binding, effects on phosphoinositide metabolism and aldosterone secretion. *Endocrinology.* 117:421–23.

207. Sen, S., Bravo, E.L., and Bumpus, F.M. 1977. Isolation of a hypertension producing compound from normal human urine. *Circ. Res.* 40, supp. 1:I5–I10.

208. Sen, S., Shainoff, J.R., Bravo, E.L., and Bumpus, F.M. 1981. Isolation of aldosterone-stimulating-factor (ASF) and its effect on rat adrenal glomerulosa cells in vitro. *Hypertension.* 3:4–10.

209. Carey, R.M., and Sen, S. 1986. Recent progress in the control of aldosterone secretion. *Rec. Prog. Horm. Res.* 42:251–95.

210. Aguilera, G., Harwood, J.P., and Catt, K.J. 1981. Somatostatin modulates effects of angiotensin II in adrenal glomerulosa zone. *Nature.* 292:262–63.

211. Miller, J.A. 1983. Somatostatin attenuates aldosterone responses to angiotensin II in normal subjects. *Proc. Univ. Otago Med. Sch.* 61:77–79.

212. Mazzocchi, G., Robba, C., Rebuffat, P., Gottardo, G., and Nussdorfer, G.G. 1985. Effect of somatostatin on the zona glomerulosa of rats treated with angiotensin II or captopril: stereology and plasma hormone concentrations. *J. Steroid Biochem.* 23:353–56.

213. Aguilera, G., Parker, D.S., and Catt, K.J. 1982. Characterization of somatostatin receptors in the rat adrenal glomerulosa zone. *Endocrinology.* 111:1376–84.

214. Maurer, R., and Reubi, J.C. 1986. Somatostatin receptors in the adrenal. *Mol. Cell. Endocr.* 45:81–90.

215. Delarue, C., Netchitailo, P., Leboulenger, F., Perroteau, I., Escher, E., and Vaudry, H. 1984. In vitro study of frog (*Rana ridibunda* Pallas) interrenal function by use of a simplified perifusion system. VII. Lack of effect of somatostatin. *Gen. Comp. Endocr.* 54:333–38.

216. Srikant, C.B., and Patel, Y.C. 1985. Somatostatin receptors in the rat adrenal

cortex: Characterization and comparison with brain and pituitary receptors. *Endocrinology*. 116:1717–23.

217. Carey, R.M., Thorner, M.O., and Ortt, E.M. 1979. Effects of metoclopramide on the renin-angiotensin-aldosterone system in man: dopaminergic control of aldosterone secretion. *J. Clin. Invest*. 63:727–35.

218. Edwards, C.R.W., Miall, P.A., Hanker, J.P., Thorner, M.O., Al-Dujaili, E.A.S., and Besser, G.M. 1975. Inhibition of the plasma aldosterone response to furosemide by bromocriptine. *Lancet*. II:903–04.

219. Norbiato, G.M., Bevilacqua, M., Raggi, D., Micossi, P., and Moroni, C. 1977. Metoclopramide increases plasma aldosterone in man. *J. Clin. Endocrinol. Metab*. 45:1313–16.

220. Aguilera, G., and Catt, K.J. 1984. Dopaminergic modulation of aldosterone secretion in the rat. *Endocrinology*. 114:176–81.

221. Campbell, D.J., Mendelsohn, F.A.O., Adam, W.R., and Funder, J.W. 1981. Metoclopramide does not elevate aldosterone in the rat. *Endocrinology*. 109:1484–91.

222. Ganguly, A. 1984. Dopaminergic regulation of aldosterone secretion: How credible? *Clin. Sci*. 66:631–37.

223. Sowers, J.R., Sharp, B., Levin, E.R., Golub, M.S., and Eggena, P. 1981. Metoclopramide, a dopamine antagonist, stimulates aldosterone secretion in rhesus monkeys but not in dogs or rabbits. *Life Sci*. 29:2171–75.

224. Dunn, M.G., and Bossman, G.H. 1981. Peripheral dopamine receptor identification: Properties of a specific dopamine receptor in the rat adrenal zona glomerulosa. *Biochem. Biophys. Res. Comm*. 99:1081–87.

225. Cuche, J.L. 1988. Dopaminergic control of aldosterone secretion. State of the art review. *Fund. Clin. Pharm*. 2:327–39.

226. Connell, J.M.C., Tonolo, G., Davies, D.L., Finlayson, J., Ball, S.G., Inglis, G., and Fraser, R. 1987. Dopamine affects angiotensin II-induced steroidogenesis by altering clearance of the peptide in man. *J. Endocr*. 113:139–46.

227. Whitehouse, B.J., Vinson, G.P., and Thody, A.J. 1982. Dopaminergic control of aldosterone: Modulation of the response of rat adrenal glomerulosa cells by pretreatment with bromocriptine or metoclopramide. *Steroids*. 39:155–63.

228. Proulx-Ferland, L., Meunier, H., Cote, J., Dumont, D., Gagne, B., and Labrie, F. 1983. Multiple factors involved in the control of ACTH and α-MSH secretion. *J. Steroid Biochem*. 19:439–45.

229. Enyedi, P., Spät, A., and Antoni, F.A. 1981. Role of prostaglandins in the control of the function of adrenal glomerulosa cells. *J. Endocr*. 91:427–37.

230. Honn, K.V., and Chavin, W. 1976. Role of prostaglandins in aldosterone production by the human adrenal. *Biochem. Biophys. Res. Comm*. 72:1319–26.

231. Saruta, T., and Kaplan, N.M. 1972. Adrenocortical steroidogenesis: The effects of prostaglandins. *J. Clin. Invest*. 51:2246–51.

232. Honn, K.V., and Chavin, W. 1976. Prostaglandin modulation of the mechanism of ACTH action in the human adrenal. *Biochem. Biophys. Res. Comm*. 73:164–70.

233. Golub, M.S., Speckart, P.F., Zia, P.K., and Horton, R. 1976. The effect of prostaglandin A_2 on renin and aldosterone in man. *Circ. Res*. 39:574–79.

234. Yoshimura, M., Takahashi, H., Takashina, R., Kajita, Y., Miyazaki, T., Ijichi, H., and Ochi, Y. 1979. Effect of prostaglandin E_1 on renin and aldosterone in hypertensive patients. *Endocrinol. Jpn*. 26:481–86.

References   **129**

235. Katz, F.H., Romfh, P., and Smith, J.A. 1975. Diurnal variation of plasma aldosterone, cortisol and renin activity in supine man. *J. Clin. Endocrinol. Metab.* 40:125–31.

236. James, V.H.T., Tunbridge, R.D.G., and Wilson, G.A. 1976. Studies on the control of aldosterone secretion in man. *J. Steroid Biochem.* 7:941–48.

237. DeForrest, J.M., Davis, J.O., Freeman, R.H., Stephens, G.A., and Watkins, B.E. 1979. Circadian changes in plasma renin activity and plasma aldosterone concentration in two-kidney hypertensive rats. *Hypertension.* 1:142–49.

238. Cugini, P., Manconi, R., Serdoz, R., Mancini, A., and Meucci, T. 1977. Influence of propranolol on circadian rhythms of plasma renin, aldosterone, and cortisol in healthy supine man. *Boll. Soc. Ital. Biol. Sper.* 53:263–69.

239. Hilfenhaus, M. 1974. In: J. Aschoff, F. Ceresa and F. Hallberg, eds. *Chronobiological Aspects of Endocrinology.* Schattauer, Stuttgart, pp. 111–15.

240. Michaelakis, A.M., and Horton, R. 1970. The relationship between plasma renin and aldosterone in normal man. *Circ. Res.* 26/27, supp. 1:I185–94.

241. Williams, G.H., Cain, J.P., Dluhy, R.G., and Underwood, R.H. 1972. Studies on the control of plasma aldosterone concentration in normal man. 1. Response to posture, acute and chronic volume depletion and sodium loading. *J. Clin. Invest.* 51:1731–42.

242. Balikian, H.M., Brodie, A.H., Dale, S.L., Melby, J.C., and Tait, J.F. 1968. Effect of posture on the metabolic clearance rate, plasma concentration and blood production rate of aldosterone in man. *J. Clin. Endocrinol. Metab.* 28:1630–40.

243. Hunter, T., and Cooper, J.A. 1985. Protein-tyrosine kinases. *Ann. Rev. Biochem.* 54:897–930.

244. Carpenter, G. 1987. Receptors for epidermal growth factor and other polypeptide mitogens. *Ann. Rev. Biochem.* 56:881–914.

245. Chabot, J-G., Walker, P., and Pelletier, G. 1986. Distribution of epidermal growth factor binding sites in the adult rat adrenal gland by light microscope autoradiography. *Acta Endocr.* 113:391–95.

246. Rasmussen, H., and Barrett, P.Q. 1984. Calcium messenger system: An integrated view. *Physiol. Rev.* 64:938–84.

247. Robison, G.A., Butcher, R.W., and Sutherland, E.W. 1971. *Cyclic AMP.* New York: Academic Press.

248. Hokin, L.E. 1985. Receptors and phosphinositide-generated second messengers. *Ann. Rev. Biochem.* 54:205–35.

249. Berridge, M.J. 1987. Inositol trisphosphate and diacylglycerol: Two interacting second messengers. *Ann. Rev. Biochem.* 56:159–93.

250. Abdel-Latif, A.A. 1986. Calcium-mobilizing receptors, polyphosphoinositides, and the generation of second messengers. *Pharm. Rev.* 38:227–72.

251. Haynes, R.C. 1958. The activation of adrenal phosphorylase by the adrenocorticotropic hormone. *J. Biol. Chem.* 233:1220–22.

252. Haynes, R.C., Koritz, S.B., and Peron, F.G. 1959. Influence of adenosine 3′,5′-monophosphate on corticoid production by rat adrenal glands. *J. Biol. Chem.* 234:1421–23.

253. Grahame-Smith, D.G., Butcher, R.W., Ney, R.L., and Sutherland, E.W. 1967. Adenosine 3′,5′-monophosphate as the intracellular mediator of the action of adrenocorticotropic hormone on the adrenal cortex. *J. Biol. Chem.* 242:5535.

254. Schimmer, B.P., Ueda, K., and Sato, G.H. 1968. Site of action of adrenocortico-trophic hormone (ACTH) in adrenal cell cultures. *Biochem. Biophys. Res. Comm.* 32:806–10.

255. Buckley, D.I., and Ramachandran, J. 1981. Characterization of corticotropin receptors on adrenocortical cells. *Proc. Nat. Acad. Sci. USA.* 78:7431–35.

256. Catalano, R.D., Stuve, L., and Ramachandran, J. 1986. Characterization of corticotropin receptors in human adrenocortical cells. *J. Clin. Endocrinol. Metab.* 62:300–04.

257. Gallo-Payet, N., and Escher, E. 1985. Adrenocorticotropin receptors in rat adrenal glomerulosa cells. *Endocrinology.* 117:38–46.

258. Buckingham, J.C., and Hodges, J.R. 1976. Hypothalamo-pituitary adrenocorti-cal function in the rat after treatment with betamethasone. *Br. J. Pharmacol.* 56:235–39.

259. Ruhmann-Wenhold, A., and Nelson, D.H. 1977. Plasma ACTH in stressed and nonstressed adrenalectomised rats. *Ann. N.Y. Acad. Sci.* 297:498–506.

260. Ramachandran, J. 1985. Corticotropin receptors, cyclic AMP and steroidogene-sis. *Endocr. Res.* 10:347–63.

261. Bost, K.L., and Blalock, J.E. 1986. Molecular characterization of a corticotropin (ACTH) receptor. *Mol. Cell. Endocr.* 44:1–9.

262. Taunton, O.D., Roth, J., and Pastan, I. 1967. ACTH stimulation of adenyl cyclase in adrenal homogenates. *Biochem. Biophys. Res. Comm.* 29:1–7.

263. Levitzki, A. 1987. Regulation of hormone-sensitive adenylate cyclase. *TIPS* 8:299–303.

264. Hildebrandt, J.D., Sekura, R.D., Codina, J., Iyengar, R., Manclark, C.R., and Birnbaumer, L. 1983. Stimulation and inhibition of adenylyl cyclases mediated by distinct regulatory proteins. *Nature.* 302:706–09.

265. Birnbaumer, L., Codina, J., Mattera, R., Cerione, R.A., Hildebrandt, J.D., Sunyer, T., Rojas, F.J., Caron, M.G., Lefkowitz, R.J., and Iyengar, R. 1985. Regulation of hormone receptors and adenylyl cyclases by guanine nucleotide binding N proteins. *Rec. Prog. Horm. Res.* 41:41–99.

266. Glynn, P., Cooper, D.M., and Schulster, D. 1979. The regulation of adenylate cyclase of the adrenal cortex. *Mol. Cell. Endocr.* 13:99–121.

267. Ross, E.M., and Gilman, A.G. 1980. Biochemical properties of hormone-sensitive adenylate cyclase. *Ann. Rev. Biochem.* 49:533–64.

268. Marie, J., and Jard, S. 1983. Angiotensin II inhibits adenylate cyclase from adrenal cortex glomerulosa zone. *FEBS lett.* 159:97–101.

269. Woodcock, E.A., and Johnston, C.I. 1984. Inhibition of adenylate cyclase in rat adrenal glomerulosa cells by angiotensin II. *Endocrinology.* 115:337–41.

270. Hausdorff, W.P., Sekura, R.D., Aguilera, G., and Catt, K.J. 1987. Control of aldosterone production by angiotensin II is mediated by two guanine nucleotide regulatory proteins. *Endocrinology.* 120:1668–78.

271. Gill, G.N., and Garren, L.D. 1969. On the mechanism of action of adrenocorti-cotropic hormone: the binding of cyclic-3′,5′-adenosine monophosphate to an adrenal cortical protein. *Proc. Nat. Acad. Sci.* 63:512–19.

272. Sala, G.B., Hayashi, K., Catt, K.J., and Dufau, M.L. 1979. Adrenocorticotropin action in isolated adrenal cells. *J. Biol. Chem.* 254:3861–65.

273. Garren, L.D., Gill, G.N., Masui, H., and Walton, G.M. 1971. On the mechanism of action of ACTH. *Rec. Prog. Horm. Res.* 27:433–78.

274. Walsh, D.A. 1978. Role of cAMP dependent protein kinase as the transducer of cAMP action. *Biochem. Pharmacol.* 27:1801–04.

275. Boyd, G.S., McNamara, B., Suckling, K.E., and Tocher, D.R. 1983. Cholesterol metabolism in the adrenal cortex. *J. Steroid Biochem.* 19:1017–27.

276. Mackie, C., Richardson, M.C., and Schulster, D. 1972. Kinetics and dose-response characteristics of adenosine 3'5'-monophosphate production by isolated rat adrenal cells stimulated with adrenocorticotropic hormone. *FEBS lett.* 23:345–48.

277. Yanagibashi, K., Kamiya, N., Lin, G., and Matsuba, M. 1978. Studies on adrenocorticotropic hormone receptor using isolated rat adrenocortical cells. *Endocrinol. Jpn.* 25:545–51.

278. Fujita, K., Aguilera, G., and Catt, K.J. 1979. The role of cyclic AMP in aldosterone production by isolated zona glomerulosa cells. *J. Biol. Chem.* 254:8567–74.

279. Albano, J.D.M., Brown, B.L., Ekins, R.P., Tait, S.A.S., and Tait, J.F. 1974. The effects of potassium, 5-hydroxytryptamine, adrenocorticotropin and angiotensin II on the concentration of adenosine 3',5'-monophosphate in suspensions of dispersed rat adrenal zona glomerulosa and zona fasciculata cells. *Biochem. J.* 142:391–400.

280. Hyatt, P.J., Tait, J.F., and Tait, S.A.S. 1986. The mechanism of the effect of K$^+$ on the steroidogenesis of rat zona glomerulosa cells of the adrenal cortex: Role of cyclic AMP. *Proc. R. Soc. Lond. B.* 227:21–42.

281. Kojima, I., Kojima, K., and Rasmussen, H. 1985. Intracellular calcium and adenosine 3',5'-cyclic monophosphate as mediators of potassium-induced aldosterone secretion. *Biochem. J.* 228:69–76.

282. Fakunding, J.L., Chow, R., and Catt, K.J. 1979. The role of calcium in the stimulation of aldosterone production by adrenocorticotropin, angiotensin II and potassium in isolated glomerulosa cells. *Endocrinology.* 105:327–33.

283. Schiffrin, E.L., Gutkowska, J., Lis, M., and Genest, J. 1982. Relative role of sodium and calcium ions in the steroidogenic response of isolated rat adrenal glomerulosa cells. *Hypertension.* 4 (suppl. II):II-36–II-42.

284. Chiu, A.T., and Freer, R.J. 1979. Angiotensin-induced steroidogenesis in rabbit adrenal: Effects of pH and calcium. *Mol. Cell. Endocr.* 13:159–66.

285. Schiffrin, E.L., Lis, M., Gutkowska, J., and Genest, J. 1981. Role of Ca^{2+} in response of adrenal glomerulosa cells to angiotensin II, ACTH, K$^+$ and ouabain. *Am. J. Physiol.* 241:E42–E46.

286. Foster, R., Lobo, M.V., Rasmussen, H., and Marusic, E.T. 1981. Calcium: its role in the mechanism of action of angiotensin II and potassium on aldosterone production. *Endocrinology.* 109:2196–201.

287. Haksar, A., and Peron, F.G. 1973. The role of calcium in the steroidogenic response of rat adrenal cells to adrenocorticotropic hormone. *Biochim. Biophys. Acta.* 313:363–371.

288. Cheitlin, R., Buckley, D.I., and Ramachandran, J. 1985. The role of extracellular calcium in corticotropin-stimulated steroidogenesis. *J. Biol. Chem.* 260:5323–27.

289. Kojima, I., Kojima, K., and Rasmussen, H. 1985. Characteristics of angiotensin II, K$^+$ and ACTH-induced calcium influx in adrenal glomerulosa cells. *J. Biol. Chem.* 260:9171–77.

290. Kojima, I., and Ogata, E. 1986. Direct demonstration of adrenocorticotropin-

induced changes in cytoplasmic free calcium with aequorin in adrenal glomerulosa cell. *J. Biol. Chem.* 261:9832–38.

291. Yanagibashi, K. 1979. Calcium ion as "second messenger" in corticoidogenic action of ACTH. *Endocrinol. Jpn.* 26:227–34.

292. Fukuda, N., Honda, M., and Hatano, M. 1988. Role of calcium ion in ACTH-induced steroidogenesis in humans. *J. Steroid Biochem.* 31:337–44.

293. Fakunding, J.L., and Catt, K.J. 1980. Dependence of aldosterone stimulation in adrenal glomerulosa cells on calcium uptake: Effects of lanthanum and verapamil. *Endocrinology.* 107:1345–53.

294. Williams, B.C., McDougall, J.G., Tait, J.F., and Tait, S.A.S. 1981. Calcium efflux and steroid output from superfused rat adrenal cells: Effects of potassium, adrenocorticotropic hormone, 5-hydroxytryptamine, adenosine 3',5'-cyclic monophosphate and angiotensins II and III. *Clin. Sci.* 61:541–51.

295. Foster, R., and Rasmussen, H. 1983. Angiotensin-mediated calcium efflux from adrenal glomerulosa cells. *Am. J. Physiol.* 245:E281–87.

296. Kojima, I., Kojima, K., and Rasmussen, H. 1985. Role of calcium fluxes in the sustained phase of angiotensin II-mediated aldosterone secretion from adrenal glomerulosa cells. *J. Biol. Chem.* 260:9177–84.

297. Apfeldorf, W.J., and Rasmussen, H. 1988. Simultaneous determination of intracellular free calcium and aldosterone production in bovine adrenal glomerulosa cells. *Cell Calcium.* 9:71–80.

298. Johnson, E.M.I., Capponi, A.M., and Vallotton, M.B. 1989. Cytosolic free calcium oscillates in single bovine adrenal glomerulosa cells in response to angiotensin II stimulation. *J. Endocr.* 122:391–402.

299. Quinn, S.J., Williams, G.H., and Tillotson, D.L. 1988. Calcium oscillations in single adrenal glomerulosa cells stimulated by angiotensin II. *Proc. Nat. Acad. Sci. USA.* 85:5754–58.

300. Kojima, I., Shibata, H., and Ogata, E. 1987. Time-dependent restoration of the trigger pool of calcium after termination of angiotensin II action in adrenal glomerulosa cells. *J. Biol. Chem.* 262:4557–63.

301. Rossier, M.F., Krause, K.H., Low, P.D., Capponi, A.M., and Vallotton, M.B. 1987. Control of cytosolic free calcium by intracellular organelles in bovine adrenal glomerulosa cells. *J. Biol. Chem.* 262:4053–58.

302. Kojima, I., Kojima, K., and Rasmussen, H. 1985. Effects of ANG II and K^+ on Ca efflux and aldosterone production in adrenal glomerulosa cells. *Am. J. Physiol.* 248:E36–E43.

303. Quinn, S.J., Cornwall, M.C., and Williams, G.H. 1987. Electrical properties of isolated rat adrenal glomerulosa and fasciculata cells. *Endocrinology.* 120:903–14.

304. Matsunaga, H., Maruyama, Y., Kojima, I., and Hoshi, T. 1987. Transient Ca^{2+}-channel current characterized by a low-threshold voltage in zona glomerulosa cells of rat adrenal cortex. *Pflugers Archiv.* 408:351–55.

305. Capponi, A.M., Lew, P.D., Jornot, L., and Vallotton, M.B. 1984. Correlation between cytosolic free calcium and aldosterone production in bovine adrenal glomerulosa cells. *J. Biol. Chem.* 259:8863–69.

306. Aguilera, G., and Catt, K.J. 1986. Participation of voltage-dependent calcium channels in the regulation of adrenal glomerulosa function by angiotensin II and potassium. *Endocrinology.* 118:112–18.

307. Cohen, C.J., McCarthy, R.T., Barrett, P.Q., and Rasmussen, H. 1988. Ca

References **133**

channels in adrenal glomerulosa cells: K^+ and angiotensin II increase T-type Ca channel current. *Proc. Nat. Acad. Sci. USA.* 85:2412–16.

308. Connor, J.A., Cornwall, M.C., and Williams, G.H. 1987. Spatially resolved cytosolic calcium response to angiotensin II and potassium in rat glomerulosa cells measured by digital imaging techniques. *J. Biol. Chem.* 262:2919–27.

309. Means, A.R., and Dedman, J.R. 1980. Calmodulin—an intracellular calcium receptor. *Nature.* 285:73–77.

310. Harper, J.F., Cheung, W.Y., Wallace, R.W., Huang, H-L., Levine, S.N., and Steiner, A.L. 1980. Localization of calmodulin in rat tissues. *Proc. Nat. Acad. Sci. USA* 77:366–70.

311. Bristow, A.F., Schulster, D., and Rodnight, R. 1981. Cyclic nucleotide-dependent phosphorylation of endogenous proteins in bovine adrenocortical cell membranes. *Biochim. Biophys. Acta.* 675:24–28.

312. Kigoshi, T., Uchida, K., and Morimoto, S. 1988. Existence of endogenous substrate proteins for Ca^{++}/calmodulin-dependent and Ca^{++}/phospholipid-dependent protein kinases in rat adrenal glomerulosa cells. *J. Steroid Biochem.* 29:277–83.

313. Balla, T., and Spät, A. 1982. The effect of various calmodulin inhibitors on the response of adrenal glomerulosa cells to angiotensin II and cyclic-AMP. *Biochem. Pharmacol.* 31:3705–07.

314. Simpson, E.R., and Williams-Smith, D.L. 1975. Effect of calcium (ion) uptake by rat adrenal mitochondria on pregnenolone formation and spectral properties of cytochrome P450. *Biochim. Biophys. Acta.* 404:309–20.

315. Koritz, S.B. 1986. The stimulation by calcium and its inhibition by ADP of cholesterol side-chain cleavage activity in adrenal mitochondria. *J. Steroid Biochem.* 24:569–76.

316. Michell, R.H. 1975. Inositol phospholipids and cell surface receptor function. *Biochim. Biophys. Acta. Biochim* 415:81–147.

317. Berridge, M.J. 1983. Rapid accumulation of inositol trisphosphate reveals that agonists hydrolyse polyphosphoinositides instead of phosphatidylinositol. *Biochem. J.* 212:849–58.

318. Kojima, I., Kojima, K., Kreuter, D., and Rasmussen, H. 1984. The temporal integration of the aldosterone secretory response to angiotensin occurs via two intracellular pathways. *J. Biol. Chem.* 259:14448–57.

319. Catt, K.J., Carson, M.C., Hausdorff, W.P., Leach-Harper, C.M., Baukal, A.J., Guillemette, G., Balla, T., and Aguilera, G. 1987. Angiotensin II receptors and mechanisms of action in adrenal glomerulosa cells. *J. Steroid Biochem.* 27:915–27.

320. Enyedi, P., Mucsi, I., Hunyady, L., Catt, K.J., and Spät, A. 1986. The role of guanyl nucleotide binding proteins in the formation of inositol phosphates in adrenal glomerulosa cells. *Biochem. Biophys. Res. Comm.* 140:941–47.

321. Cockroft, S. 1987. Polyphosphoinositide phosphodiesterase: regulation by a novel guanine nucleotide binding protein, G_p. *TIBS.* 12:75–78.

322. Rossier, M.F., Capponi, A.M., and Vallotton, M.B. 1988. Inositol trisphosphate isomers in angiotensin II-stimulated adrenal glomerulosa cells. *Mol. Cell. Endocr.* 57:163–68.

323. Balla, T., Guillemette, G., Baukal, A.J., and Catt, K.J. 1987. Metabolism of inositol 1,3,4-trisphosphate to a new tetrakisphosphate isomer in angiotensin-stimulated adrenal glomerulosa cells. *J. Biol. Chem.* 262:9952–55.

324. Farese, R.V., Larson, R.E., and Davis, J.S. 1984. Rapid effects of angiotensin

II on polyphosphoinositide metabolism in the rat adrenal glomerulosa. *Endocrinology.* 114:302–04.

325. Enyedi, P., Buki, B., Mucsi, I., and Spät, A. 1985. Polyphosphoinositide metabolism in adrenal glomerulosa cells. *Mol. Cell. Endocr.* 41:105–12.

326. Ganguly, A., Chiou, S., West, L.A., and Davis, J.S. 1986. Aldosterone secretagogues and inositol trisphosphate as intracellular mediator. *J. Hypertens.* 4, supp. 6:S361–63.

327. Storey, D.J., Shears, S.B., Kirk, C.J., and Michell, R.H. 1984. Stepwise enzymatic dephosphorylation of inositol 1,4,5-trisphosphate to inositol in liver. *Nature.* 312:374–76.

328. Irvine, R.F., Letcher, A.J., Heslop, J.P., and Berridge, M.J. 1986. *Nature.* 320:631–34.

329. Hansen, C.A., Mah, S., and Williamson, J.R. 1986. Formation and metabolism of inositol 1,3,4,5-tetrakisphosphate in liver. *J. Biol. Chem.* 261:8100–03.

330. Balla, T., Baukal, A.J., Guillemette, G., Morgan, R.O., and Catt, K.J. 1986. Angiotensin-stimulated production of inositol trisphosphate isomers and rapid metabolism through inositol 4-monophosphate in adrenal glomerulosa cells. *Proc. Nat. Acad. Sci. USA.* 83:9323–27.

331. Guillemette, G., Baukal, A.J., Balla, T., and Catt, K.J. 1987. Angiotensin-induced formation and metabolism of inositol polyphosphates in bovine adrenal glomerulosa cells. *Biochem. Biophys. Res. Comm.* 142:15–22.

332. Vilgrain, I., Cochet, C., and Chambaz, E.M. 1984. Hormonal regulation of a calcium-activated, phospholipid-dependent protein kinase in bovine adrenal cortex. *J. Biol. Chem.* 259:3403–06.

333. Kraft, A.S., and Anderson, W.R. 1983. Phorbol esters increase the amount of Ca^{++}-phospholipid dependent protein kinase associated with plasma membrane. *Nature.* 301:621–23.

334. Capponi, A.M., Rossier, M., Lang, U., Lew, P.D., and Vallotton, M.B. 1986. Comparison of the signal transduction mechanisms for angiotensin II in adrenal zona glomerulosa and vascular smooth muscle cells. *J. Hypertens.* 4 (suppl. 6):S419 20.

335. Lang, U., and Vallotton, M.B. 1987. Angiotensin II but not potassium induces subcellular redistribution of protein kinase C in bovine adrenocortical cells. *J. Biol. Chem.* 262:8047–50.

336. Spät, A. 1988. Stimulus-secretion coupling in angiotensin-stimulated adrenal glomerulosa cells. *J. Steroid Biochem.* 29:443–53.

337. Vinson, G.P., Laird, S.M., Whitehouse, B.J., and Illuson, J.P. 1989 Specific effects of agonists of the calcium messenger system on secretion of "late-pathway" steroid products by intact tissue and dispersed cells of the rat adrenal zona glomerulosa. *J. Mol. Endocr.* 2:157–65.

338. Berridge, M.J., and Irvine, R.F. 1984. Inositol trisphosphate, a novel second messenger in cellular signal transduction. *Nature.* 312:315–21.

339. Baukal, A.J., Guillemette, G., Rubin, R., Spät, A., Catt, K.J. 1985. Binding sites for inositol trisphosphate in the bovine adrenal cortex. *Biochem. Biophys. Res. Comm.* 133:532–38.

340. Guillemette, G., Balla, T., Baukal, A.J., Spät, A., and Catt, K.J. 1987. Intracellular receptors for inositol 1,4,5-trisphosphate in angiotensin II target tissues. *J. Biol. Chem.* 262:1010–15.

341. Balla, T., Szebeny, M., Kanyar, B., and Spät, A. 1985. Angiotensin II

and FCCP mobilizes calcium from different intracellular pools in adrenal glomerulosa cells; analysis of calcium fluxes. *Cell Calcium.* 6:327–42.

342. Carafoli, E. 1987. Intracellular calcium homeostasis. *Ann. Rev. Biochem.* 56:395–433.

343. Metz, S.A. 1988. Arachidonic acid and its metabolites: Evolving roles as transmembrane signals for insulin release. *Prostaglandins, Leukotrienes and Essential Fatty Acids-Reviews.* 32:187–202.

344. Needleman, P., Turk, J., Jakschik, B.A., Morrison, A.R., and Lefkowith, J.B. 1986. Arachidonic acid metabolism. *Ann. Rev. Biochem.* 55:69–102.

345. Berridge, M.J. 1981. Phosphatidylinositol hydrolysis. *Mol. Cell. Endocr.* 24:115.

346. Laychock, S.G., Franson, R.C., Weglicki, W.B., and Rubin, R.P. 1977. Identification and partial characterization of phospholipases in isolated adrenocortical cells. *Biochem. J.* 164:753–56.

347. Schrey, M.P., and Rubin, R.P. 1979. Characterization of a calcium-mediated activation of arachidonic acid turnover in adrenal phospholipids by corticotropin. *J. Biol. Chem.* 254:11234–41.

348. Kojima, I., Kojima, K., and Rasmussen, H. 1985. Possible role of phospholipase A_2 action and arachidonic acid metabolism in angiotensin II-mediated aldosterone secretion. *Endocrinology.* 117:1057–66.

349. Bergstrom, S. 1967. Prostaglandins; members of a new hormonal system. *Science.* 157:382–91.

350. Laychock, S.G., and Rubin, R.P. 1975. ACTH-induced prostaglandin biosynthesis from ^{3}H arachidonic acid by adrenocortical cells. *Prostaglandins.* 10:529–40.

351. Shaw, J.E., and Ramwell, P.W. 1967. Prostaglandin release from the adrenal gland. In: S. Bergstrom, and B. Samuelsson, eds. "Prostaglandins" Proceedings of the second Nobel Symposium, Stockholm. 1966 pp. 293–99.

352. Chanderbhan, R., Treadwell, C.R., Hodges, V.A., and Vahouny, G.V. 1977. ACTH-induced synthesis of prostaglandin E_2 in adrenal cortical cells. *Fed. Proc.* 36:674.

353. Swartz, S.L., and Williams, G.H. 1983. Role of prostaglandins in adrenal steroidogenesis. *Endocrinology.* 113:992–96.

354. Campbell, W.B., and Gomez-Sanchez, C.E. 1980. Role of prostaglandins in angiotensin-induced steroidogenesis. *Hypertension.* 2:471.

355. Laychock, S.G., Warner, W., and Rubin, R.P. 1977. Further studies on the mechanisms controlling prostaglandin biosynthesis in the cat adrenal cortex: The role of calcium and cyclic AMP. *Endocrinology.* 100:74–81.

356. Segre, G., Bianchi, E., Bruni, G., and Dalpra, P. 1983. Indomethacin and sulindac inhibition of ACTH-stimulated cortisol release from adrenal glands in vitro. *Pharm. Res. Comm.* 15:347–60.

357. Gallant, S., and Brownie, A.C. 1973. The in vivo effect of indomethacin and prostaglandin E2 on ACTH and dibutyryl cyclic AMP induced steroidogenesis in hypophysectomized rats. *Biochem. Biophys. Res. Comm.* 55:831–36.

358. Vukoson, M.B., Kramer, R.E., Pope, M., Greiner, J.W., and Colby, H.D. 1976. Failure of indomethacin to affect adrenal responsiveness to ACTH in vitro. *Horm. Metab. Res.* 8:325–26.

359. Gerber, J.G., and Nies, A.S. 1979. The failure of indomethacin to alter ACTH-induced hyperaemia or steroidogenesis in the anaesthetized dog. *Br. J. Pharmacol.* 67:217–20.

360. Jones, D.B., Marante, D., Williams, B.C., and Edwards, C.R.W. 1987. Adrenal synthesis of corticosterone in response to ACTH in rats is influenced by leukotriene A_4 and by lipoxygenase intermediates. *J. Endocr.* 112:253–58.

361. Nadler, J.L., Naterjan, R., and Stern, N. 1987. Specific action of the lipoxygenase pathway in mediating angiotensin II-induced aldosterone synthesis in isolated adrenal glomerulosa cells. *J. Clin. Invest.* 80:1763–69.

362. Stone, D., and Hechter, O. 1954. Studies on ACTH action in perfused bovine adrenals: The site of action of ACTH in corticosteroidogenesis. *Arch. Biochim. Biophys.* 51:457–69.

363. Aguilera, G., and Catt, K.J. 1979. Loci of action of regulators of aldosterone biosynthesis in isolated glomerulosa cells. *Endocrinology.* 104:1046–52.

364. Bisgaier, C.L., Chanderbhan, R., Hinds, R.W., and Vahouny, G.V. 1985. Adrenal cholesterol esters as substrate source for steroidogenesis. *J. Steroid Biochem.* 23:967–74.

365. Beckett, G.J., and Boyd, G.S. 1977. Purification and control of bovine adrenal cholesterol ester hydrolase and evidence for the activation of the enzyme by phosphorylation. *Eur. J. Biochem.* 72:223–33.

366. Brown, M.S., Kovanen, P.T., and Goldstein, J.L. 1979. Receptor-mediated uptake of lipoprotein-cholesterol and its utilization for steroid synthesis in the rat adrenal cortex. *Rec. Prog. Horm. Res.* 35:215–49.

367. Gwynne, J.T., and Strauss, J.F. 1982. The role of lipoproteins in steroidogenesis and cholesterol metabolism in steroidogenic glands. *Endocr. Rev.* 3:299–329.

368. Fidge, N., Leonard-Kanevsky, M., and Nestel, P. 1984. The hormonal stimulation of high-density lipoprotein binding, internalisation and degradation by cultured rat adrenal cortical cells. *Biochim. Biophys. Acta.* 793:180–86.

369. Gwynne, J.T., and Hess, B. 1980. The role of high-density lipoproteins in rat adrenal cholesterol metabolism and steroidogenesis. *J. Biol. Chem.* 255:10875–83.

370. Verschoor-Klootwyk, A.H., Verschoor, L., Azhar, S., and Reaven, G.M. 1982. Role of exogenous cholesterol in regulation of adrenal steroidogenesis in the rat. *J. Biol. Chem.* 257:7666–71.

371. Rainey, W.E., Shay, J.W., and Mason, J.I. 1986. ACTH induction of 3-hydroxy-3-methylglutaryl coenzyme A reductase, cholesterol biosynthesis and steroidogenesis in primary cultures of bovine adrenocortical cells. *J. Biol. Chem.* 261: 7322–26.

372. Ferguson, J.J. 1963. Protein synthesis and adrenocorticotropin responsiveness. *J. Biol. Chem.* 238:2754–59.

373. Farese, R.V. 1964. Inhibition of the steroidogenic effect of ACTH and incorporation of amino acid into rat adrenal protein by chloramphenicol. *Biochim. Biophys. Acta.* 87:701–03.

374. Garren, L.D., Ney, R.L., and Davis, W.W. 1965. Studies on the role of protein synthesis in the regulation of corticosterone production by adrenocorticotropic hormone in vivo. *Proc. Nat. Acad. Sci. USA.* 53:1443–50.

375. Garren, L.D., Gill, G.N., Masui, H., and Walton, G.M., 1971. On the mechanism of action of ACTH. *Rec. Prog. Horm. Res.* 27:433–78.

376. Davis, W.W., and Garren, L.D. 1968. On the mechanism of action of adrenocorticotropic hormone. *J. Biol. Chem.* 243:5153–157.

377. Koritz, S.B., and Kumar, A.M. 1970. On the mechanism of action of the adrenocorticotrophic hormone. *J. Biol. Chem.* 245:152–59.

References

378. Bell, J.J., and Harding, B.W. 1974. The acute actions of adrenocorticotropic hormone on adrenal steroidogenesis. *Biochim. Biophys. Acta.* 348:285–98.

379. Davis, W.W., and Garren, L.D. 1966. Evidence for the stimulation by adrenocorticotropic hormone of the conversion of cholesterol esters to cholesterol in the adrenal in vivo. *Biochem. Biophys. Res. Comm.* 24:805–10.

380. Simpson, E.R., McCarthy, J.L., and Peterson, J.A. 1978. Evidence that the cycloheximide-sensitive site of adrenocorticotropic hormone action is the mitochondria. *J. Biol. Chem.* 253:3135–39.

381. Crivello, J.F., and Jefcoate, C.R. 1978. Mechanism of corticotropin action in rat adrenal cells. The effects of inhibitors of protein synthesis and of microfilament formation on corticosterone synthesis. *Biochim. Biophys. Acta.* 542:315–29.

382. Crivello, J.F., and Jefcoate, C.R. 1980. Intracellular movement of cholesterol in rat adrenal cells: Kinetics and effects of inhibitors. *J. Biol. Chem.* 255:8144–51.

383. Vahouny, G.V., Dennis, P., Chanderbhan, R., Fiskum, G., Noland, B.J., and Scallen, T.J. 1984. Sterol carrier protein $_2$ (SCP$_2$)-mediated transfer of cholesterol to mitochondrial inner membranes. *Biochem. Biophys. Res. Comm.* 122:509–15.

384. Mori, M., and Marsh, J.M. 1982. The site of luteinising hormone stimulation of steroidogenesis in mitochondria of the rat corpus luteum. *J. Biol. Chem.* 257:6178–83.

385. Privalle, C.T., Crivello, J.F., and Jefcoate, C.R. 1983. Regulation of intramitochondrial cholesterol transfer to side-chain cleavage cytochrome P450 in rat adrenal gland. *Proc. Natl. Acad. Sci. USA.* 80:702–06.

386. Kimura, T. 1986. Transduction of ACTH signal from plasma membrane to mitochondria in adrenocortical steroidogenesis. Effects of peptide, phospholipid and calcium. *J. Steroid Biochem.* 25:711–16.

387. Privalle, C.T., McNamara, B.C., Dhariwal, M.S., and Jefcoate, C.R. 1987. ACTH control of cholesterol side-chain cleavage at adrenal mitochondrial cytochrome P450$_{scc}$. Regulation of intramitochondrial cholesterol transfer. *Mol. Cell. Endocrin.* 53:87–101.

388. Kan, K.W., Ritter, M.C., Ungar, F., and Dempsey, M.E. 1972. The role of a carrier protein in cholesterol and steroid hormone synthesis by adrenal enzymes. *Biochem. Biophys. Res. Comm.* 48:423–29.

389. Chanderbhan, R., Noland, B.J., Scallen, T.J., and Vahouny, G.V. 1982. Sterol carrier protein$_2$: Delivery of cholesterol from adrenal lipid droplets to mitochondria for pregnenolone synthesis. *J. Biol. Chem.* 257:8928–34.

390. Conneely, O.M., Headon, D.R., Olson, C.D., Ungar, F., and Dempsey, M.E. 1984. Intramitochondrial movement of adrenal sterol carrier protein with cholesterol in response to corticotropin. *Proc. Nat. Acad. Sci. USA.* 81:2970–74.

391. Vahouny, G.V., Chanderbhan, R., Noland, B.J., Irwin, D., Dennis, P., Lambeth, J.D., and Scallen, T.J. 1983. Sterol carrier protein$_2$: Identification of adrenal sterol carrier protein$_2$ and site of action for mitochondrial cholesterol utilisation. *J. Biol. Chem.* 258:11731–37.

392. Trzeciak, W.H., Simpson, E.R., Scallen, T.J., Vahouny, G.V., and Waterman, M.R. 1987. Studies on the synthesis of sterol carrier protein-2 in rat adrenocortical cells in monolayer culture. Regulation by ACTH and dibutyryl cyclic 3'5'-AMP. *J. Biol. Chem.* 262:3713–17.

393. Krueger, R.J., and Orme-Johnson, N.R. 1983. Acute adrenocorticotropic

hormone stimulation of adrenal corticosteroidogenesis: Discovery of a rapidly induced protein. *J. Biol. Chem.* 258:10159–67.

394. Pon, L.A., Hartigan, J.A., and Orme-Johnson, N.R. 1986. Acute ACTH regulation of adrenal corticosteroid biosynthesis: Rapid accumulation of a phosphoprotein. *J. Biol. Chem.* 261:13309–16.

395. Pedersen, R.C., and Brownie, A.C. 1983. Cholesterol side-chain cleavage in the rat adrenal cortex: Isolation of a cycloheximide-sensitive activator peptide. *Proc. Nat. Acad. Sci. USA.* 80:1882–86.

396. Solano, A.R., Neher, R., and Podesta, E.J. 1984. Rat adrenal cycloheximide-sensitive factors and phospholipids in the control of acute steroidogenesis. *J. Steroid Biochem.* 21:111–16.

397. Yanagibashi, K., Ohno, Y., Kawamura, M., and Hall, P.F. 1988. The regulation of intracellular transport of cholesterol in bovine adrenal cells: Purification of a novel protein. *Endocrinology.* 123:2075–82.

398. Vinson, G.P. 1987. The stimulation of steroidogenesis by corticotrophin: the role of intracellular regulatory peptides and proteins. *J. Endocr.* 114:163–65.

399. Purvis, J.L., Canick, J.A., Mason, J.I., Estabrook, R.W., and McCarthy, J.L. 1973. Lifetime of adrenal cytochrome P450 as influenced by ACTH. *Ann. N.Y. Acad. Sci. USA.* 212:319–42.

400. Simpson, E.R., Mason, J.I., John, M.E., Zuber, M.X., Rodgers, R.J., and Waterman, M.R. 1987. Regulation of the biosynthesis of steroidogenic enzymes. *J. Steroid Biochem.* 27:801–05.

4

Effects of
Corticosteroids

4.1 Mode of action

4.1.1 Cellular aspects

The intracellular events consequent on tissue stimulation by steroid hormones have been the object of intensive study over the last 25 years. This surge of activity stemmed from the findings of Jensen and others [1], that target organs specifically retain administered radioactive steroid, whereas nontarget organs do not. This, in turn, led to the hypothesis that steroids, like other hormones, exert their actions through their recognition and binding by specific receptors, which are only expressed in hormone sensitive target tissues. The interaction of a hormone with its receptor, to form a hormone-receptor complex, is the first critical step in the tissue response to hormonal stimulation. It leads to "activation" of the receptor, an incompletely defined process which results from a structural or conformational change in the protein. This may include, or be the result of, loss of association with a 90 kDa heat shock protein, and the formation of a receptor homodimer. Hormonal responses to steroids are then generated by the binding of the activated hormone receptor complex to DNA enhancer sites which then bring about the activation of specific genes, generating increased mRNA and protein synthesis [2-4] (Fig. 4.1).

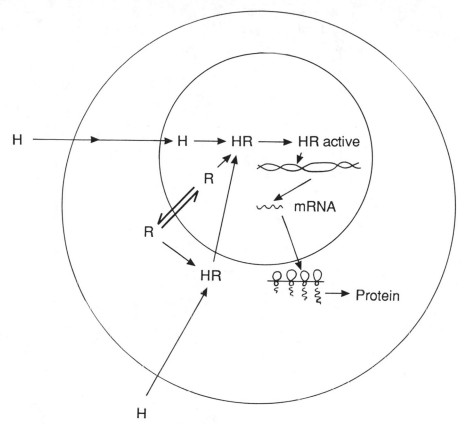

FIGURE 4.1 A general interpretation of the cellular events involved in the activation of target cell function by glucocorticoid stimulation. The site of receptor activation is not firmly established. H = hormone, R = receptor, HR = hormone-receptor complex.

Formation of the hormone-receptor complex   It is generally reasonable to assume that, in acute dose-related responses to hormones, the formation of the hormone receptor complex is the rate limiting step. In the reaction

$$H_f + R_u \underset{k_2}{\overset{k_1}{\rightleftharpoons}} HR$$

where H_f = free hormone, R_u = unoccupied receptor, and HR = the hormone-receptor complex, the kinetics of the reaction are governed by the two-rate constants, k_1 and k_2. The equilibrium constants, k_1/k_2, and k_2/k_1, are designated K_a and K_d, and are known, respectively, as the association constant and the dissociation constant. From the equation

$$k_1 [H_f] [R_u] = k_2 [HR]$$

it will be seen that, at equilibrium

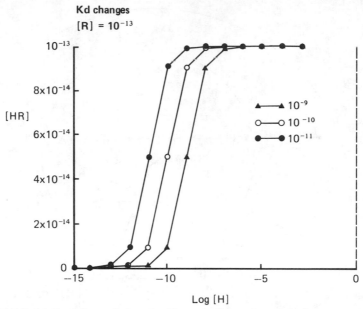

FIGURE 4.2 (a) The effect of changing K_d on the formation of the hormone receptor (HR) complex. These computer generated curves illustrate responses in a system in which the receptor concentration is 10^{-13} moles per l, and the K_d is 10^{-9}, 10^{-10}, or 10^{-11} moles per l.

$$\frac{[\text{HR}]}{[\text{H}_f][\text{R}_u]} = \frac{k_1}{k_2} = K_a \qquad \text{Eq. (1)}$$

and

$$\frac{[\text{H}_f][\text{R}_u]}{[\text{HR}]} = \frac{k_2}{k_1} = K_d$$

Therefore, when receptor occupancy is half maximal, and consequently $[\text{HR}] = [\text{R}_u]$, clearly $[\text{H}_f] = K_d$. In other words, K_d, which is expressed in concentration units, in practical terms signifies the concentration of free hormone required to give half maximal binding of hormone in any given system. Therefore, K_a, (the reciprocal of the dissociation constant, K_d) is expressed in liters per mole, and is a dilution constant.

The dissociation constant, K_d, is therefore a valuable expression of the avidity with which the receptor binds hormone, and thus indicates the sensitivity of the response. The effect of changing K_d on the formation of HR is illustrated in Figure 4.2a. This is not necessarily the same as the biological response, since, in many systems, maximal responses may be

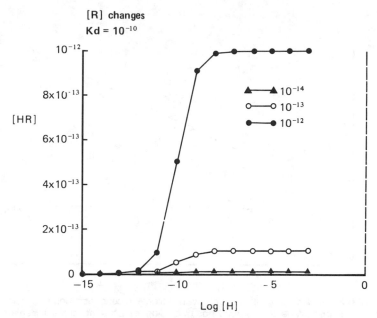

FIGURE 4.2 (b) The effect of changing receptor concentration on the formation of the hormone receptor (HR) complex. In these curves, the K_d is constant at 10^{-10} moles per l, while receptor concentration varies between 10^{-12} and 10^{-14} moles per l.

In both of these figures, data are presented as semilog plots, the conventional form for dose response data. This has the effect of seemingly exaggerating the effect of receptor concentration compared with K_d. However, a corollary is that, in general, the amplitude of response is often easier to evaluate with precision than the sensitivity, that is, minimum effective dose. The biological response does not necessarily precisely reflect the formation of HR since full responses may often be achieved when only a proportion of the total receptor population available is occupied.

achieved with occupancy of only a proportion of the receptors available. The effect of changing total receptor concentration, which clearly also has a bearing on the amount of HR formed, is also shown (Fig. 4.2b).

Methods for characterization and localization of receptors to steroid hormones   Steroid hormone receptors have mainly been characterized and localized by the use of three methodologies [2,3,5]. These are

1. Saturation kinetics, and the Scatchard plot, usually on cell fractions obtained by ultracentrifugation.
2. Physico-chemical fractionation. Originally, sucrose density gradient fractionation of hormone-receptor complexes was used, with characterization of the receptor in terms of its sedimentation coefficient in Svedberg units (S) [6]. Hormone receptor complexes also have been fractionated, with better resolution, for example, on DEAE cellulose columns [7,8], or by isoelectric focussing polyacrylamide gel electrophoresis.

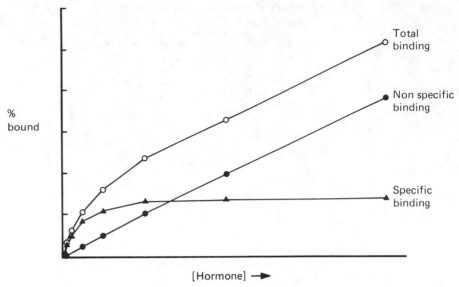

FIGURE 4.3 (a) Saturation of receptor with increasing concentrations of radio-labeled hormone. In most preparations two kinds of binding are present, the saturable specific binding to receptor, and the nonsaturable binding to other cellular components. These can be discriminated by the addition in duplicate experiments of excess amounts of unlabeled hormone, which competes with and displaces labeled hormone from the receptor, leaving only nonspecific binding in evidence. Subtraction of values for nonspecific binding from those for total binding then produces values for specific binding, in which the saturation effect is clearly revealed.

3. More recently, monoclonal antibodies have been developed to many receptor types [9,10].

These three methodologies give different information about the receptors. Saturation analysis, and the transformation of the data into the form of the Scatchard plot (Fig. 4.3a,b) gives information about the concentration of receptors, usually in a homogenate, nuclear or soluble fraction of the target tissue under study, and the avidity of the binding, in terms of the K_a (or K_d). The Svedberg unit, on the other hand, reflects the molecular weight and conformation of the molecule, and is of importance in the study of receptor transformations, which in some cases (e.g., the estrogen receptor) but not others (e.g., the glucocorticoid receptor) are thought to be associated with their activation. More rigorous fractionation techniques such as isoelectric focussing also give information on the existence of receptor isoforms, although the relationship between the information gained by these different methods is not always clear. Monoclonal antibodies to receptors can be used to give information on the location of the receptors within the cell as seen in histological preparations, and can also be used for quantitative assessment of receptors in subcellular components.

Data obtained in these different ways is not always consistent, and may

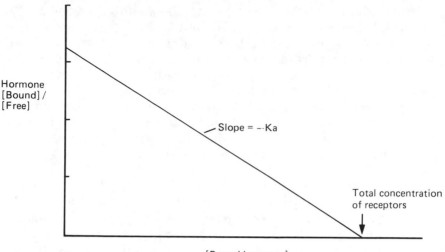

Hormone [Bound] / [Free]

Slope = --Ka

Total concentration of receptors

[Bound hormone]

FIGURE 4.3 (b) Transformation of saturation data into the linear Scatchard plot. If [HR] is taken to be equal to the concentration of specifically bound steroid $[H_b]$, and the total (occupied and unoccupied) receptor concentration is [T], the unoccupied receptor concentration (R_u) is given by $[T - H_b]$. Equation 1 (see text) can then be rewritten

$$\frac{[H_b]}{[H_f][T - H_b]} = \frac{k_1}{k_2} = K_a$$

Rearranging this

$$\frac{[H_b]}{[H_f]} = K_a \cdot [T] - K_a \cdot [H_b]$$

Thus in the Scatchard plot, in which $[H_b]/[H_f]$ is plotted against $[H_b]$, the value of the slope of the curve is equal to $-K_a$, and the intercept on the x axis gives the total concentration of hormone receptor, that is, when $[H_b]/[H_f] = 0$, $[H_b] = [T]$.

lead, without careful interpretation, to conclusions which are partially or completely erroneous. This certainly seems to have been the case in studies on the intracellular localization of receptors, which we will discuss.

Intracellular localization of steroid hormone receptors The earlier data, obtained by using radioactively labeled steroid to bind to the receptor, led, in the 1960s, to the two-stage model of steroid hormone action [11,12]. Broadly, this held that unoccupied steroid receptors were located in the cytoplasm of the cell, as indicated by their appearance in the cytosolic supernatant in studies on subcellular fractions obtained by conventional ultracentrifugation methods. Steroid hormones enter the cell, which they may do by a process of simple diffusion (though see Chap. 2), and bind to the receptor, which then undergoes the structural changes associated with activation. Activation facilitates the transfer of the hormone receptor complex to the nucleus.

In this model, therefore, free (unoccupied) receptors are found only in the cytosol. However, towards the end of the 1970s, it became apparent that some of the data on which the two-stage theory was based was artefactual. In particular, partitioning of receptors may depend on the free water content of the nuclear and cytosolic fractions [13], and enucleation experiments showed that unoccupied glucocorticoid, progesterone, and estrogen receptors were associated with the nucleoplasts in a rat pituitary cell line [14]. Other findings led to similar conclusions, and it was finally with the development of the monoclonal antibody labeling techniques that it was shown that estrogen receptors appear to be located only in the nucleus, and none are seen in the cytoplasm [15].

Whether this is true under all conditions, and in all tissues, cannot be decided. Certainly there are differences between the classes of steroid, and, coexisting with the enucleation data referred to, there is other good evidence that the glucocorticoid receptor is distributed between more than one compartment of the cells of its target organs. As in the case of studies on the estrogen and progesterone receptors, the initial biochemical data suggested that unoccupied glucocorticoid receptors occurred in the cytoplasm. However, in the case of the glucocorticoid receptor, this concept was confirmed by subsequent studies in which labeling with polyclonal and monoclonal antibodies showed that it may occur in both cytoplasmic and nuclear compartments, in agreement with the two-stage theory [16–18].

Given the close homology between the receptors of the different steroid types (see below), it is difficult to see how the mechanisms of action can be very different. It is perhaps likely that the unoccupied receptor is always partitioned between nucleus and cytoplasm (if only in a perinuclear cytoplasmic shell), and the only differences between the receptor types in this respect are those of partition coefficients. The general concept of glucocorticoid receptor localization, and action, are shown in Figure 4.1.

Structures of steroid hormone receptors   In recent years, the amino acid sequences of the major classes of steroid receptors have been deduced from the structures of the exons of the genes which code for them. This has been achieved for both glucocorticoid and mineralocorticoid receptors. While the steroid receptors differ in length, their structures show striking homologies with each other, and also with the receptors for 1,25-dihydroxycholecalciferol, thyroid hormone, and the ecdysteroids (hormones controlling molting in insects) [19,20]. Studies suggest that receptors are arranged in domains which are linked by "hinge" regions (Figs. 4.4 and 4.5). The DNA binding domain, of about 70 amino acid residues, is the most highly conserved. The steroid binding domains also show considerable homology, and there is a degree of conservation in the end domains. In the DNA binding region, the conservation of cysteines and basic and hydrophobic residues is seemingly of particular importance because, as in other proteins which bind to nucleic acids, pairs of cysteines and histidines may bind a zinc ion, through coordinate bonding, forming one or more "zinc fingers" in the protein which

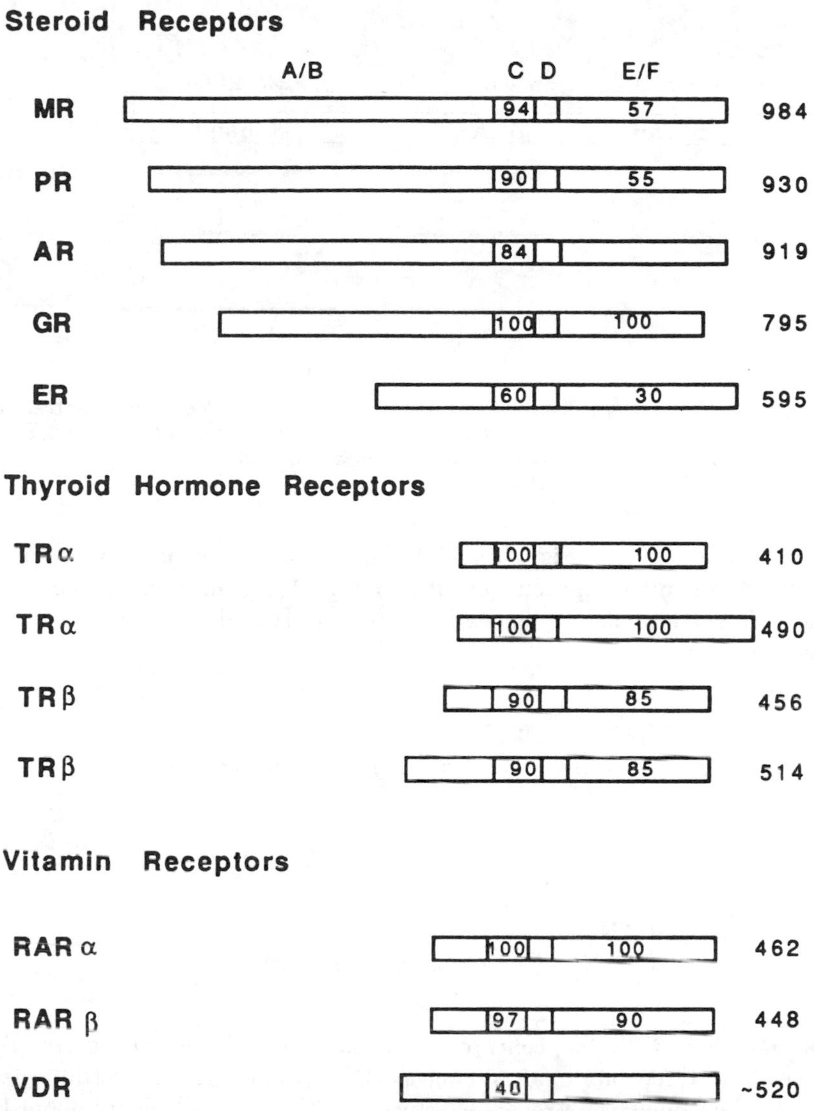

Steroid Receptors

	A/B	C D	E/F	
MR		94	57	984
PR		90	55	930
AR		84		919
GR		100	100	795
ER		60	30	595

Thyroid Hormone Receptors

TR α	100	100	410
TR α	100	100	490
TR β	90	85	456
TR β	90	85	514

Vitamin Receptors

RAR α	100	100	462
RAR β	97	90	448
VDR	40		~520

FIGURE 4.4 Schematic diagram of the structure of nuclear receptors. The receptors, which are divided into regions A through F are as follows: mineralocorticoid (MR), progesterone (PR), androgen (AR), estrogen (ER), vitamin D (VDR), thyroid (TR), and retinoic acid (RAR). The number of amino acid sequences for each receptor in humans is shown on the right. They have been compared in region C, the DNA binding domain, and region E, the ligand binding domain, and the percent homologies where known are shown for the steroid receptors relative to GR, thyroid receptors relative to TRα, and retinoic acid receptors relative to RARα. (Reproduced with permission from ref. 20.)

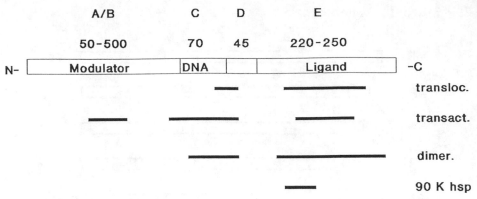

FIGURE 4.5 Functional organization of nuclear receptors: transloc. = nuclear translocation region; transact. = transactivation domain; dimer = dimerization domain; 90 K hsp = 90 kDa heat shock protein binding region. See ref. 22.

insert into the DNA (Fig. 4.6) [21]. The general structure of a steroid hormone dependent operon, and its functional organization, is shown in Fig. 4.7. Palindromic sequences in the glucocorticoid response elements, thus

      5'-AGAACACAGTGTTCT-3'
      3'-TCTTGTGTCACAAGA-5'   (perfect palindrome)

or

      5'-GGTACANNNTGTTCT-3'
      3'-CCATGTNNNACAAGA-5'   (consensus)

are consistent with the concept that, like other steroid receptors, the glucocorticoid receptor binds to nuclear DNA as a dimer, and that dimerization is a consequence of receptor activation. That this is brought about by the loss of association with the 90 kDa heat shock protein is consistent with the observation that the heat shock protein binding domain forms part of the dimerization domain (Fig. 4.5) [22]. Other aspects of the nature of steroid hormone receptor activation are still unclear.

In the case of the glucocorticoid receptor (unlike the estrogen receptor), activation is not associated with a change in the sedimentation coefficient, but may be associated with a change in surface charge, and conformation [22,23]. In the case of the estrogen receptor, it is possible that phosphorylation and dephosphorylation sequences play a part in the activation and deactivation processes, and in the regulation of the availability of the receptor within the cell [24].

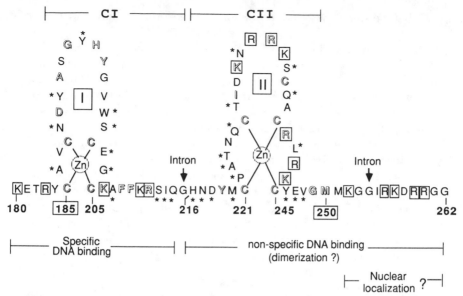

FIGURE 4.6 Hypothetical structure of human ER region C (amino acids 180–262) showing possible arrangement of "zinc finger" region of receptor protein, which may insert into the DNA of the control region of steroid responsive genes. (Figure reproduced with permission from ref. 21.)

The glucocorticoid receptor is a monomer of molecular weight $\sim$ 94 000 [23]. It is present in concentrations of about 25 000 molecules per cell in target organs, and its K_d is in the nanomolar range [23,25]. The human mineralocorticoid receptor cDNA has been cloned and has been found to encode a 107 000 molecular weight protein [26]. The relationships and

FIGURE 4.7 General form of glucocorticoid sensitive operon. The glucocorticoid responsive elements (GREs) are enhancerlike, from their position and orientation independent properties, and are hence called "hormone responsive enhancer elements." After the hormone receptor complex (HR) binds to the GRE, two different transcription factors, already present in the nucleus, are recruited to the core promoter, and form an initiation complex with RNA polymerase I.

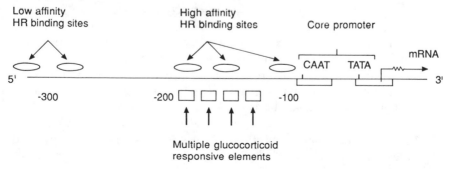

homologies between these and other nuclear receptors is shown in Figure 4.4.

In the kidney, the diversity of corticosteroid receptor types has led to the introduction of an additional nomenclature. The type I receptor, which binds aldosterone with high affinity (K_d 0.1 to 1 nmol/1) is present in the rat at concentrations of 1 to 10 fmol/mg protein, and can be regarded as the mineralocorticoid receptor. The type II receptor binds dexamethasone with high affinity (K_d 1 to 10 nmol/1), and is present in higher concentrations (by a factor of 10) than the mineralocorticoid receptor (type I). A further binding protein, termed type III, is corticosterone specific (i.e., it does not bind dexamethasone), and is thus corticosteroid binding globulinlike in nature (see Chap. 2). It is not clear whether it is indeed a true receptor [27,28]. It has been suggested that there may be different types of corticosteroid receptors in different tissues, which may account for the diversity of the observed responses [25], but the molecular basis of such heterogeneity has not been established, and may depend on other factors.

4.1.2 Classification of corticosteroid action

In defining the function of an endocrine gland, the classical starting point is to investigate the changes produced by its removal. Relatively early in the history of adrenal physiology, when adrenalectomy first became possible, it was realized that the cortex shared with the islets of Langerhans the distinction that its proper functioning was essential for life [29]. However, explanation of the metabolic disturbances of body function which result after adrenalectomy or in the presence of excess hormone is not simple and, even now, the role of the adrenal cortical hormones in normal physiology is by no means fully understood. The situation is made complicated by the number and diversity of corticosteroid effects which have been described, and which are difficult to fit into a simple conceptual framework.

Removal of the adrenals causes numerous changes in the body, which are almost all related to the deficit of corticosteroids (rather than catecholamines). Sodium is lost from the body, blood volume decreases and hypotension occurs. Plasma levels of sodium fall and levels of potassium rise, and there is muscular weakness and reduced cardiac contractility. Also, fasting plasma glucose levels are low, and liver and muscle glycogen stores are depleted. Changes are seen in the nervous system, and depressed apathetic behavior is characteristic. There is also a decreased ability to withstand disturbance of many kinds, both physical (such as infection or starvation) or psychological (such as fear or emotional stress). These changes can be reversed by administration of corticosteroids, and it has become traditional to separate their actions into two general classes.

Mineralocorticoid actions are those associated with changes in water and electrolyte metabolism.

Glucocorticoid actions are those associated with changes in intermediary metabolism and resistance to stress.

While this has proved a convenient classification from a pragmatic point of view, it has also proved to be an oversimplification, as will emerge from the discussion that follows. There are five commonly recognized classes of steroid hormone in mammals: androgen, estrogen, progestogen, mineralocorticoid, and glucocorticoid. Historically, the distinction is based on what the steroids do, rather than on their structure. There is a consensus on the definition of mineralocorticoids as agents which either change the ratio of sodium to potassium in urine or which modulate unidirectional sodium transport in epithelia. For glucocorticoid activity there is less agreement, reflecting the diverse effects of these hormones in the organism. The original descriptions were related to the ability to modulate intermediary metabolism with an assay based on glycogen deposition in the liver. Currently, bioassays for glucocorticoids may often be based on their anti-inflammatory actions (see Sec. 4.3.3). Alternative approaches to defining the activity of a steroid are to examine its interaction with the binding of a glucocorticoid or a mineralocorticoid to tissue receptors. However, the classification of corticosteroid receptors is not straightforward, for instance as previously noted in the kidney, aldosterone binds not only to specific high-affinity sites (Type I: believed to mediate the mineralocorticoid responses) and to lower affinity sites which also bind dexamethasone (Type II: believed to mediate the glucocorticoid responses) but also to further sites which bind corticosterone and aldosterone but not dexamethasone (these are referred to as Type III binding sites by some authors [25,28]). The sites may be characterized on the basis of the hierarchy of binding affinity they show for the different steroids.

Type   I   Aldosterone > DOC > corticosterone > cortisol > dexamethasone
Type   II   Dexamethasone > corticosterone > cortisol = aldosterone
Type III   Corticosterone > cortisol >> aldosterone and dexamethasone

High-affinity dexamethasone and aldosterone-binding sites have been detected in the brain together with a further type, termed "nonclassical corticosterone-preferring sites." These are prominent in the hippocampus and show much higher affinity for corticosterone than for dexamethasone and binding characteristics very similar to those of the renal Type I receptor [28,30].

In general, when investigating the effects of an unknown steroid it is desirable to compare results obtained with different methods before judging the significance of any particular estimate of activity. On occasions different assays have given discordant results, a fact which has been used in support

TABLE 4.1 Activities of corticosteroids

Steroid	Relative activity in assay				
	MR	**Na+**	**GLY**	**GR**	**AI**
Aldosterone	1	1	0.15	1	—
DOC	0.8	0.03	0.02	1	—
Corticosterone	0.2	0.004	0.36	2	0.3
Cortisol	0.1	0.001	1	1	1
18-OH-DOC	0.015	0.004	—	0.02	—
Dexamethasone	0.05	0.001	17–250	10	25–169
9α-fluorocortisol	1*	0.15	6–13	10*	8–15

MR = mineralocorticoid receptor assay based on competition to ^{3}H-aldosterone binding sites in rat kidney
Na+ = bioassay, normally based on change of urinary Na+/K+ in the adrenalectomized rat
GLY = bioassay, based on glycogen deposition in the adrenalectomized rat liver
GR = glucocorticoid receptor assay, based on competition with ^{3}H-dexamethasone binding sites in rat or sheep(*) kidney
AI = bioassay based on anti-inflammatory activity
Sources: Ref. 26, 31, 32, 42, 106, 107, 144

of the argument that different facets of corticosteroid action may be modulated by further receptors. This is illustrated by the effects of corticosteroids on the cardiovascular system for which a "hypertensinogenic" class of steroid action has been proposed on the basis of the ability of some steroids to raise blood pressure without showing significant "classical" glucocorticoid or mineralocorticoid activity ([32] and see Sec. 4.2.8). Table 4.1 shows a comparison of effects of some corticosteroids assessed by a variety of methods.

A further complication is caused by the fact that few steroids are absolutely specific in their effects: naturally occurring glucocorticoids such as cortisol show mineralocorticoid activity, and aldosterone shows glucocorticoid activity. Several factors must be taken into account when judging whether these "crossover" actions will make any significant contribution to the physiology of the animal. First, and most obvious, is the circulating levels of the steroids in the plasma: since that of aldosterone is 2 to 3 orders of magnitude lower than that of cortisol, it is very unlikely that it exerts any significant glucocorticoid effect in vivo. Second, it is held to be the free rather than protein bound steroid which leaves the circulation and has access to the cellular receptors, and which is biologically active. The affinity and capacity of the binding of the different corticosteroids by albumin and transcortin varies and must also be considered. In practice, in most species, the levels of free cortisol and corticosterone approximate to 1 percent of total circulating levels (see Chap. 2). A further degree of specificity may be introduced by interactions in the whole animal which enable mineralocorticoid target tissues to show a pronounced selectivity in their response to aldosterone in the face of very low levels of the hormone. It has been suggested that this is achieved by the presence in tissues such as kidney and colon of the enzyme 11β-hydroxysteroid dehydrogenase which converts cortisol to cortisone and corticosterone to 11-dehydrocorticosterone. These 11-oxo derivatives do not bind to the mineralocorticoid receptors, with the result that the parent compounds are effectively prevented from expressing their effect. Thus, the target tissue specificity may be said to be enzyme rather than receptor mediated. In contrast, in the hippocampus where no 11β-hydroxysteroid dehydrogenase activity is detected, the Type I receptor will bind corticosterone [33,34]. Finally, it should be remembered that the blood contains at least ten steroids at concentrations sufficient for biological activity and that the presence of one may modify the action of another. For example, it is well known that progesterone can act as a mineralocorticoid antagonist [35], and some steroids such as 5α-dihydrocortisol have been claimed to act as amplifiers of aldosterone action [36].

4.2 Effects of corticosteroids on electrolyte and water metabolism and the cardiovascular system

4.2.1 General aspects

Aldosterone and other steroids with mineralocorticoid activity increase the reabsorption of sodium in the kidney and in secretory epithelia, thus reducing the sodium content of urine, saliva, sweat, gastric juice, and faeces [37]. Effectively, increased amounts of sodium ions (Na^+) are exchanged for potassium or hydrogen ions (K^+ or H^+) in the kidney tubules under the influence of aldosterone, leading to decreased Na^+ excretion (antinatriuresis), increased K^+ excretion (kaliuresis) and increased urine acidity. This aspect is exploited in the classical bioassay for mineralocorticoid activity used in the isolation of aldosterone [38], in which the change in urinary Na^+/K^+ ratio in the adrenalectomized rat is used as an index of potency of the administered steroid. (Table 4.1; Fig. 4.8). It is also possible that aldosterone affects the uptake of K^+ and Na^+ into nonepithelial cells in the body such as arterial smooth muscle and brain [39].

With prolonged mineralocorticoid administration, extracellular fluid (ECF) volume expansion occurs as a result of Na^+ retention, and renin production is depressed. However, after a few days, the kidney "escapes" from the Na^+ retaining action of the mineralocorticoid but not from the effect on potassium secretion (see Sec. 4.2.3). Thus, one of the main effects of exposure of an organism to excessive quantities of mineralocorticoids is a decrease in total body K^+, reflected in hypokalaemia. Magnesium and calcium excretion is also increased during prolonged mineralocorticoid

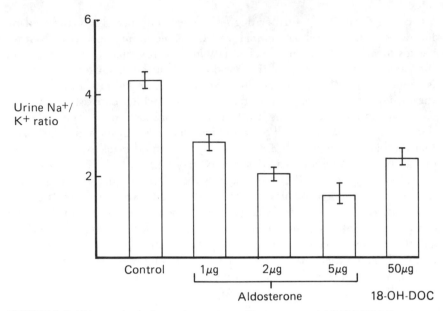

FIGURE 4.8 Urinary electrolyte responses to aldosterone and 18-OH-DOC in adrenal-ectomized male rats. Urine was collected for 3 hours following the injection of steroid or vehicle, and the results are shown as urinary Na^+/K^+ ratio. (Figure drawn from data obtained by Dr. H. Vazir.)

administration, but these effects are not usually seen following acute administration and may be secondary to ECF volume expansion [40,41].

Aldosterone is recognized to be the most potent mineralocorticoid secreted by the adrenal cortex, although DOC and 18-OH-DOC (which are produced in significant amounts in some species, notably the rat) may also make a significant contribution [42]. Cortisol and corticosterone may also manifest appreciable mineralocorticoid activity particularly when secretion is elevated (Table 4.1).

Glucocorticoid activity can also affect kidney function, in part through an action on glomerular filtration rate (GFR), [43]. In addition, one of the classical signs of adrenocortical insufficiency is the inability to excrete a water load. This was used as the basis of an early diagnostic test, and seems to be mainly related to glucocorticoid deficiency [44], (see Sec. 4.2.5).

4.2.2 Effects of mineralocorticoids on sodium transport

Probably, the most important action of aldosterone in mammals is to increase reabsorption of Na^+ in the kidney. It only affects a very small proportion of the filtered Na^+ ($< 0.5\%$), nevertheless, this is essential for life and the losses must be replaced. Thus, provision of 0.9 percent NaCl as drinking fluid is required for the survival of adrenalectomized animals.

Much of the current knowledge of mineralocorticoid action on the kidney

ALDOSTERONE   DEXAMETHASONE

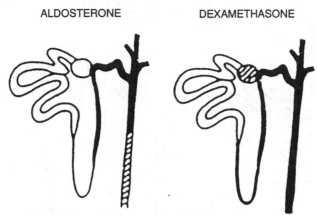

FIGURE 4.9 Specific nuclear binding of aldosterone and dexamethasone along the length of the nephron. Dark zones are those showing maximum specific binding, and hatched zones those showing lower levels of specific binding. (Figure modified and reproduced with permission from ref. 47.)

has been derived from studies using animals pretreated with large doses of hormone for several days. Although this has provided much useful data, some of the changes observed may have been only secondary to changes in overall Na^+/K^+ balance in the body and not a reflection of the primary short term effect of the hormone. Nevertheless, there is now general agreement that the major sites of aldosterone action in the nephron are in the distal tubule and collecting duct [45,46]. A high concentration of aldosterone receptors has been found in these areas, with a greater density in the cortical than the medullary portion of the collecting duct (Fig. 4.9). A lesser degree of specific nuclear binding is found in the cortical part of the ascending limb of the Loop of Henle, and none occurs in the proximal tubule [47]. This correlates with the conclusion drawn from recent studies that physiological concentrations of mineralocorticoids do not affect electrolyte transport in the proximal tubule [48]. Glucocorticoid receptors have also been found in the distal nephron, but it is not clear what effects they might be mediating (Fig. 4.9; [49] and see Sec. 4.2.5).

In the last 15 years, techniques have been available for the study of the acute actions of mineralocorticoids in the mammalian kidney [50]. Prior to this, amphibian epithelia were used as model systems, and the urinary bladder of the toad, which responds to aldosterone by increasing the active transport of Na^+ has received much attention. It has provided a particularly useful experimental model for the study of aldosterone action because Na^+ transport through its single-cell-thick epithelium, which consists largely of one-cell type, can be studied in the Ussing chamber [51]. In this apparatus, the bladder is mounted as a sheet separating two identical physiological

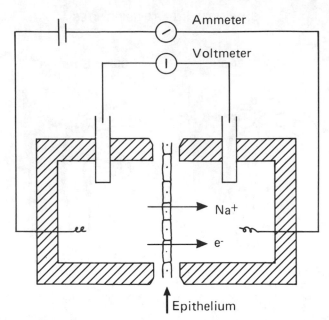

FIGURE 4.10 (a) Ussing apparatus for study of ion transport. A sheet of isolated epithelium is mounted in the chamber so that it separates two solutions of identical composition. The electrical potential generated by transport of Na$^+$ is opposed by an external current source. When these two are exactly equal, the voltage measured at the meter is zero, and the current measured (SSC) by the ammeter is proportional to the rate of transport of Na$^+$.

solutions, and an electric potential is generated as a result of sodium transport from the mucosal side (normally in contact with urine) to the serosal side (normally in contact with interstitial fluid). An external current, the short-circuit current (SCC), can be applied to the chamber in sufficient magnitude to nullify the spontaneous potential difference, and this can be used to quantify the rate of sodium transport, which is increased on exposure to mineralocorticoids (Fig. 4.10a,b). Early studies with this apparatus revealed a lag time between exposure of the toad bladder to aldosterone and the increase in SCC. In addition, the involvement of both RNA and protein synthesis was suggested by the observations that the effects of aldosterone could be blocked by both actinomycin D and puromycin [52]. Other indications that the mode of action of aldosterone was similar to that of other steroid hormones came from the fact that the cellular uptake of aldosterone preceded its biological effect, that aldosterone could be shown to bind preferentially to cell nuclei, and that uptake of labeled uridine into epithelial RNA was increased by aldosterone [53,54]. Further exploration of the system has been concerned with the role of the newly synthesized protein in the response to aldosterone. Sodium transport in the collecting duct of the kidney, amphibian bladder, and other aldosterone responsive epithelia depends on asymmetrical properties of the cells which possess

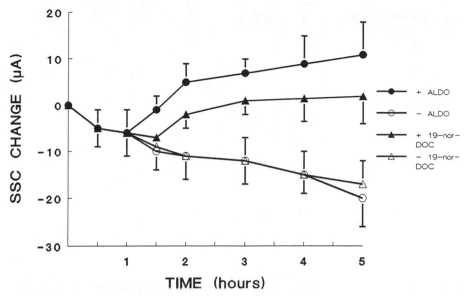

FIGURE 4.10 (b) Difference in short-circuit current (SSC) occurring over 5 hours in toad bladders mounted in an Ussing chamber after addition of aldosterone and 19-nor-DOC or diluent (control) to the serosal (blood) side. (Figure reproduced with permission from ref. 110.)

amiloride-sensitive sodium channels in the surface (termed apical in the kidney) exposed to urine or tubular fluid and ATPase on the surface exposed to interstitial fluid (termed serosal or basolateral). It is believed that sodium diffuses down its electrochemical gradient into the cell from the tubular fluid and is transported out into the interstitial fluid at the basolateral surface by the energy requiring sodium pump (Na^+/K^+ ATPase; Fig. 4.11). With this model in mind it is evident that there are at least three ways in which the aldosterone-induced protein might act.

1. By increasing the permeability of the apical membrane to Na^+.
2. By increasing the activity of the basolateral membrane Na^+/K^+ ATPase.
3. By increasing the availability of ATP to drive the pump.

Over the years there has been much debate over the existence and relative importance of these different mechanisms. The opposing views may, however, be reconciled if aldosterone is seen to control Na^+ and K^+ transport in a pleiotropic fashion, with actions at all three loci, and stimulation of the synthesis of a number of proteins.

The hypothesis that the action of aldosterone on Na^+ transport is mediated by a primary action on the permeability of the mucosal membrane was first suggested by Crabb in 1961 [55] on the basis of its effects on toad bladder. The concept was supported by the finding that the number of amiloride-

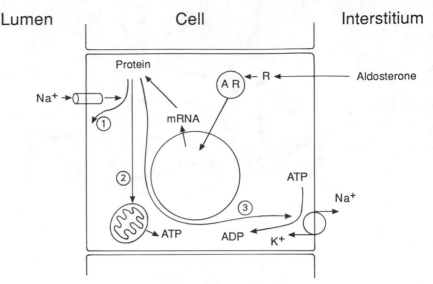

FIGURE 4.11 A scheme for mineralocorticoid action in an epithelial cell. Aldosterone (A) binds to a specific receptor (R), the complex is activated and translocates to the nucleus. As a result of interaction with chromatin sites, the synthesis of specific mRNA is stimulated.

Three possible sites of action of the resultant protein(s) are indicated and include (1) the recruitment of amiloride-sensitive sodium channels, (2) the stimulation of mitochondrial enzymes involved in ATP production, and (3) the activation of Na$^+$/K$^+$ ATPase. The protein may also be involved in remodeling of the apical membrane.

binding sites (reflecting the number of sodium channels) is increased in aldosterone-treated toad bladders [56]. More recently, analysis of voltage-current relationships after aldosterone treatment in kidney tubules also showed that the apical membrane Na$^+$ permeability was increased [57]. This action of aldosterone could be the result of the insertion of new channels into the apical membrane or of the opening of previously silent channels. The fact that trypsinisation of the membrane resulted in a decrease in both basal and aldosterone stimulated sodium transport favors the latter [58], although it is likely that synthesis of new channel protein is also involved in the longer term responses. It appears that the mechanisms which are responsible for the change between the conductive and non-conductive states of the sodium channels may involve membrane remodeling with alterations in lipids and proteins, for example, by methylation [59–63].

An increase in Na$^+$ transport through epithelia obviously requires increased activity of the sodium pump in the basolateral surfaces of the cells. A source of continuing controversy has been whether this is simply secondary to the increased entry of sodium into the cells or requires induction of new catalytic units [64–66]. In general, it seems that the evidence supports the idea that the long-term effects of aldosterone on sodium transport involve the synthesis of new ATPase, and increased synthesis of the mRNAs coding for the α and β subunits has been detected

[67]. In contrast, the effects seen in the short term, which are blocked by amiloride, have been explained by stimulation of the pump through increased entry of Na^+ into the cells [63–66].

Research into the effects of aldosterone on the generation of ATP has focussed on the enzymes involved in the tricarboxylic acid cycle, and in particular citrate synthase. The activity of this enzyme, which may be a rate-limiting enzyme in ATP synthesis in the kidney, is decreased after adrenalectomy and restored by aldosterone [68,69].

4.2.3 Escape from mineralocorticoid actions

It is well known that prolonged administration of mineralocorticoid hormones induces only a transient period of sodium retention, and after 2 to 3 days a natriuresis occurs which returns sodium balance to normal, despite the continued administration of the steroid (Fig. 4.12). This phenomenon has been termed "mineralocorticoid escape," and has been the subject of much investigation over the last 35 years [70,71]. One of the favored hypotheses to explain escape has been that it occurs as a result of compensatory mechanisms activated by the expansion of extracellular fluid volume [72]. The recent isolation of atrial natriuretic peptides (ANP), which are secreted in response to volume expansion has added substance to these ideas, and it has been possible to show that levels of ANP are elevated during escape from the sodium retaining effects of both aldosterone and DOC [73,74]. There are several mechanisms by which the peptide could counteract the effects of aldosterone since it has been shown to increase GFR and inhibit tubular reabsorption of sodium as well as inhibiting the release of renin (thus decreasing the formation of angiotensin II and preventing its antinatriuretic effects). The overall contribution that ANP makes to mineralocorticoid escape remains to be determined, however, since there are also data which suggest that changes in renal hemodynamics can play an important part in determining sodium excretion [75].

4.2.4 Effects on potassium and acid secretion

Precise control of the potassium ion concentration in extracellular fluid is essential to normal vertebrate physiology, and a number of mechanisms exist to buffer both chronic and acute increases in K^+ intake. Ingested potassium can be partially disposed of in the intracellular pool as a temporary means of avoiding sudden increases in plasma K^+; hormones such as insulin and epinephrine are important mediators of this process, and it is probable that aldosterone also contributes [76,77]. Long-term homeostasis of body K^+ requires that the excess be excreted in the urine: there is little evidence that the amount lost in feces is variable even though mineralocorticoids are known to increase the fecal potassium content [76,78]. Although primarily thought of as a sodium conserving hormone, the importance of aldosterone in the regulation of potassium metabolism in the whole animal, particularly

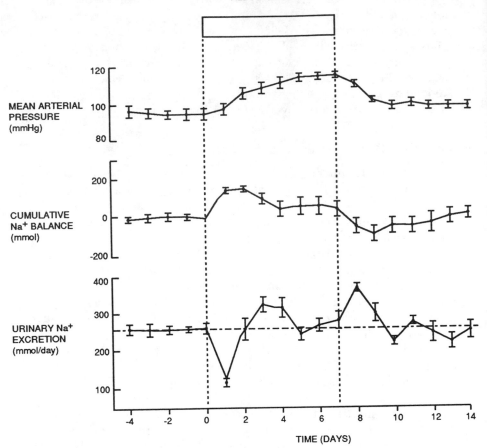

FIGURE 4.12 Effect of aldosterone infusion (14μg/kg per day) for 6 days on urinary sodium (Na+) excretion, Na+ balance and mean arterial pressure in dogs maintained on a constant Na+ intake of 269 mmol per day. Note that sodium retention returns to control values after 2 days—the "escape" phenomenon. (Figure reproduced with permission from ref. 75.)

in the long term, is increasingly recognized. After a period of adaptation to a high K^+ intake, a series of changes takes place in the kidney which lead to the enhanced capacity to excrete an acute load of K^+. This response is associated with an increase in aldosterone levels in the plasma and increased secretion of K^+ in the distal tubules and cortical collecting duct. It has been demonstrated that increased aldosterone levels act in concert with the increase in K^+ intake to insure the development of this response [79,80] (Fig. 4.13). On a shorter time scale, ambient aldosterone levels have been shown to play a part in determining potassium excretion following an acute load, although demonstration of this may be difficult because of the interaction of the several factors involved [81,82].

The cellular mechanisms have not been particularly well understood until

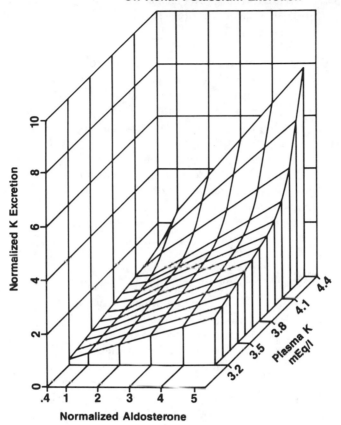

**Effects Of Aldosterone And Plasma Potassium
On Renal Potassium Excretion**

Normalized K Excretion

Plasma K mEq/l

Normalized Aldosterone

FIGURE 4.13 Three dimensional representation of the relationship between plasma potassium (K) concentration, potassium excretion, and dose in adrenalectomized dogs receiving aldosterone treatment. The data are normalized so that 50 µg aldosterone/day = 1.0, and 22 mmol K^+/day = 1.0. (Figure reproduced with permission from ref. 79.)

recently, probably because there has been no readily accessible experimental model in which to pursue the studies: aldosterone does not affect K^+ transport in the toad bladder. In whole animal experiments, usually involving relatively long-term treatment with large doses of mineralocorticoids, increased potassium secretion is seen in association with increased Na^+ reabsorption. This is believed to be explained by active peritubular uptake being stimulated through increased ATPase activity, and also by the increased membrane conductance and transepithelial potential difference, which facilitates the passive movement through the luminal membrane [45,83]. Recent studies have demonstrated a primary effect of aldosterone in increasing the permeability of the cortical collecting tubule to both sodium and potassium [84]. Since the secretion of K^+ and H^+ are closely linked in some systems, it is suggested that one of the actions of aldosterone is to activate a Na^+/H^+ countertransport mechanism resulting in increased H^+ secretion and intracellular alkalinization. This leads in turn to increased K^+ conductance of the apical membrane (with recruitment of new channels) and K^+ secretion [85,86].

The roles played by mineralocorticoids in the control of acid-base balance are not clearly defined, although it is well known that hypokalemic alkalosis occurs in states of mineralocorticoid excess and acidosis in deficiency. There have been suggestions that the increase in H^+ secretion produced by mineralocorticoids is dependent on a relative deficit of K^+, and therefore, occurs essentially as a substitution of one ion for the other [87,88]. However, there is also evidence that aldosterone may actively maintain renal H^+ secretion, and that the effect is direct rather than secondary to changes in electrolyte balance (see above and [89,90]). The development of metabolic alkalosis with mineralocorticoid excess can thus be seen as an exaggeration of the normal physiological role of the hormone. An action to increase renal NH_4^+ secretion has also been detected [91].

4.2.5 Effects of glucocorticoids on the kidney

The effect of glucocorticoids on renal hemodynamics is well known: adrenal-ectomy reduces both glomerular filtration rate and renal plasma flow, ACTH and glucocorticoids (but not mineralocorticoids) increase both parameters [92]. Micropuncture studies showed significantly increased glomerular blood flow and filtration in methylprednisolone treated rats, which were explained by decreases in both afferent and efferent arteriolar resistance. It seems possible that these changes might be due to an action of glucocorticoids on vascular smooth muscle [93]. The increase in renal potassium excretion seen after glucocorticoid treatment does not seem to be a direct tubular effect, but rather secondary to the hemodynamic changes [86].

Adrenocortical insufficiency is associated with defects both in the ability to dilute and to concentrate the urine, reflecting glucocorticoid effects both on vasopressin action and secretion [44,94–97]. Glucocorticoid pretreatment of intact animals seems to attenuate or eliminate the vasopressin response to stimuli such as hemorrhage. This suggests the possibility of a glucocorticoid modulation of vasopressin secretion, although the physiological significance of this system in normal physiology is not clear. In glucocorticoid-deficient rats, plasma vasopressin levels are elevated; they are returned to normal by treatment with corticosterone, as is the ability to excrete a water load [95] (Fig. 4.14). It seems unlikely that the elevated levels of vasopressin can be solely explained by the hyperosmotic, hypovolemic conditions after adrenalectomy, and that there is a significant contribution from the loss of the central glucocorticoid negative feedback [96].

Glucocorticoids may also interact with vasopressin action in the kidney. It was originally suggested that they were necessary for the maintenance of the full impermeability to water of the distal nephron in order to explain

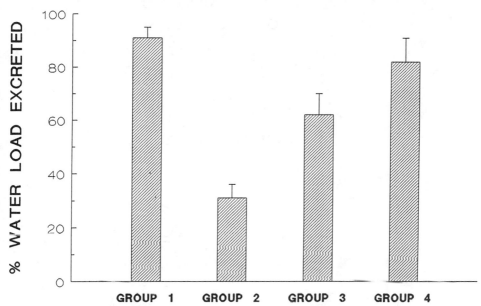

FIGURE 4.14 Effect of glucocorticoid deficiency and corticosterone replacement on the excretion of a water load in rats. Group 1 = Untreated controls, Group 2,3,4 = Hypophysectomized animals: Group 2 = No replacement, that is, glucocorticoid deficient, Group 3 = Corticosterone replacement at 300 μg/day, and Group 4 at 600 μg/day. (Figure modified and reproduced with permission from ref. 95.)

the inability to excrete a water load, but no direct effects were demonstrated in isolated tubule preparations. In fact, a *reduced* responsiveness of renal tubules to vasopressin in terms of both water permeability and cyclic AMP generation has been observed after adrenalectomy. This defect was restored to normal by dexamethasone, and perhaps explains the reduced ability to concentrate the urine in adrenal insufficiency [96].

4.2.6 Other steroids with mineralocorticoid activity

A number of steroids with mineralocorticoid activity have attracted attention over the years as possible causative agents in hypertensive disease. This has arisen because it is possible to identify a subgroup of patients with essential hypertension who show a syndrome suggestive of exposure to excess mineralocorticoid, "low-renin hypertension." It has not been possible to show that known mineralocorticoids such as aldosterone and DOC are produced in sufficient quantity to account for the metabolic changes. This has led to the hypothesis that another (possibly as yet unidentified) compound is present [98]. This has seemed a particularly plausible argument in the

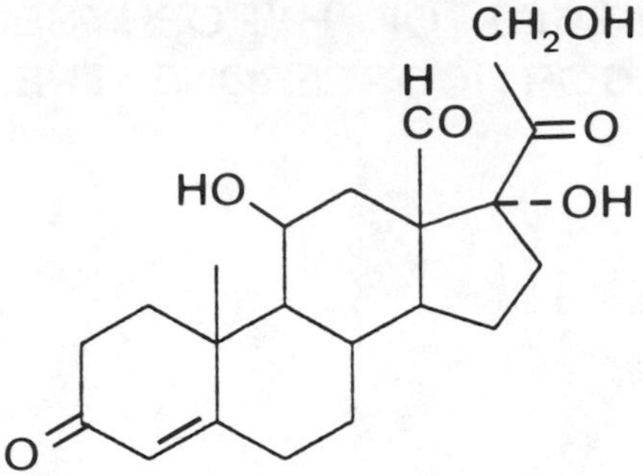

18-oxo-cortisol

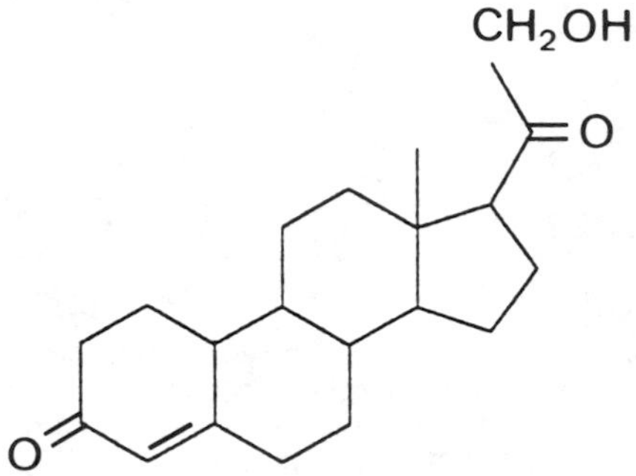

19-nor-DOC

FIGURE 4.15 Structures of 18-oxo-cortisol and 19-nor-DOC. For other structures, see Fig. 2.3.

case of steroids such as those now described, which possess mineralocorticoid activity and yet whose secretion is controlled by ACTH. However, none has yet been unequivocally demonstrated to be involved in essential hypertension.

18-OH-DOC was isolated and characterized in 1961, and shown to be a major product of the rat adrenal gland, rivaling corticosterone in the quantities produced [42,99]. It has been shown to exert weak antinatriuretic activity in the adrenalectomized rat, with a potency of approximately 1/250 that of aldosterone (Table 4.1), and to increase sodium transport in the isolated toad bladder [100–102]. Some authors have found that it possesses a greater capacity to raise blood pressure than would be predicted from its antinatriuretic activity, and it may be involved in the pathogenesis of some forms of experimental hypertension in the rat [103].

18-Oxo-cortisol ((Fig. 4.15), which could also be called 17-hydroxy aldosterone) has been isolated from the urine of patients with hyperaldosteronism and shows weak mineralocorticoid and glucocorticoid activity [104–106]. In sheep it produces a significant increase in blood pressure which is associated with the classical mineralocorticoid effects of hypokalemia and transient urinary sodium retention, with no evidence of glucocorticoid effects. It is more effective in raising blood pressure than an "equivalent" dose (in terms of biological activity) of aldosterone [107]. In contrast, 18-OH-cortisol was found to show no significant glucocorticoid or mineralocorticoid activity [108].

19-Nor-DOC is produced peripherally (mainly in the kidney) in the rat from an adrenocortical precursor, presumably 19-OH-DOC. It possesses mineralocorticoid activity when tested in both the toad bladder and rat assay systems [109,110]. It is capable of raising blood pressure in rats at doses that are approximately 10 to 25 times larger than those of aldosterone, and its production has been linked with some forms of hypertension [111,112].

4.2.7 Salt appetite

Regulation of the body sodium content requires that, in the face of shortage of this ion, both excessive losses are prevented and additional sodium is consumed. The drive to consume sodium is stimulated under conditions of sodium deficiency and it would seem an effective economy of mechanism if the renin-angiotensin system and aldosterone were also to be involved in the stimulation of salt appetite. There is some support for this concept in that both mineralocorticoids and angiotensin II in the brain have been shown to increase saline consumption in the rat (probably by independent mechanisms). However, it seems unlikely that these are the only factors involved, because such maneuvers have not been found to be effective in other rodents [113]. The Type I high-affinity-binding sites for aldosterone found in the brain may be involved in mediating the appetite response [114]. In some species, ACTH induces a large increase of NaCl intake, and here it seems likely that glucocorticoids are playing a part in mediating the process [115].

4.2.8 Effects of corticosteroids on the cardiovascular system

Hypotension is characteristic of adrenal insufficiency and it is said that it can be corrected very rapidly (within 1 to 2 minutes) by administration of cortisol. From this it has been concluded that glucocorticoids can affect cardiovascular function by mechanisms which are independent of their relatively slow effects on body fluid redistribution [116]. The action seems to involve a reversal of the reduced myocardial effectiveness and arteriolar tone which accompanies glucocorticoid deficiency, when vascular smooth muscle becomes unresponsive to catecholamines. Glucocorticoids both potentiate responses to catecholamines and inhibit synthesis of prostacyclin (see Sec. 4.3.3) which may contribute to their effects on blood pressure [117–119].

Conversely, hypertension is a well-known consequence of excess cortico-steroid secretion, be it cortisol, as in Cushing's syndrome, or aldosterone, as in Conn's syndrome. It is perhaps surprising, therefore, that current knowledge of the mechanisms involved is so incomplete. The classical view is that corticosteroid induced hypertension can be ascribed to the sodium retaining activity of the compounds, which leads to expansion of extracellular fluid (ECF) volume. As a result, cardiac output is increased in the initial stages of the hypertensive process, subsequently "vascular autoregulation" occurs with an increase in peripheral resistance which maintains the elevated blood pressure [120] [Fig. 4.16(a)]. Increasingly, this is seen as only a partial explanation and currently much effort is directed towards devising and testing other models. It has long been known that the blood pressure increases that can be produced by treatment with a mineralocorticoid are attenuated by sodium restriction and potentiated by sodium loading [121]. In fact, a widely used model of experimental hypertension is provided by deoxycorticosterone acetate (DOCA) treated, sodium loaded animals in which the sodium retention is compounded by unilateral nephrectomy! [122]. In contrast, blood pressure increases produced by treatment with glucocorticoids such as cortisol and dexamethasone are independent of sodium status [123,124], and must therefore, involve fundamentally different mechanisms. A further complication that should be borne in mind in any discussion of the hypertensive actions of steroids, is that there is a great deal of species variation in the responses, so that proposing a universal mechanism even for one class of steroid may be premature [123].

Mineralocorticoid-induced hypertension The rise in arterial pressure pro-duced by mineralocorticoid treatment in the salt sensitized animal has been almost universally found to be associated with an increase in peripheral resistance but not necessarily blood volume or cardiac output [125]. This has led to a search for mechanisms by which mineralocorticoids in combination with Na^+ might increase vascular resistance. Some studies carried out in experimental animals have suggested an important role for the sympathetic nervous system in the development and maintenance of the hypertension. The evidence offered in support of this possibility include the increased levels of circulating catecholamines and norepinephrine turnover, and the prevention of DOCA-induced hypertension by chemical sympathectomy [126–128]. However, in humans the data suggest that mineralocorticoid hypertension is associated with decreased, rather than increased sympathetic activity, so that this concept remains conjectural [129]. A recent suggestion has been that aldosterone might exert a hyperten-sive effect by a direct action on the central nervous system [130]. Another line of evidence favors the idea that vasopressin might play an important role in mineralocorticoid hypertension, particularly in the rat where plasma vasopressin levels are increased during DOCA treatment [131]. Other ideas are based on mineralocorticoid-induced changes in membrane permeability of vascular smooth muscle, leading to an abnormal cation turnover, depolar-

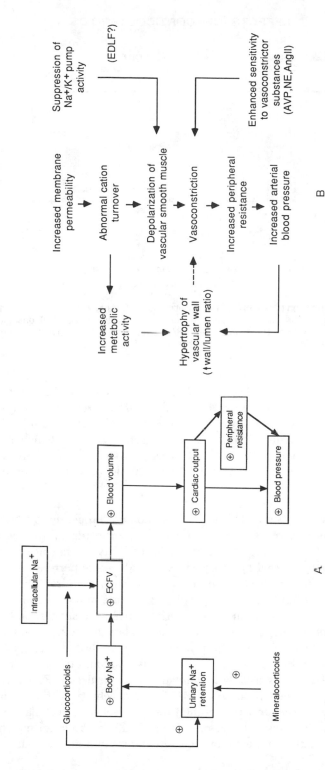

FIGURE 4.16 (a) Proposed mechanisms by which corticosteroids increase blood pressure. Sodium retention (or shifts of sodium from the intracellular space) leads to expansion of ECF volume, increased cardiac output and; eventually, increased peripheral resistance. ECFV = extracellular fluid volume, + = increase.
(b) A possible sequence of events in mineralocorticoid-induced development of increased peripheral resistance and hypertension. EDLF = endogenous-digitalis-like factor, NE = norepinephrine, ANG II = angiotensin II. (Figure reproduced with permission from ref. 134.)

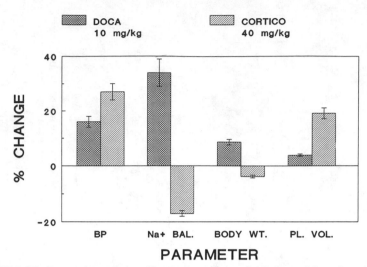

FIGURE 4.17 Comparison of the effects of treatment for 5 days with a mineralocorticoid (DOC acetate, DOCA) and a glucocorticoid (corticosterone, cortico) on blood pressure (BP), cumulative sodium balance (Na⁺ BAL.), body weight, and plasma volume (PL. VOL.) in rats. (Figure drawn from data obtained by Dr. H. Vazir.)

ization and vasoconstriction [132,133]. The observed enhanced sensitivity to pressor agents and hypertrophy of the vascular wall have been incorporated into this scheme as part of a unifying hypothesis [134] [Fig. 4.16(b)]. The proposed circulating endogenous-digitalis-like factor (EDLF), which it is suggested, is released from the central nervous system (or possibly the adrenal) in response to increases in ECF volume may also make a contribution [135,136].

Glucocorticoid hypertension   It has been common until recently to ascribe the blood pressure increases produced by glucocorticoids to the same mechanism as that proposed for mineralocorticoids. This was in spite of the knowledge, available since 1952, that cortisone-induced hypertension was not affected by sodium restriction and developed much more rapidly than the mineralocorticoid-induced variety [124]. One attempt at reconciling these facts within the conventional framework has been the suggestion that glucocorticoid-induced hypertension occurs secondary to shifts of sodium from the intracellular to the extracellular space leading to increased intravascular volume [Fig. 4.16(a)], [137]. This takes account of the fact that glucocorticoid treatment is usually accompanied by natriuresis (Fig. 4.17). Alternatively, it has been proposed that increased activity of the renin-angiotensin system, or potentiation of the pressor effects of catecholamines may underlie the increases in blood pressure [137–139]. The latter may be caused by a direct action of the steroids on vascular smooth muscle, and the finding of glucocorticoid receptors in the cells may provide corroborative evidence for this [140]. However, there is doubt as to whether the enhanced responsiveness is sufficient to account for the full extent of the blood pressure increases produced by glucocorticoids. Current research effort is still directed towards elucidation of the cause of the increase in peripheral resistance which seems, as in mineralocorticoid hypertension, to be at the center of the phenomenon, and which also may involve alterations in membrane ion transport [141].

ACTH-induced hypertension Increases in blood pressure can be induced by relatively small doses of ACTH in humans and several animal species including the sheep and rat. In humans, it seems possible that the effects of ACTH can be explained simply in terms of hypersecretion of cortisol [142]. In contrast, in the sheep this is not true and more than 10 years of effort have been expended in attempting to dissect the mechanisms underlying the cardiovascular responses [143]. ACTH-induced hypertension in sheep shows some characteristics of mineralocorticoid hypertension (initial sodium retention, hypernatremia, and hypokalemia), and some characteristics of glucocorticoid hypertension (blood pressure increases develop rapidly and are not dependent on sodium retention). However, it was not possible to reproduce the effects of infusions of ACTH using infusions of appropriate amounts of the known biologically active corticosteroids produced by the sheep adrenal (cortisol, aldosterone, DOC, etc.). Further investigation revealed that the presence of two otherwise biologically inactive steroids, 20α-hydroxy-4-pregnene-3-one and 17α-20α-dihydroxy-4-pregnene-3-one was necessary for an increase in blood pressure to be seen. On the basis of this and further studies using synthetic steroids, it was proposed that a new class of corticosteroid action—"hypertensinogenic"—should be established [33,144]. It is not yet clear what role, if any, these receptors might play in human essential hypertension.

4.3 Glucocorticoid effects

4.3.1 Effects of corticosteroids on intermediary metabolism

The typical metabolic effects of glucocorticoids in the whole animal include a stimulation of protein catabolism and hepatic glycogenesis and gluconeogenesis, with decreased glucose uptake and utilization in peripheral tissues. These effects are minimal in the fed state, but are very obvious during fasting when plasma glucose levels and liver glycogen levels are increased with glucocorticoid treatment. They are brought about by actions on protein, carbohydrate, and fat metabolism occurring mainly in liver, muscle, and adipose tissue; together these contribute to the maintenance of blood glucose levels. Thus, the effects of glucocorticoids, although very much secondary to those of insulin in the control of metabolism, essentially oppose its actions

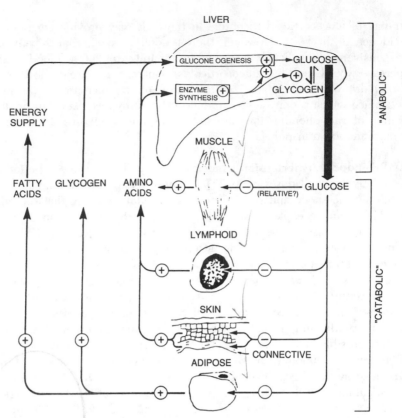

FIGURE 4.18 Actions of glucocorticoids in promoting gluconeogenesis. These are shown as decreased glucose uptake and catabolism in muscle, skin, lymphoid and adipose tissue, and anabolic actions in the liver with increased synthesis of gluconeogenic enzymes and glycogen. + = stimulation; − = inhibition. (Figure reproduced with permission from ref. 145.)

and can be seen as providing a protection against long term glucose deprivation (Fig. 4.18), [145].

Protein metabolism Glucocorticoids increase protein catabolism in the periphery resulting in increased levels of amino acids in plasma, and, as a consequence, administration of excessive amounts of glucocorticoids results in muscle wasting (Fig. 4.19). The action seems to involve both decreased protein synthesis and stimulation of proteolysis [145,146]. Suppression of protein synthesis is demonstrated by the decreased incorporation of amino acids into rat muscle following glucocorticoid treatment, and the degree of response is greater in "fast-twitch" muscle than in the "slow-twitch" variety [147,148]. Increased proteolysis has been demonstrated by studying plasma leucine and alanine dynamics in humans during a cortisol infusion, which indicated that there was also increased de novo synthesis of alanine, a

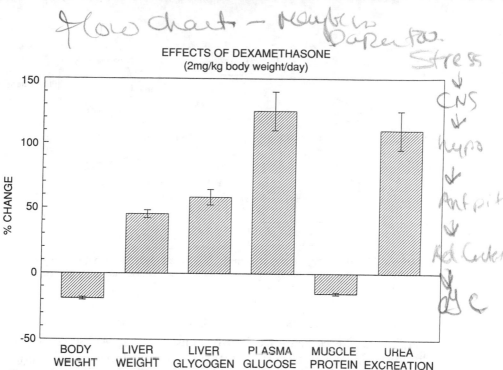

FIGURE 4.19 The effects of treatment with dexamethasone (2mg/kg body weight/day for 7 days) on metabolic parameters in rats. Note the decrease in body weight and loss of muscle protein, accompanied by an increase in the relative weight of the liver (calculated as % body weight). (Figure drawn from data obtained by Dr. N.I. Onyezili.)

potential gluconeogenic substrate [149]. The proteolytic effects of glucocorticoids are antagonized by insulin [150].

Peripheral glucose metabolism   Short-term glucocorticoid treatment leads to modest increases in plasma glucose concentration, long-term treatment adversely affects glucose tolerance and may lead to fasting hyperglycemia [151]. In the whole animal there is a reduction in glucose uptake and clearance from the blood, which is consistent with in vitro studies which show a direct inhibitory effect of glucocorticoids on glucose oxidation in muscle and adipose tissue, and on glucose transport in adipocytes [151–154]. The antagonism of the action of insulin in these tissues probably involves both alterations in insulin's binding to its receptor and in postreceptor events [154,155].

Hepatic glucose metabolism   Glucocorticoids have two major effects on carbohydrate metabolism in the liver: they increase the rate of hepatic gluconeogenesis during fasting and they also increase the deposition of glycogen. These two actions occur independently, and may oppose each other [156,157]. Adrenalectomy has relatively little effect on the concentration of glycogen in the liver, whereas glucocorticoids greatly increase it, particularly

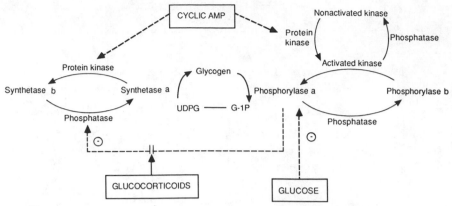

FIGURE 4.20 The control of glycogen metabolism in the liver. + = activation, − = inhibition. ‖ = action of glucocorticoids to dissociate synthetase phosphatase from phosphorylase *a* inhibition.

after fasting. This response has long been used as the basis of an assay for glucocorticoid activity (see Table 4.1). In essence, glucocorticoids permit glycogen synthesis to occur in the liver even when plasma glucose levels are low, so that glycogen metabolism is freed from the tight control normally exerted by glucose [157]. The rate of glycogen synthesis in the liver is proportional to the concentration of synthetase *a,* and the activity of this enzyme is normally controlled by glucose levels. This mechanism involves inhibition of phosphorylase *a* by high concentrations of glucose, which allows the activation of synthetase *b* by synthetase phosphatase. During glucocorticoid treatment, synthetase phosphatase is dissociated from the inhibition exerted by phosphorylase *a* (Fig. 4.20), thus allowing glycogen synthesis to occur under conditions when it would normally be inhibited. There is evidence that this dissociation is brought about by a protein factor synthesized under the influence of glucocorticoids [157,158].

Glucocorticoids also increase the liver's capacity for gluconeogenesis, partly by stimulating enzyme synthesis and partly by increasing hepatic responsiveness to catecholamines and glucagon. The activity of many steps in the gluconeogenic pathway are increased by glucocorticoids, and total hepatic protein synthesis is increased. Stimulation of the transaminase enzymes is very obvious, although influences on steps distal to this stage may be of greater functional importance, but there is relatively little information on this aspect [145,151,157]. The permissive role of glucocorticoids in increasing the sensitivity of the liver to the gluconeogenic actions of glucagon and catecholamines is better understood. This is revealed in adrenalectomized animals where the responses to these hormones are impaired and may be restored by cortisol administration [159]. The ability of glucagon to stimulate gluconeogenesis depends on the phosphorylation of pyruvate kinase and other enzymes by cyclic AMP-dependent protein kinases. The absence of glucocorticoids reduces the ability of cyclic AMP to effect this activation, rather than reducing the amount of cyclic AMP formed [160].

Effects on fat metabolism   The principal effect of glucocorticoid in adipose tissue is increased lipolysis, partly caused by direct effects of the hormones on lipase but also because of enhanced activity of lipolytic hormones (such as glucagon) and decreased glucose uptake [161,162]. Paradoxically, in humans states of glucocorticoid excess are characterized by deposition of fat in the trunk and abdomen, and weight gain. This phenomenon is not well understood since the redistribution of body fat is not so evident in experimental animals, although the epididymal fat pad in rats may be preserved preferentially. Since the lipolytic actions of glucocorticoids are opposed by insulin, which is elevated in hypercorticoid states, it has been suggested that the insulin response predominates in areas of high-fat mass, while the direct steroid lipolytic response predominates in other areas [163].

Interactions with other hormones   As mentioned, in many respects gluco-corticoids oppose the actions of insulin, and the two hormones are often found to act reciprocally. Thus, the effects of cortisol are not only most obvious in the fasted state, but may also be revealed when insulin secretion is deficient. Net hepatic glucose production may be unchanged during acute glucocorticoid treatment, although the actions of insulin to decrease glucose output by the liver are impaired [150]. Elevated insulin levels are seen during chronic glucocorticoid excess, which are probably explained in part by the glucocorticoid-induced hyperglycemia. The combination of elevated insulin and glucose levels suggest a state of insulin resistance, which may involve a direct antagonism by glucocorticoids of insulin action [164].

Glucocorticoids may interact with other hormones in a "permissive" fashion, that is they show the ability to amplify their action, without having a clear primary role. These synergistic interactions often occur with hormones that activate adenylate cyclase such as glucagon and epinephrine, as for example, in the control of gluconeogenesis mentioned [165,166].

4.3.2 Effects of corticosteroids on growth and wound-healing

One of the most immediately conspicuous effects of treatment with glucocorticoid hormones is the inhibition of somatic growth. This action was noted as early as 1940 when cortisone first became available for experimentation [167]. Inhibition of growth was a serious complication in children treated chronically with corticosteroids, although recently the problems have been alleviated by the use of alternate day therapy (see Chap. 7). It has been claimed that the dose of steroid required to produce significant inhibition of growth can be as low as that equivalent to two to three times the average daily secretion [167,168]. The growth-inhibitory effects of glucocorticoids are not confined to the skeleton, and inhibition of DNA synthesis is seen in skeletal, muscle, kidney, and liver in weanling animals. In the case of the last tissue, there is an apparent paradox, in that the liver weight actually increases in proportion to body weight in the adult animal treated with high doses of cortisol, an effect which is secondary to the stimulation of gluconeogenesis (Fig. 4.19 and see previous text). Tissues which renew themselves by cell proliferation such as the gut mucosa, testis, and spleen are relatively resistant to the effects of glucocorticoids [167]. Glucocorticoid treatment can also lead to decreased linear bone growth and osteoporosis (decreased density of bone).

The idea that glucocorticoids may be interfering with normal growth hormone (GH) release in response to stimuli acting through the central nervous system is supported by several clinical studies. In contrast, it has been demonstrated that glucocorticoids enhance the response of the pituitary to GH-releasing hormone (GHRH), and in vitro increase GH release as well as GH gene transcription [169,170]. Since a reduction in thymidine incorporation and cell proliferation can be seen in isolated tissues treated with glucocorticoids, the possibility of an effect to antagonize the actions of GH at the tissue level has also been explored [167]. Serum insulinlike growth factor (IGF) levels are not suppressed in patients with glucocorticoid excess, although the steroids have been found to modulate the secretion of IGF by fetal rat hepatocytes in culture [169,171]. In vitro, growth suppression is seen in tumor cell lines, and depends on the presence of glucocorticoid receptor. Cell lines with defective receptors, which are resistant to glucocorticoids, have been described and extensively studied. Recently, a unifying theory which attributes these actions of glucocorticoids to a reduction in growth factor production and activity has begun to emerge. For example, it has been found that glucocorticoid treatment of a smooth muscle cell line leads to decreased growth and decreased production of a platelet-derived growth factor (PDGF)-like molecule and in an epithelial cell line the production of epithelial growth factor (EGF) was found to be decreased [171–173].

Glucocorticoids have a number of effects on calcium and bone, which may be involved in the development of osteoporosis. There is decreased calcium absorption in the gut, and increased calcium loss in the urine, leading to negative calcium balance [174]. Increased bone loss and decreased bone accretion occur which are in part caused by the direct action of glucocorticoids, as well as their actions in concert with parathyroid hormone and 1,25-dihydroxycholecalciferol [145].

The ability to repair wounds and regenerate damaged tissue is essential for the survival of the animal. The process may be divided into three stages: an inflammatory phase (discussed in Sec. 4.3.3) followed by phases of proliferation and remodeling. Glucocorticoids both delay the process of wound-healing and alter the properties of wound tissue by interfering at all stages in the process. Central to the repair process is the growth of connective tissue and proliferation of fibroblasts, in which there is very convincing evidence demonstrating inhibitory effects of glucocorticoids on collagen formation and synthesis of glycosaminoglycans both in vitro and in vivo [175]. However, the study of fibroblast growth has generated many contradictory results, with increased proliferation being found in the presence of glucocorticoids in some experimental models. Nevertheless, there is some suggestion that, here again, decreased growth may be linked to decreased production of factors such as interleukin I and colony stimulating factor [176]. Remodeling in some tissues is associated with the activity of proteases such as plasminogen activator and collagenase, and these enzymes appear to be modulated by glucocorticoids. Dexamethasone has been found to reduce the activity of plasminogen activator by inducing the synthesis of a protein inhibitor [177].

It is worth pointing out, however, that the absence of glucocorticoid has little effect on growth if the individual is clinically well in other respects, thus this aspect of glucocorticoid action is usually considered to be "unphysiological." Paradoxically, however, removal of the adrenals can prevent the development of obesity in a number of animal models. The critical action of glucocorticoids here may lie in the stimulation of food intake, and promotion of the funneling of the ingested calories into fat stores [178].

4.3.3 Actions of glucocorticoids on the inflammatory response

The anti-inflammatory effects of glucocorticoids were discovered by Hench and Kendall in 1949 and have provoked much research and controversy ever since, particularly on the question of how to rationalize these actions within a physiological context (see Sec. 4.3.6). Nevertheless, suppression of inflammation with synthetic glucocorticoids constitutes a very important use for these drugs in medical practice, even though there is no correction of the underlying disease process, simply amelioration of the symptoms (see Chap. 7). In order to understand current knowledge on the ways in which glucocorticoids suppress inflammation, it is useful to have some understanding of the processes involved in the response to tissue injury, which may be started by a variety of different stimuli, including bacterial infection, mechanical, chemical or thermal injury, or anoxia. Alternatively, it may arise as part of an autoimmune response, when the process may become self-perpetuating and there is no phase of repair. Most forms of acute and chronic inflammation involve activity of the immune system in recognizing the foreign material, and amplifying the response by initiating the production of intercellular mediators. The early events are characterized by vasodilation at the site of injury, which leads to reddening of the skin. The vascular endothelium increases in permeability and tissue fluid accumulates, blood platelets and leukocytes attach to the damaged endothelial wall, and activation of blood coagulation with formation of a thrombus may occur. Leukocytes penetrate the tissue and proceed to migrate towards the cells where the inflammatory response was generated. This amoeboid movement is known as chemotaxis. The repair process requires removal of fibrin, dead leukocytes, and foreign material by macrophages, and in the case of a wound, tissue proliferation and remodeling follow. In chronic inflammation, when the initial stimulus is not eliminated (possibly because it is an autoantibody) the reaction persists. There is copious exudate and the damaged region may be invaded by fibrous tissue—the granulation reaction [179]. There are many mediators of these responses whose actions may be interdependent, and they include components of the complement system, eicosanoids, histamine (and serotonin in some species), cytokines, and kinins [180]. For example, bradykinin is released during tissue damage and causes vascular changes which may be secondary to the release of arachidonic acid and formation of prostaglandins, thromboxane A_2 and leukotrienes through the cyclooxygenase and lipoxygenase branches of the biosynthetic pathway (Fig. 4.21). These metabolites have been detected in tissue extracts, and several lines of evidence link these metabolites to the inflammatory response [179–181].

Glucocorticoids possess a number of properties that contribute to their anti-inflammatory activity. They decrease the release of mediators of the vascular response and reduce the accumulation of fluid, and through their inhibitory actions on connective tissue, they decrease the severity of the granulation reaction. There is increasing evidence that these actions of glucocorticoids are mediated by inhibition of the production (and possibly the action) of many of the intercellular mediators of inflammation. Generally, when these effects have been investigated in cell culture systems, glucocorticoids have been found to act in a steroid specific fashion, at concentrations approximating those found in physiological fluids [182,183].

It has been proposed that the glucocorticoid-induced inhibition of arachidonic acid release is achieved by the synthesis of proteins (collectively termed lipocortin) which inhibit cellular phospholipases. Thus, their effect is to reduce the activity of the enzymes which degrade membrane phospholipids and release arachidonic acid, and, in contrast to nonsteroidal anti-inflammatory drugs, reduce the production of both prostaglandins and leukotrienes (Fig. 4.21). Several laboratories have been involved in the purification and characterization of lipocortin from various sources, and during this time the proteins were also referred to as lipomodulin and macrocortin [184,185]. Regulation of lipocortin may involve phosphorylation by tyrosine kinases, resulting in reduction of the arachidonic acid release response [186]. Further, glucocorticoid-induced proteins have been detected which inhibit histamine release from mast cells but have no effect on phospholipase activity. These are called vasocortin or vasoregulin [187,188]. Glucocorticoids have also been found to inhibit the production of many of the cytokines that have been implicated in the inflammatory responses, for example, interleukin 2, macrophage-activating factor, colony-stimulating factor and γ-interferon [189].

The structure of human lipocortin was obtained in 1986 using the techniques of molecular biology which showed that it was a highly polar, linear polypeptide with molecular mass of about 37,000 and with a potential

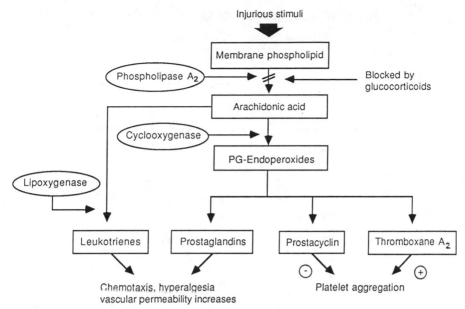

FIGURE 4.21 Involvement of intercellular mediators in the inflammatory response. Possible actions of the products are indicated. + = stimulation; − = inhibition.

site for glycosylation [185]. This molecule is now known as lipocortin I, since it appears to form part of a family of structurally related proteins that bind to phospholipid in a calcium-dependent manner. At least six lipocortinlike molecules are now recognized, of which five have a molecular weight of approximately 35,000, while the sixth has a mass of 68,000. Each has two domains, a small N-terminal domain and a core domain that is formed by a 4- or 8-fold repeat of a 70 amino acid conserved segment (Fig. 4.22), [190]. These proteins have also been pursued by other workers in investigations of intracellular signaling mechanisms, with the result that the nomenclature is very confused and they may also be referred to as annexins and calpactins [191]. For example, the sequences of human lipocortin I and II were found to be identical with those of calpactins II and I, which were known to be relatively abundant membrane cytoskeletal proteins and substrates for tyrosine kinases and also called p35 and p36 (Table 4.2) [192]. Although the common structure of lipocortins I to VI suggests a common function for the proteins, their cellular and tissue distribution differs and may indicate specific roles for each [190]. Although recombinant lipocortin I has been found to inhibit eicosanoid production and the inflammatory process in vivo [185,193], the physiological significance of the observations has been questioned because some authors have claimed that the inhibition of phospholipase A_2 activity in vitro may be caused by substrate depletion rather than a specific effect on the enzyme itself [194,195]. The mechanism by which lipocortin inhibits inflammation awaits further definition, neverthe-

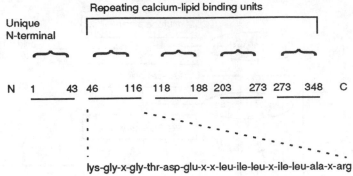

INTERNAL STRUCTURE OF LIPOCORTIN 1

lys-gly-x-gly-thr-asp-glu-x-x-leu-ile-leu-x-ile-leu-ala-x-arg

FIGURE 4.22 The internal structure of lipocortin I. The molecule contains a unique N-terminal domain and a domain consisting of four repeat units each of which begin with the same 17 amino acid sequence. The latter is conserved among the lipocortins and each repeated unit seems to be a calcium binding site. (Figure reproduced with permission from ref. 185.)

TABLE 4.2 Nomenclature of lipocortins and structurally related proteins which bind phospholipids in a calcium dependent manner

Lipocortin	Relative Mass	Apparent Identity
Lipocortin I	35,000	Calpactin II
		p35
		Chromobindin 9 or 16
Lipocortin II	38,000	Calpactin I
		p36
		Chromobindin 8
		Protein I
Lipocortin III	35,000	Calcimedin 35
Lipocortin IV	33,000	Chromobindin 4
		Endonexin
		Protein II
		32 kDa Calectrin
Lipocortin V	33,000	Chromobindin 5 or 7
		Endonexin II
		Anticoagulant protein (PAP)
Lipocortin VI	68,000	Protein III
		Chromobindin 20
		67 kDA Calectrin
		Calcimedin 68
		p68

Based on data from ref. 190 and 195

less, the potential importance of these studies is obvious, since they could lead to methods by which anti-inflammatory effects could be produced without the undesirable side effects of existing glucocorticoid drugs. Some success has already been achieved in this area with the report that novel anti-inflammatory peptides have been produced which are derived from lipocortin I [196]. It is also of interest that antilipocortin I autoantibodies have been detected in the sera from patients with chronic rheumatoid arthritis and systemic lupus erythmatosis, and that high titers of antibody were associated with resistance to the anti-inflammatory effects of glucocorticoids [185].

4.3.4 Effects of steroids on the immune system

The immunosuppressive characteristics of glucocorticoids have also been widely exploited clinically in a variety of situations (see Chap. 7) including the treatment of lymphoid neoplasia. Here, glucocorticoids are exerting their most profound effect on the immune system, that of killing cells. This cytotoxic effect has been intensively studied, and is dramatically illustrated in rats by the effect of a single injection of dexamethasone on the thymus glands of immature animals: after 48 hours there is a 50 percent reduction in thymus weight, and an 80 percent reduction in thymocyte population [197,198]. There are, however, marked species differences in sensitivity to steroids, and in humans and guinea pig steroids have relatively little effect on the immune system, whereas rats and mice are particularly sensitive to their effects [198,199]. For instance, in humans only a transient decrease in lymphocyte number is seen following a single dose of glucocorticoid, and this can be attributed to a movement of cells out of the vascular compartment [199]. In experimental animals, a variety of metabolic disturbances have been detected in glucocorticoid treated thymocytes, including reduced transport of metabolic precursors and decreased protein and nucleic acid synthesis [200]. However, the magnitude of these perturbations is relatively small and does not seem sufficient to account for the remarkable increase in lymphocyte mortality. An alternative approach has been to suggest that the actions of glucocorticoids involves the induction of a nucleolytic gene product [201]. The multiple alterations in lymphocyte metabolism could then be seen as resulting from disorganization of gene transcription and inappropriate protein biosynthesis. In favor of this view are the facts that glucocorticoids mediate DNA degradation in lymphocytes and induction of nuclease activity, both these effects being blocked by glucocorticoid antagonists [197].

The general immune suppression that is seen with glucocorticoid treatment involves more than a reduction of lymphocyte number, and interactions at most stages in the system have been recorded. As noted in Sec. 4.2.3, glucocorticoids impair the ability of cells to produce a number of soluble factors including interleukin I and II, and this reduces the proliferative effects of antigens on lymphocytes [199]. The effects of glucocorticoids on

B-cell function are complex, at low or moderate doses the number of cells and titer of antibody may actually be increased. However, at high doses they suppress antibody production and suppress multiplication of activated B-cells [202,203]. In addition, glucocorticoids appear to inhibit production of certain components of the complement system [204].

4.3.5 Behavioral effects of corticosteroids

That corticosteroids might affect central nervous system function and behavior has been a possibility since Cushing first noted that psychiatric symptoms often accompanied physical signs in cases of the syndrome which bears his name. The commonest abnormality is depression, which varies widely in severity between individuals, but which usually remits when adrenal function returns to normal. Conversely, depressed patients may have many of the biochemical characteristics of Cushing's syndrome, in particular elevated plasma cortisol levels, so that differential diagnosis of the two conditions can cause problems (see Chap. 5 and [205]). While experimental studies have shown that excess of glucocorticoids can influence many aspects of behavior including both mood and learning, as yet there is relatively little understanding of what part corticosteroids play in the regulation of normal brain function [206].

Corticosteroids readily enter the brain, and several receptor systems have been described in nervous tissue. As well as the classical glucocorticoid Type II receptors (see Sec. 4.1.1), a binding component which resembles the kidney Type I mineralocorticoid receptor has been detected in rat brain [207]. It seems to be expressed in two subtypes, one of which shows a marked preference for corticosterone (over aldosterone) and which resides mainly in the neurons of the limbic system. This shows stringent specificity in responsiveness to corticosterone in the modulation of steroid-dependent adaptive behavior, and neurotransmission. It has been suggested that these receptors may be involved in mediating tonic responses which are entrained to diurnal variations in ACTH. In contrast, the Type II receptor whose binding affinity is such that it would only be occupied at higher levels of corticosterone secretion, may be involved in mediating the response to stress [206]. The Type-I mineralocorticoid receptor is thought to mediate the effect of aldosterone on salt appetite [114].

Steroids are also known to produce effects on the central nervous system that are distinct from the relatively slowly induced and long-lasting effects as mentioned. These are actions which result in rapid suppression of CNS excitability, categorized as hypnotic and anxiolytic. Progesterone, DOC and several of their metabolites are potent inducers of sedation and anesthesia, and this fact has been exploited in the development of the steroid anesthetic, alphaxalone [208]. The short latency for these actions makes it unlikely that they are mediated by the "classical" intracellular receptors through effects on the genome. It has recently been found that the $5a$-reduced metabolites of DOC and progesterone appear to act in a fashion similar to the barbiturate anesthetics, and that their sedative actions may be attributed to their enhancement of the action of the inhibitory neurotransmitter gamma-aminobutyric acid (GABA). This potentiation seems to be brought about by a direct interaction with the GABA A receptor-chloride ion channel complex (Fig. 4.23), [208]. It remains to be seen whether other steroid effects are brought by a similar direct membrane effect.

4.3.6 Stress and the role of corticosteroids

The association between the adrenal gland and stress is sufficiently well known to have entered popular scientific mythology, and it is ironic that there is still no clear understanding of how corticosteroids protect against stress. The original finding that the pituitary-adrenocortical axis is activated under stressful conditions dates back to experiments reported by Selye in 1936, which were followed by the proposal of the "general adaptation syndrome" as mechanism to explain a number of diseases including arthritis. Stress was defined as almost any threat to the organism, whether physical such as injury or infection or psychological such as fear, pain, or anxiety. According to this theory, glucocorticoid in general afforded protection (in some undefined way) against the effects of stress, however, prolonged exposure to the stress and elevated levels of the hormones could lead to breakdown in some systems of the body and "diseases of adaptation" [209]. Although evoking a great deal of interest at the time, this view has not been

FIGURE 4.23 Diagrammatic representation of the GABA A-receptor-chloride channel complex indicating a possible site of action of the active steroids with the barbiturate recognition site (BARB); BDZ = benzodiazepine recognition site. (Figure reproduced with permission from ref. 207.)

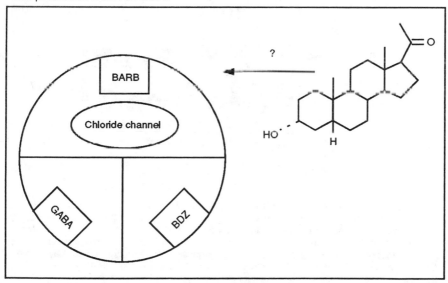

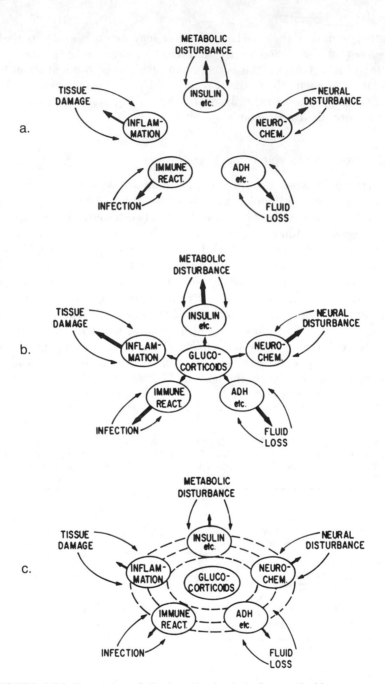

FIGURE 4.24 Illustration of the hypothesis that glucocorticoids suppress defense reactions in stress and prevent them from overshooting. In (a) the effects of stress in the form of metabolic or neural disturbance, fluid loss, infection or tissue damage are illustrated. In (b) glucocorticoids are shown in their conventional role of potentiators of the reactions to stress, in contrast (c) suggests that glucocorticoids may instead attenuate the responses of the immune and inflammatory systems, and so on, to stress. (Figure reproduced with permission from ref. 214.)

substantiated by subsequent research. Nevertheless, it remains true that stress leads to secretion of ACTH and increased production of cortisol and/ or corticosterone, while if the level of adrenocortical function is inadequate, an animal may not survive even a relatively mild disturbance. It has also been the practice to administer high doses of glucocorticoids prior to surgery to patients who may have some degree of adrenal suppression, even though this may cause delayed wound healing and unwanted immunosuppression [210]. The problem remains, however, of explaining what aspect of glucocorticoid action confers on them the ability to increase resistance to stress. The traditional view that has formed is that glucocorticoid may be useful in maintaining vascular reactivity to catecholamines, since most of the stressful stimuli that increase ACTH secretion also activate the adrenal medulla and the sympathetic nervous system [211]. Attempts also have been made to relate the effects of glucocorticoid on fat and protein metabolism to increased resistance to stress [211–213]. Recently, however, views on this subject have been challenged by Munck's suggestion that glucocorticoid response to stress exists not as part of the body's defense mechanisms against stress but rather as a counterregulatory system to buffer responses of other systems [183,214]. This is based on the idea that the immunosuppressive and anti-inflammatory actions (discussed in Secs. 4.3.3 and 4.3.4), often mediated through suppression of a range of intracellular mediators, are a normal part of glucocorticoid physiology, rather than a pharmacological side effect (Fig. 4.24). Thus, the attenuation by glucocorticoids of those of the body's defense mechanisms mediated by agents such as prostaglandins and lymphokines, could be seen as preventing an overshoot of the response to stress and thereby helping maintain vascular integrity [215].

The actions of glucocorticoids in the fetus can also be rationalized as anticipatory responses to the stress of birth. By their action through Type II receptors, they coordinate a number of developmental events which are essential for the transition from intrauterine to extrauterine life. For example, glucocorticoids have been found to stimulate the synthesis of pulmonary surfactant, the induction of some gut and hepatic enzymes and to play a part in the control of the transition from fetal to adult hemoglobin [216,217].

References

1. Jensen, E.V., and Jacobson, H.I. 1962. Basic guides to the mechanism of estrogen action. *Recent Progr. Hormone Res.* 18:387–414.

2. King, R.J.B., and Mainwaring, W.I.P. 1974. *Steroid-cell Interactions.* London: Butterworths.

3. Walters, M.R. 1985. Steroid hormone receptors and the nucleus. *Endocrine Rev.* 6:512–43.

4. Rousseau, G.G. 1984. Control of gene expression by glucocorticoid hormones. *Biochem. J.* 234:1–12.

5. Leake, R.E., and Habib, F. 1987. Steroid hormone receptors: Assay and

characterization. In: B. Green and R.E. Leake, eds. *Steroid Hormones: a Practical Approach.* Oxford and Washington: IRL Press, pp. 67–92.

6. Jensen, E.V., DeSombre, E.R., and Jungblut, P.W. 1967. Estrogen receptors in hormone responsive tissues and tumors. In: R.W. Wissler, T.L. Dao, and S. Wood, eds. *Endogenous Factors Influencing Host-tumors Balance.* Chicago: University of Chicago Press, pp. 15–30.

7. Gustafsson, J.A., Carlstedt-Duke, J., Wrange, O., Okret, A., and Wikstrom, A.-C. 1986. Functional analysis of the purified glucocorticoid receptor. *J. Steroid Biochem.* 24:63–68.

8. Wrange, O., Carlstedt-Duke, J., and Gustafsson, J.A. 1979. Purification of the glucocorticoid receptor from rat liver cytosol. *J. Biol. Chem.* 254:9284–90.

9. Greene, G.L., Nolan, C., Engler, J.P., and Jensen, E.V. 1980. Monoclonal antibodies to human estrogen receptor. *Proc. Nat. Acad. Sci., U.S.A.* 77:5115–19.

10. Okret, S., Wikstrom, A.C., Wrange, O., Andersson, B., and Gustafsson, J.A. 1984. Monoclonal antibodies against the rat liver glucocorticoid receptor. *Proc. Nat. Acad. Sci., U.S.A.* 81:1609–13.

11. Jensen, E.V., Suzuki, T., Kawashima, T., Stumpf, W.E., Jungblut, P.W., and DeSombre, E.R. 1968. A two-step mechanism for the interaction of estradiol with the rat uterus. *Proc. Nat. Acad. Sci., U.S.A.* 59:632–38.

12. Gorski, J., Toft, D., Shyamala, G., Smith, D. and Notides, A. 1968. Hormone receptors: Studies on the interaction of estrogen with the uterus. *Recent Progr. Hormone Res.* 24:45–80.

13. Sheridan, P.J., Buchanan, J.M., Anselmo, V.C., and Martin, P.M. 1979. Equilibrium: The intracellular distribution of steroid receptors. *Nature, London.* 282:579–82.

14. Welshons, W.V., Krummel, B.M., and Gorski, J. 1985. Nuclear localization of unoccupied receptors for glucocorticoids, estrogens and progesterone in GH_3 cells. *Endocrinology.* 117:2140–47.

15. King, W.J., and Greene, G.L. 1984. Monoclonal antibodies localise oestrogen receptor in the nuclei of target cells. *Nature, London.* 307:745–47.

16. Govindan, M.V. 1980. Immunofluorescence microscopy of the intracellular translocation of glucocorticoid-receptor complexes in rat hepatoma (HTC) cells. *Exp. Cell Res.* 127:293–97.

17. Papamichail, M., Tsokos, G., Tsawdaroglou, N., and Sekeris, C.E. 1980. Immunocytochemical demonstration of glucocorticoid receptors in different cell types and their translocation from the cytoplasm to the cell nucleus in the presence of dexamethasone. *Exp. Cell Res.* 125:490–93.

18. Fuxe, K., Wikstrom, A.-C., Okret, S., Agnati, L.F., Harfstrand, A., Yu, Z.-Y., Granholm, L., Zoli, M., Vale, W., and Gustafsson, J.-A. 1985. Mapping of the glucocorticoid receptor immunoreactive neurons in the rat tel- and diencephalon using a monoclonal antibody against rat liver glucocorticoid receptor. *Endocrinology.* 117:1803–12.

19. King, R.J.B. 1987. Structure and function of steroid receptors. *J. Endocr.* 114:341–49.

20. Parker, M.G. 1988. The expanding family of hormone receptors. *J. Endocr.* 119:175–77.

21. Green, S., Kumar, V., Theulaz, I., Wahli, W., and Chambon, P. 1988. The N-terminal DNA-binding "zinc finger" of the oestrogen and glucocorticoid receptors determines target gene specificity. *EMBO J.* 7:3037–44.

22. Beato, M. 1989. Gene regulation by steroid hormones. *Cell.* 56:335–44.

23. Gustafsson, J.-A., Carlstedt-Duke, J., Wrange, O., Akret, S., and Wikstrom, A.-C. 1986. Functional analysis of the glucocorticoid receptor. *J. Steroid Biochem.* 24:63–68.

24. Raymoure, W.J., McNaught, R.W., Greene, G.L., and Smith, R.G. 1986. Receptor interconversion model of hormone action. *J. Biol. Chem.* 261:17018–25.

25. Funder, J.W., and Barlow, J.W. 1980. Heterogeneity of glucocorticoid receptors. *Circulation Res.* 46: supp. I:I-83–I-87.

26. Arriza, J.L., Weinberger, C., Cerelli, C., Glaser, T.M., Handelin, B.L., Housman, D.E., and Evans, R.M. 1987. Cloning of human mineralocorticoid receptor complementary DNA: Structural and functional kinship with the glucocorticoid receptor. *Science.* 237:268–75.

27. Marver, D. 1984. Evidence of corticosteroid action along the nephron. *Am. J. Physiol.* 246:F111–23.

28. Marver, D. 1985. The mineralocorticoid receptor. In: G. Litwak, ed. *Biochemical Actions of Hormones.* London: Academic Press, vol. XII, pp. 385–429.

29. Tait, J.F., and Tait, S.A.S. 1979. Recent perspectives on the history of the adrenal cortex. *J. Endocrinol.* 83:3P–24P.

30. Funder, J.W., and Sheppard, K. 1987. Adrenocortical steroids and the brain. *Ann. Rev. Physiol.* 49:397–411.

31. Wolff, M.E. 1981. Anti-inflammatory steroids. In: M.E. Wolff, ed. *Burger's Medicinal Chemistry,* 4th Ed. Part III, New York: Wiley, pp. 1273–1316.

32. Coghlan, J.P., Denton, D.A., Fan, J.S.K., MacDougall, J.P., and Scoggins, B.A. 1976. Hypertensive effect of 17α,20α-dihydroxyprogesterone and 17α,-hydroxyprogesterone in the sheep. *Nature.* 263:608–09.

33. Stewart, P.M., Valentino, R., Wallace, A.M., Burt, D., Shackleton, C.H.L., and Edwards, C.R.W. 1987. Mineralocorticoid activity of liquorice: 11-beta-hydroxysteroid dehydrogenase deficiency comes of age. *Lancet.* 2:821–24.

34. Funder, J.W., Pearce, P.T., Smith, R., and Smith, A.I. 1988. Mineralocorticoid action: Target tissue specificity is enzyme, not receptor, mediated. *Science.* 242:583–85.

35. Landau, R.L., and Lugibihl, K. 1958. Inhibition of the sodium-retaining influence of aldosterone by progesterone. *J. Clin. Endocrinol. Metab.* 18:1237–45.

36. Adam, W.R., Funder, J.W., Mercer, J., and Ulick, S. 1978. Amplification of the action of aldosterone by 5α-dihydrocortisol. *Endocrinology.* 103:465–71.

37. Laragh, J.H., and Sealey, J.E. 1973. The renin-angiotensin-aldosterone hormonal system and regulation of sodium, potassium and blood pressure homeostasis. In: J. Orloff and R.W. Berliner, eds. *Handbook of Physiology, Renal Physiology.* Washington D.C.: American Physiological Society, Sec. 8, pp. 831–908.

38. Simpson, S.A.S., and Tait, J.F. 1952. A quantitative method for the bioassay of the effect of adrenal cortical steroids on mineral metabolism. *Endocrinology.* 50:150–61.

39. Worcel, M., and Moura, A.-M. 1987. Arterial effects of aldosterone and antimineralocorticoid compounds—mechanism of action. *J. Steroid Biochem.* 27:865–69.

40. Horton, R., and Biglieri, E.G. 1962. Effect of aldosterone on the metabolism of magnesium. *J. Clin. Endocrinol. Metab.* 22:1187–90.

41. Leman, J.J., Jr., Piering, W.F., and Lennon, E.J. 1970. Studies of the acute effects of aldosterone and cortisol on the interrelationship between renal sodium, calcium and magnesium excretion in normal man. *Nephron.* 7:117–30.

42. Nichols, M.G., Brown, W.C.B., Hay, G.D., Mason, P.A., and Fraser, R. 1979. Arterial levels and mineralocorticoid activity of 18-hydroxy-11-deoxycorticosterone in the rat. *J. Steroid Biochem.* 10:67–70.

43. Baylis, C., and Brenner, B.M. 1978. Mechanism of the glucocorticoid-induced increase in glomerular filtration rate. *Amer. J. Physiol.* 234:F166–70.

44. Schwartz, M.J., Keil, L.C., and Reid, I.A. 1983. Role of vasopressin in blood pressure regulation during adrenal insufficiency. *Endocrinology.* 112:234–38.

45. Schwartz, G.B., and Burg, M.B. 1978. Mineralocorticoid effect on cation transport by cortical collecting tubules in vitro. *Amer. J. Physiol.* 235:F576–85.

46. Horisberger, J.D., and Diezi, J. 1983. Effects of mineralocorticoids on Na^+ and K^+ excretion in the adrenalectomised rat. *Amer. J. Physiol.* 245:F89–99.

47. Bonvalet, J.-P. 1987. Binding and action of aldosterone, dexamethasone, 1-25 $OH_2 D_3$ and estradiol along the nephron. *J. Steroid Biochem.* 27:953–61.

48. Knepper, M.A., and Burg, M.B. 1981. Increased fluid absorption and cell volume in isolated rabbit proximal straight tubules in vivo after DOCA administration. *Amer. J. Physiol.* 241:F502–08.

49. Lee, S.M.K., Chekal, M.A., and Katz, A.I. 1983. Corticosterone binding sites along the rabbit nephron. *Amer. J. Physiol.* 244:F504–09.

50. Morel, F., and Doucet, A. 1986. Hormonal control of kidney functions at the cell level. *Physiol. Rev.* 66:377–497.

51. Ussing, H.H., and Zerahn, K. 1951. Active transport of sodium as the source of electric current in the short-circuited isolated frog skin. *Acta Physiol. Scand.* 23:110–27.

52. Leaf, A., Anderson, J., and Page, L.P. 1958. Active sodium transport by the isolated toad bladder. *J. General Physiol.* 41:657–68.

53. Edelman, I.S., Bogoroch, R., and Porter, G.A. 1963. On the mechanism of action of aldosterone on sodium transport: The role of protein synthesis. *Proc. Natl. Acad. Sci. USA.* 50:1169–77.

54. Porter, G.A., Bogoroch, R., and Edelman, I.S. 1963. On the mechanism of action of aldosterone on sodium transport: The role of RNA synthesis. *Proc. Natl. Acad. Sci. USA.* 52:1326–33.

55. Crabbe, J. 1961. Site of action of aldosterone on the bladder of the toad. *Nature.* 200:787–88.

56. Cuthbert, A.W., and Shum, W.K. 1975. The effects of vasopressin and aldosterone on amiloride binding sites in toad bladder epithelial cells. *Proc. Roy. Soc. London, Series B.* 189:543–75.

57. Palmer, L.G., Edelman, I.S., and Lindemann, B. 1980. Current voltage analysis of apical sodium transport in toad urinary bladder: Effect of inhibitors of transport and metabolism. *J. Membrane Biol.* 57:59–71.

58. Garty, H., and Edelman, I.S. 1983. Amiloride-sensitive trypsinization of apical sodium channels. Analysis of hormonal regulation of sodium transport in toad bladder. *J. General Physiol.* 81:785–803.

59. Yorio, T., and Bentley, P.J. 1978. Phospholipase A and the mechanism of action of aldosterone. *Nature.* 271:79–81.

60. Goodman, D.B.P., Wong, M., and Rasmussen, H. 1975. Aldosterone-induced membrane phospholipid fatty acid metabolism in the toad urinary bladder. *Biochemistry.* 14:2803–09.

61. Lien, E.L., Goodman, D.B.P., and Rasmussen, H. 1976. Effects of inhibitors of protein and RNA synthesis on aldosterone stimulated changes in phospholipid fatty acid metabolism in the toad urinary bladder. *Bioch. Biophys. Acta.* 421:210–17.

62. Sariban-Sohraby, S., Burg, M., Wiesmann, W.P., Chiang, P.K., and Johnson, J.P. 1984. Methylation increases sodium transport into A6 membrane vesicles: possible mode of aldosterone action. *Science.* 225:745–46.

63. Garty, H., and Benos, D.J. 1988. The amiloride-blockable sodium channel. *Physiol. Rev.* 68:309–73.

64. Verrey, F., Schaerer, E., Zoerkler, P., Paccolat, M.P., Geering, K., Kraehen-buhl, J.P., and Rossier, B.C. 1987. Regulation by aldosterone of Na$^+$, K$^+$ ATP-ase, messenger RNAs, protein synthesis and sodium transport in cultured kidney cells. *J. Cell. Biol.* 104:1231–37.

65. Jorgensen, P.L. 1972. The role of aldosterone in the regulation of Na$^+$/K$^+$-ATPase activity in the rat kidney. *J. Steroid Biochem.* 3:181–91.

66. Garg, L.C., Knepper, M.A., and Burg, M.B. 1981. Mineralocorticoid effects on Na-K-ATPase in individual nephron segments. *Amer. J. Physiol.* 240:F536–44.

67. Petty, K.J., Kokko, J.P., and Marver, D. 1981. Secondary effect of aldosterone on Na-K-ATPase activity in the rabbit cortical collecting tubule. *J. Clin. Invest.* 68:1514 21.

68. Kirsten, E., Kirsten, F., Leaf, A., and Sharp, G.W.G. 1968. Increased activity of enzymes of the tricarboxylic acid cycle in response to aldosterone in the toad bladder. *Pflugers Archiv.* 314:213–25.

69. Law, P.Y., and Edelman, I.S. 1978. Induction of citrate synthase by aldosterone in the rat kidney. *J. Mol. Biol.* 41:41–64.

70. Daughady, W.H., and MacBryde, C.M. 1950. Renal and adrenal mechanisms of salt conservation. The excretion of urinary formaldehyodogenic steroids and 17-ketogenic steroids during salt deprivation and desoxycorticosterone administration. *J. Clin. Invest.* 25:591–601.

71. Davis, J.O., and Howell, D.S. 1953. Comparative effect of ACTH, cortisone and DOCA on renal function, electrolyte excretion and water exchange in normal dogs. *Endocrinology.* 52:245–55.

72. Knox, F.G., Burnett, J.C., Kohan, D.E., Spielman, W.S. and Strand, J. C. 1980. Escape from the sodium-retaining effects of mineralocorticoids. *Kidney Internat.* 17:263–76.

73. Metzler, C.H., Gardner, D.G., Keil, L.C., Baxter, J.D., and Ramsay, D.R. 1987. Increased synthesis and release of atrial peptide during DOCA escape in conscious dogs. *Amer. J. Physiol.* 252:R188–92.

74. Granger, J.P., Burnett, J.C., Romero, J.C., Opgenorth, T.J., Salazar, J., and Joyce, M. 1987. Elevated levels of atrial natriuretic peptide during aldosterone escape. *Amer. J. Physiol.* 252:R878–82.

75. Hall, J.E., Granger, J.P., Smith, M.J. Jr., and Premen, A.J. 1984. Role of renal haemodynamics and arterial pressure in aldosterone escape. *Hypertension.* 6 Suppl. I: I.183–I.192.

76. Young, D.B. 1988. Quantitative analysis of aldosterone's role in potassium regulation. *Amer. J. Physiol.* 255:F811–22.

77. Brown, R.S. 1986. Extrarenal potassium homeostasis. *Kidney Internat.* 30:116–27.

78. Charron, R.C., Leme, C.E., Wilson, D.R., Ing, T.S., and Wong, O.M. 1969. The effect of adrenal steroids on stool composition, as revealed by in vivo dialysis of faeces. *Clin. Science.* 37:151–67.

79. Young, D.B., and Paulsen, A.W. 1983. Interrelated effects of aldosterone and plasma potassium on potassium excretion. *Amer. J. Physiol.* 244:F28–34.

80. Stanton, B., Pan, L., Deetjen, H., Guckian, V., and Giebisch, G. 1987. Independent effects of aldosterone and potassium on induction of potassium adaptation by the kidney. *J. Clin. Invest.* 79:198–206.

81. Adam, W.R., Ellis, A.G., and Andrews, B.A. 1987. Aldosterone is a physiologically significant kaliuretic hormone. *Amer. J. Physiol.* 252:F1048–52.

82. Field, M.J., and Giebisch, G.H. 1985. Hormonal control of renal potassium excretion. *Kidney Internat.* 27:379–87.

83. Wiederholt, M., Schoormans, W., Fischer, F., and Behn, C. 1973. Mechanism of action of aldosterone on potassium transfer in the rat kidney. *Pflugers Arch.* 345:117–23.

84. Stokes, J.B. 1985. Mineralocorticoid effect on K^+ permeability of the rabbit cortical collecting tubule. *Kidney Internat.* 28:640–45.

85. Oberleithner, H., Weigt, M., Westphale, H.-J., and Wang, W. 1987. Aldosterone activates Na^+/H^+ exchange and raises cytoplasmic pH in target cells of the amphibian kidney. *Proc. Natl. Acad. Sci. USA.* 84:1464–68.

86. Wang, W.H., Henderson, R.M., Geibel, J., White, S., and Giebisch, G. 1989. Mechanism of aldosterone induced increase of K^+ conductance in early distal renal collecting tubule cells of the frog. *J. Membrane. Biol.* 111:277–89.

87. Mills, J.N., Thomas, S., and Williamson, K.S. 1961. The effect of intravenous aldosterone on urinary electrolyte excretion of the recumbent human subject. *J. Physiol.* 156:415–23.

88. Burnell, J.M., Teubner, E.J., and Simpson, D.P. 1974. Metabolic acidosis accompanying potassium deprivation. *Amer. J. Physiol.* 227:329–33.

89. Hulter, H.N., Sebastian, A., Sigala, J.F., Licht, J.H., Glynn, R.D., Schambelan, M., and Biglieri, E.G. 1980. Pathogenesis of renal hyperchloremic acidosis resulting from dietary potassium restriction in the dog. Role of aldosterone. *Amer. J. Physiol.* 238:F79–91.

90. Higashihara, E., Carter, N.W., Pucacco, L., and Kokko, J.P. 1984. Aldosterone effects on papillary collecting pH profile of the rat. *Amer. J. Physiol.* 246:F725–31.

91. Welbourne, T.C., and Francoeur, D. 1977. Influence of aldosterone on renal NH_3 production. *Amer. J. Physiol.* 233:E56–60.

92. Uete, T., and Venning, E.H. 1962. Interplay between various adrenocortical steroids with respect to electrolyte excretion. *Endocrinology.* 71:768–78.

93. Baylis, C., and Brenner, B.M. 1978. Mechanism of the glucocorticoid-induced increase in glomerular filtration rate. *Amer. J. Physiol.* 234:F166–70.

94. Raff, H. 1987. Glucocorticoid inhibition of neurohypophysial vasopressin secretion. *Amer. J. Physiol.* 252:R635–44.

95. Mandell, I.R., DeFronzo, R.A., Robertson, G.L., and Forrest, J.N. 1980. Role of plasma arginine vasopressin in the impaired water diuresis of isolated glucocorticoid deficiency in the rat. *Kidney Internat.* 17:186–95.

96. Boykin, J., De Torrente, A., Erickson, A., Robertson, G., and Schrier, R.W.

1978. Role of plasma vasopressin in impaired water excretion of glucocorticoid deficiency. *J. Clin. Invest.* 62:738–44.

97. Schwartz, M.J., and Kokko, J.P. 1980. Urinary concentrating defect of adrenal insufficiency. *J. Clin. Invest.* 66:234–42.

98. Spark, R.F., and Melby, J.C. 1971. Hypertension and low plasma renin activity: presumptive evidence for mineralocorticoid excess. *Ann. Int. Med.* 75:831–36.

99. Birmingham, M.K., and Ward, P.J. 1961. The identification of the Porter-Silber chromogen secreted by the rat adrenal. *J. Biol. Chem.* 236:1661–67.

100. Kagawa, C.M., and Pappo, R. 1962. Renal electrolyte effects of synthetic 18-hydroxylated steroids in adrenalectomised rats. *Proc. Soc. Exp. Biol. Med.* 109:982–85.

101. Porter, G.A., and Kimsey, J. 1971. Assessment of the mineralocorticoid activity of 18-hydroxy-11-deoxycorticosterone in the isolated toad bladder. *Endocrinology.* 89:353–57.

102. Feldman, D., and Funder, J.W. 1973. The binding of 18-hydroxydeoxycorticosterone and 18-hydroxycorticosterone to mineralocorticoid receptors in the rat kidney. *Endocrinology.* 92:1389–95.

103. Melby, J.C., Dale, S.L., Grekin, R.J., Gaunt, R., and Wilson, T.E. 1972. 18-Hydroxy-11-deoxycorticosterone (18-OH-DOC) secretion in human and experimental hypertension. *Recent Prog. Horm. Res.* 28:287–351.

104. Ulick, S., Land, M., and Chu, M.D. 1983. 18-oxo-cortisol, a naturally occurring mineralocorticoid agonist. *Endocrinology.* 113:2320–22.

105. Gomez-Sanchez, C.E., Gomez-Sanchez, E.P., Smith, J.S., Ferris, M.W., and Foecking, M.F. 1985. Receptor binding and biological activity of 18-oxo-cortisol. *Endocrinology.* 116:6–10.

106. Spence, C.D., Coghlan, J.P., Denton, D.A., Gomez-Sanchez, C., Mills, E.H., Whitworth, J.A., and Scoggins, B.A. 1987. Blood pressure and metabolic effects of 18-oxo-cortisol in sheep. *J. Steroid Biochem.* 28:441–43.

107. Hall, C.E., and Gomez-Sanchez, C.E. 1986. Hypertensive potency of 18-oxo-cortisol in the rat. *Hypertension.* 8:317–22.

108. Gomez-Sanchez, C.E., Gomez-Sanchez, E.P., Smith, J.S., Ferris, M.W., and Foecking, M.F. 1984. Receptor binding and biological activity of 18-hydroxy-cortisol. *Endocrinology.* 115:462–66.

109. Gomez-Sanchez, C.E., Holland, O.B., Murray, B.A., Lloyd, H.A., and Milewich, L. 1979. 19 Nor-deoxycorticosterone, a potent mineralocorticoid isolated from the urine of rats with regenerating adrenals. *Endocrinology.* 105:708–11.

110. Perrone, R.D., Schwartz, J.H., Bengele, H.H., Dale, S.L., Melby, J.C., and Alexander, E.A. 1981. Mineralocorticoid activity of 19-nor-DOC and 19-OH-DOC in toad bladder. *Amer. J. Physiol.* 241:E406–09.

111. Gorsline, J., and Morris, D.J. 1985. The hypertensinogenic activity of 19-nor-deoxycorticosterone in the adrenalectomised spontaneously hypertensive rat. *J. Steroid Biochem.* 23:535–36.

112. Ehlers, M.E., Griffing, G.T., Wilson, T.E., and Melby, J.C. 1987. Elevated urinary 19-nor-deoxycorticosterone glucuronide in Cushing's syndrome. *J. Clin. Endocrinol. Metab.* 64:926–30.

113. Fregly, M.J., and Rowland, N.E. 1986. Hormonal and neural mechanisms of sodium appetite. *News Physiol. Sci.* 1:51–54.

114. Weisinger, R.S.W., Coghlan, J.P., Denton, D.A. et al. 1980. ACTH-elicited sodium appetite. *Amer. J. Physiol.* 239:E45–50.

References   **189**

115. Denton, D.A. 1984. *The Hunger for Salt*. Berlin: Springer-Verlag.

116. Bondy, P.K. 1985. Disorders of the adrenal cortex. In: J.D. Wilson and D.W. Foster, eds. *Williams Textbook of Endocrinology*, 7th Ed. Philadelphia: W.B. Saunders, pp. 816–90.

117. Lefer, A.M. 1975. Corticosteroids and circulatory function. In: H. Blaschko, G. Sayers and A.D. Smith, eds. *Handbook of Physiology*, Vol. 6: Adrenal Gland, Washington D.C.: American Physiological Society, pp. 191–207.

118. Schomig, A., Luth, B., Dietz, R., and Gross, F. 1976. Changes in vascular smooth muscle sensitivity to vasoconstrictor agents induced by corticosteroids, adrenalectomy and differing salt intake in rats. *Clin. Sci. Mol. Med.* 51:61s–63s.

119. Axelrod, L. 1983. Inhibition of prostacyclin production mediates the permissive effects of glucocorticoids on vascular tone. *Lancet.* 1:904–06.

120. Guyton, A.C., Coleman, T.G., and Granger, J.H. 1972. Circulation: overall regulation. *Ann. Rev. Physiol.* 34:13–46.

121. Selye, H., Hall, C.E., and Rowley, E.M. 1943. Malignant hypertension produced by treatment with desoxycorticosterone acetate and sodium chloride. *Canadian Med. Ass. J.* 49:88–92.

122. Carretero, O.A., and Romero, J.C. 1977. Production and characteristics of experimental hypertension in animals. In: J. Genest, E, Koiw, and O. Kuchel, eds. *Hypertension*. New York: McGraw Hill, pp. 485–507.

123. Scoggins, B.A., Coghlan, J.P., Denton, D.A., Mills, E.H., Nelson, M.A., Spence, C.D., and Whitworth, J.A. 1984. How do adrenocortical steroid hormones produce hypertension? *Clin. Exp. Hypertension—Theory and Practice.* A6:315–84.

124. Knowlten, A.I., Loeb, E.N., Stoerk, H.C., White, J.P., and Heffernan, J.F. 1952. Induction of hypertension in normal and adrenalectomized rats given cortisone acetate. *J. Experimental Med.* 96:187–205.

125. Schalekamp, M.A.D.H., Wenting, G.J., and Man in 't Veld. 1981. Pathogenesis of mineralocorticoid hypertension. In: E.G. Biglieri and M. Schambelan, eds. *Clinics in Endocrinology and Metabolism*, Vol. 10 No. 3. *Endocrine Hypertension*. London: W.B. Saunders, pp. 397–418.

126. Reid, J.L., Zivin, J.A., and Kopin, I.J. 1975. Central and peripheral adrenergic mechanisms in the development of deoxycorticosterone-saline hypertension in rats. *Circulation Res.* 37:569–77.

127. Beilin, L.J., and Ziakas, G. 1972. Vascular reactivity in post-deoxycorticosterone hypertension in rats and its relation to irreversible hypertension in man. *Clin. Sci.* 42:579–90.

128. Songu-Mize, E., Bealer, S.L.L., and Caldwell, R.W. 1982. Effect of AV3V lesions on development of DOCA-salt hypertension and vascular Na^+—pump activity. *Hypertension.* 4:575–80.

129. Distler, A., Philipp, T., Luth, B., and Wucherer, G. 1979. Studies on the mechanism of mineralocorticoid-induced blood pressure in man. *Clin. Science.* 57:303s–05s.

130. Gomez-Sanchez, E.P. 1986. Intracerebroventricular infusion of aldosterone induces hypertension in rats. *Endocrinology.* 118:819–23.

131. Mohring, J., Mohring, B., Petri, M., and Haack, D. 1977. Vasopressor role of ADH in the pathogenesis of malignant DOC hypertension. *Amer. J. Physiol.* 232:F260–69.

132. Moreland, R.S., Lamd, F.S., Webb, R.C., and Bohr, D.F. 1984. Functional

evidence for increased sodium permeability in aortae from DOCA hypertensive rats. *Hypertension.* 6 Supp. I:I88–94.

133. Gurwitz, E.T., and Jones, A.W. 1982. Altered ion transport and its reversal in aldosterone hypertensive rat. *Amer. J. Physiol.* 243:H927–33.

134. Bravo, E.L., 1986. Aldosterone and other adrenal steroids. In: A. Zanchetti and R.C. Tarazi, eds. *Handbook of Hypertension*, Vol. 8. *Pathophysiology of Hypertension—Regulatory Mechanisms*, Amsterdam: Elsevier, pp. 603–24.

135. De Wardner, H.E. 1985. Concept of natriuretic hormone. *Physiol. Rev.* 65:658–759.

136. Pernollet, M.G., Ali, R.M., Meyer, P., and Devynck, M.-A. 1986. Are the circulating digitalis-like compounds of adrenal origin? *J. Hypertension.* 4 Supp. 6:S382–84.

137. Haack, D., Mohring, J., Mohring, B., Petri, M., and Hackenthal, E. 1977. Comparative study on development of corticosterone and DOCA hypertension in rats. *Amer. J. Physiol.* 233:F403–11.

138. Krakoff, L.R., Selvadurai, R., Sutter, E. 1975. Effect of methylprednisolone upon arterial pressure and the renin-angiotensin system in the rat. *Amer. J. Physiol.* 228:613–17.

139. Waeber, B., Gavras, H., Bresnahan, M.R., Gavras, I., and Brunner, H.R. 1983. Role of vasoconstrictor systems in experimental glucocorticoid hypertension in rats. *Clin. Science.* 65:255–61.

140. Kornel, L., Ramsay, C., Kanmarlapudi, N., Travers, T., and Packer, W. 1982. Evidence for the presence in arterial walls of intra-cellular-molecular mechanism for action of mineralocorticoids. *Clin. Exp. Hypertension—Theory and Practice.* A4:1561–82.

141. Grunfeld, J.-P., Eloy, L., Moura, A.-M., Ganeval, D., Ramos-Frendo, B., and Worcel, M. 1985. Effects of antiglucocorticoids on glucocorticoid hypertension in the rat. *Hypertension.* 7:292–99.

142. Whitworth, J.A., Gordon, D., and Scoggins, B.A. 1987. Dose-response relationships for adrenocorticotrophin-induced hypertension in man. *Clin. Exp. Pharmacol. and Physiol.* 14:65–71.

143. Scoggins, B.A., Coghlan, J.P., and Denton, D.A. 1984. ACTH-induced hypertension in sheep. In: W. deJong, ed. *Handbook of Hypertension*, Vol. 4, *Experimental and Genetic Models of Hypertension.* Amsterdam: Elsevier, pp. 107–34.

144. Coghlan, J.P., Butkus, A., Denton, D.A., McDougall, J.G., Scoggins, B.A., and Whitworth, J.A. 1979. Blood pressure and metabolic effects of 9-alpha-fluoro-hydrocortisone in sheep. *Clin. Exp. Hypertension.* 1:629–48.

145. Baxter, J.D., and Tyrell, J.B. 1987. The adrenal cortex. In: P. Felig, J.D. Baxter, A.E. Broadus, and L.A. Frohman, eds. *Endocrinology and Metabolism*, 2nd ed. New York: McGraw-Hill, pp. 511–650.

146. Bondy, P.K., Ingle, D.J., and Meeks, R.C. 1954. Influence of adrenal cortical hormones upon the plasma levels of amino acids in eviscerated rats. *Endocrinology.* 55:354–60.

147. Kaplan, S.A., and Naghreda-Shimizu, C.S. 1962. Effect of cortisol on amino acids in skeletal muscle and plasma, *Endocrinology.* 72:267–72.

148. Odedra, B.R., and Millward, D.J. 1982. Effect of corticosterone treatment on muscle protein turnover in adrenalectomized rats and diabetic rats maintained on insulin. *Biochem. J.* 204:663–72.

149. Simmons, P.S., Miles, J.M., Gerich, J.E., and Haymond, M.W. 1984.

Increased proteolysis. An effect of increases in plasma cortisol within the physiologic range. *J. Clin. Invest.* 73:412–20.

150. Clerc, D., Wick, H., and Keller, U. 1986. Acute cortisol excess results in unimpaired insulin action on lipolysis and branched chain amino acids, but not on glucose kinetics and C-peptide concentrations in man. *Metabolism.* 35:404–10.

151. Cahill, G.F. 1971. Action of adrenal cortical steroids on carbohydrate metabolism. In: N.P. Christy, ed. *The Human Adrenal Cortex.* New York: Harper Row, pp. 205–39.

152. Munck, A. 1971. Glucocorticoid inhibition of glucose uptake by peripheral tissues, old and new mechanisms, molecular mechanisms and physiological significance. *Perspectives Biol. Med.* 14:265–89.

153. Livingston, J.N., and Lockwood, D.H. 1975. Effect of glucocorticoids on the glucose transport system of isolated fat cells. *J. Biol. Chem.* 250:8353–60.

154. Olefsky, J. 1975. Effect of dexamethasone in insulin binding, glucose transport and glucose oxidation of isolated rat adipocytes. *J. Clin. Invest.* 56:1499–08.

155. Shamoon, H., Soman, V., and Sherwin, R.S. 1980. The influence of acute physiological increments of cortisol on fuel metabolism and insulin binding to monocytes in normal humans. *J. Clin. Endocrinol. Metab.* 50:495–501.

156. Long, C.N.H., Katzin, B., and Fry, E.G. 1940. The adrenal cortex and carbohydrate metabolism. *Endocrinology.* 26:309–44.

157. Hers, H.G. 1985. Effects of glucocorticoids on carbohydrate metabolism. *Agents Act.* 17:248–62.

158. Laloux, M., Stalmans, W., and Hers, H.G. 1983. On the mechanism by which glucocorticoids cause the activation of glycogen synthase in mouse and rat livers. *Europ. J. Biochem.* 136:175–81.

159. Exton, J.H., Friedmann, H., Wong, E.H.A., Brineaux, J.P., Corbin, J.D., and Park, C.R. 1972. Interaction of glucocorticoids with glucagon and epinephrine in the control of gluconeogenesis and glycogenolysis in liver and of lipolysis in adipose tissue. *J. Biol. Chem.* 247:3579–88.

160. Chan, T.M. 1984. The permissive effects of glucocorticoid on hepatic gluconeogenesis. *J. Biol. Chem.* 259:7426–32.

161. Fain, J. 1968. Effects of dibutyryl-3,5′-AMP, theophylline and norepinephrine on the lipolytic activity of growth hormone and glucocorticoids in fat cells. *Endocrinology.* 82:125–36.

162. Rudman, D., and Di Girolamo, M. 1971. Effect of adrenal cortical steroids on lipid metabolism. In: N.P. Christy, ed. *The Human Adrenal Cortex.* New York: Harper Row, pp. 241–72.

163. Lamberts, S.W.J., Timmermans, H.A.T., Kramer-Blankenstijn, M., and Birkenhager, J.C. 1975. The mechanism of the potentiating effect of glucocorticoids on catecholamine-induced lipolysis. *Metabolism.* 24:681–89.

164. Lenzen, S., and Bailey, C.J. 1984. Thyroid hormones, gonadal and adrenocortical steroids and the function of the islets of Langerhans. *Endocrine Rev.* 5:411–34.

165. Malbon, C.C., Rapiejko, P.J., and Watkins, D.C. 1988. Permissive hormone regulation of hormone sensitive effector systems. *Trends Pharmacol. Sci.* 9:33–36.

166. Ingle, D.J. 1954. Permissibility of hormone action. A review. *Acta Endocrinol.* 17:172–86.

167. Loeb, J.N. 1976. Corticosteroids and growth. *New England J. Med.* 295:547–52.

168. Underwood, L.E., and Van Wyk, J.J. 1985. Normal and aberrant growth. In: J.D. Wilson and D.W. Foster, eds. *Williams Textbook of Endocrinology*, 7th ed. Philadelphia: W.B. Saunders, pp. 155–205.

169. Wehrenberg, W.B., Baird, A., and Ling, N. 1983. Potent interaction between glucocorticoids and growth hormone-releasing factor in vivo. *Science.* 221:556–58.

170. Robins, D.M., Peak, I., Seeburg, P.H., and Axel, R. 1982. Regulated expression of human growth hormone genes in mouse cells. *Cell.* 29:623–31.

171. Richman, R.A., Benedict, M.R., Florini, J.R., and Toly, B.A. 1985. Hormonal regulation of somatomedin secretion by fetal rat hepatocytes in primary cultures. *Endocrinology.* 116:180–88.

172. Kurokawa, R., Kyakumoto, S., and Ota, M. 1988. Glucocorticoid regulates secretion of epidermal growth factor in the human salivary gland adenocarcinoma cell line. *J. Endocrinol.* 116:451–55.

173. Norris, J.S., Cornett, L.E., Hardin, J.W., Kohler, P.O., MacLeod, S.L., Srivastava, A., Syms, A.J., and Smith, R.G. 1984. Autocrine regulation of growth, II Glucocorticoids inhibit transcription of c-sis oncogene-specific RNA transcripts. *Biochem. Biophys. Res. Comm.* 122:124–28.

174. Singer, F.R. 1987. Metabolic bone disease. In: P. Felig, J.D. Baxter, A.E. Broadus and L.A. Frohman, eds. *Endocrinology and Metabolism*, 2nd ed. New York: McGraw-Hill, pp. 1454–99.

175. Smith, T.J. 1982. Dexamethasone regulation of glycosaminoglycan synthesis in cultured human skin fibroblasts. *J. Clin. Invest.* 74:2157–63.

176. Durant, S., Duval, D., and Homo-Delarche, F. 1986. Factors involved in the control of fibroblast proliferation by glucocorticoids: a review. *Endocrine Rev.* 7:254–68.

177. Jenkins, N., and Ellison, J.D. 1986. Corticosteroids suppress plasminogen activation in the bovine Sertoli cell. *J. Endocrinol.* 108:R1–R3.

178. Dallman, M.F. 1984. Viewing the ventromedial hypothalamus from the adrenal gland. *Amer. J. Physiol.* 246:R1–R12.

179. Bowman, W.C., and Rand, M.J. 1980. *Textbook of Pharmacology.* 2nd ed. Oxford, U.K.: Blackwell Scientific Publications, pp. 13.15–13.17.

180. Regoli, D. 1987. Kinins. In: *Inflammation—mediators and mechanisms. Brit. Med. Bull.* 43/2:270–84

181. Salmon, J.A., and Riggs, G.A. 1987. Prostaglandins and leukotrienes as inflammatory mediators. In: *Inflammation—mediators and mechanisms. Brit. Med. Bull.* 43/2:225–96.

182. Bowen, D.L., and Fauci, A.S. 1988. Adrenal corticosteroids. In: J.I. Gallin, I.M. Goldstein, and R. Snyderman, eds. *Inflammation, Basic Principles and Clinical Correlates.* New York: Raven Press, pp. 877–95.

183. Munck, A., Guyre, A.P., and Holbrook, N.J. 1984. Physiological functions of glucocorticoids in stress and their relation to pharmacological actions. *Endocrine Rev.* 5:25–44.

184. Flower, R.J. 1986. The mediators of steroid action. *Nature.* 320:20–21.

185. Flower, R.J. 1988. Lipocortin and the mechanism of action of the glucocorticoids. *Brit. J. Pharm.* 94:987–1015.

186. Hirata, F., Stracke, M.L., and Schiffman, E. 1987. Regulation of prostaglandin

formation by glucocorticoids and their second messenger, lipocortins. *J. Steroid Biochem.* 27:1053–56.

187. Carnuccio, R., Di Rosa, M., Guerrasio, B., Iuvone, T., and Sautebin, L. 1987. Vasocortin, a novel glucocorticoid-induced anti-inflammatory protein. *Brit. J. Pharmacol.* 90:443–45.

188. Oyannagui, Y., and Suzuki, S. 1985. Vasoregulin, a glucocorticoid-inducible vascular permeability protein. *Agents Act.* 17:270–77.

189. Kelso, A., and Munck, A. 1984. Glucocorticoid inhibition of lymphokine secretion by alloreactive T lymphocyte clones. *J. Immunology.* 133:784–89.

190. Pepinsky, R.B., Tizard, R., Mattaliano, R.J., Sinclair, L.K., Miller, G.T., Browning, J.L., Chow, E.P., Burne, C., Huang, K.-S., Pratt, D., Watcher, L., Hession, C., Frey, A.Z., and Wallner, B.P. 1988. Five distinct calcium and phospholipid binding proteins share homology with lipocortin I. *J. Biol. Chem.* 263:10799–811.

191. Crumpton, M.J., and Dedman, J.R. 1990. Protein terminology tangle. *Nature.* 345:212.

192. Haigler, H.T., Fitch, J.M., Jones, J.M., and Schlaepfer, D.D. 1989. Two lipocortin-like proteins, endonexin II and anchorin CII, may be alternate splices of the same gene. *Trends Biochem. Sci.* 14:48–50.

193. Cirino, G., Flower, R.J., Browning, J.L., Sinclair, L.K., and Pepinsky, R.B. 1987. Recombinant human lipocortin 1 inhibits thromboxane release from guinea pig isolated perfused lung. *Nature.* 328:270–72.

194. Hollenberg, M.D. 1988. Protein tyrosine kinase substrates: Rosetta stones or structural elements? *Trends Pharmacol. Sci.* 9:63–66.

195. Davidson, F.F., Dennis, E.A., Powell, M., and Glenney, J.R. 1987. Inhibition of phospholipase A_2 by "lipocortins" and calpactins. An effect of binding to substrate phospholipid. *J. Biol. Chem.* 262:1698–705.

196. Miele, L., Cordella-Miele, E., Facchiano, A., and Mukherjee, A.B. 1988. Novel anti-inflammatory peptides from the region of highest similarity between uteroglobin and lipocortin I. *Nature.* 335:726–30.

197. Compton, M.M., Caron, L.M., and Cidlowski, J.A. 1987. Glucocorticoid action on the immune system. *J. Steroid Biochem.* 27:201–08.

198. Claman, H.N. 1972. Corticosteroids and lymphoid cells. *New England J. Med.* 287:388–97.

199. Cupps, T.R., and Fauci, A.S. 1982. Corticosteroid-mediated immunoregulation in man. *Immunological Rev.* 65:133–55.

200. Munck, A., and Crabtree, G.R. 1981. Glucocorticoid-induced lymphocyte death. In: I.D. Bowen and R.A. Lockshin, eds. *Cell Death in Biology and Pathology.* Chapman and Hall: New York, pp. 329–59.

201. Gasson, J.C., and Bourgeois, S. 1983. A new determinant of glucocorticoid sensitivity in cell lines. *J. Cell Biol.* 96:409–15.

202. Cupps, T.R., Gerrard, T.L., Falkoff, R.J.M., Whalen, G., and Fauci, A.S. 1985. Effects of in vitro corticosteroids on B cell activation, proliferation and differentiation. *J. Clin. Invest.* 75:754–61.

203. Parillo, J.E., and Fauci, A.S. 1979. Mechanisms of glucocorticoid action on immune processes. *Ann. Rev. Pharmacol. Toxicol.* 19:179–201.

204. Hammerschmidt, D.E., White, J.G., Craddock, P.R., and Jacob, H.S. 1979. Corticosteroids inhibit complement-induced granulocyte aggregation, a possible mechanism for their efficacy in shock states. *J. Clin. Invest.* 63:798–803.

205. Anonymous Editorial. 1986. Depression and Cushing's Syndrome. *Lancet.* 2:550–51.

206. McEwen, B.S., De Kloet, E.R., and Rostene, W. 1986. Adrenal steroid receptors and actions in the nervous system. *Physiological Rev.* 66:1121–88.

207. Lambert, J.J., Peters, J.A., and Cottrell, G.A. 1987. Actions of synthetic and endogenous steroids on the GABA A receptor. *Trends Pharmacol. Sci.* 8:224–27.

208. Majewska, M.D., Harrison, N.L., Schwartz, R.D., Barker, J.L., and Paul, S.M. 1986. Steroid hormone metabolites are barbiturate-like modulators of the GABA receptor. *Science.* 232:1004–07.

209. Selye, H. 1946. The general adaptation syndrome and the diseases of adaptation. *J. Clin. Endocrinol. Metab.* 6:117–230.

210. Udelsman, R., Ramp, J., Gallucci, W.T., Gordon, A., Lipford, E., Norton, J.A., Loriaux, D.L., and Chrousos, G.P. 1987. Adaptation during surgical stress. A reevaluation of the role of glucocorticoids. *J. Clin. Invest.* 77:1377–81.

211. Ramsey, E.R., and Goldstein, M.S. 1957. The adrenal cortex and the sympathetic nervous system. *Physiological Rev.* 37:115–95.

212. Sayers, G. 1950. The adrenal cortex and homeostasis. *Physiological Rev.* 30:241–320.

213. Hofmann, F.G. 1971. Role of the adrenal cortex in homeostasis and growth. In: N.P. Christy, ed. *The Human Adrenal Cortex.* New York: Harper Row, pp. 303–16.

214. Munck, A., and Guyre, P.M. 1986. Glucocorticoid hormones in stress, physiological and pharmacological actions. *News Physiol. Sci.* 1:69–72.

215. Gaillard, R.-C., and Al-Dalmuji, S. 1987. Stress and the pituitary-adrenal axis. In: A. Grossman, ed. *Neuroendocrinology of Stress, Baillere's Clinical Endocrinology and Metabolism* Vol. 1: No. 2. London: Baillere Tindall, pp. 319–54.

216. Winter, J.S.D. 1985. The adrenal cortex in the fetus and neonate. In: D.C. Anderson and J.S.D. Winter, eds. *Adrenal Cortex.* London: Butterworth, pp. 32–56.

217. Mendelson, C.R., and Boggaram, V. 1989. Regulation of pulmonary surfactant protein synthesis in fetal lung: A major role of glucocorticoids and cyclic AMP. *Trends Endocrinol. Metab* 1:20–25.

5

Disorders of the Adrenal Cortex I: Hyperfunction

The diseases that are caused by abnormal function of the adrenal cortex can be divided into three general categories: those caused by excess corticosteroid production, those caused by a deficiency in corticosteroid production, and those in which the pattern of products is altered because of a deficiency of an enzyme in the biosynthetic pathway. This last type of disorder is especially important in the adrenal cortex, and can appear in a number of forms depending on the location and severity of the lesion. Impaired adrenal function can also occur as a consequence of the destruction of the adrenal cortex (when it is known as Addison's disease) and results in deficient production of glucocorticoids and/or mineralocorticoids. These disorders are discussed in Chap. 6.

The symptoms of glucocorticoid excess are collectively termed Cushing's syndrome, while excess mineralocorticoid production is most commonly associated with Conn's syndrome.

5.1 Cushing's syndrome

The clinical characteristics of Cushing's syndrome arise when the body is chronically exposed to higher than normal levels of glucocorticoids. The principal signs and symptoms of the first case were described by Harvey Cushing in 1912 "as obesity, hypertrichosis, and amenorrhea" with the suggestion that the syndrome may have a pituitary, pineal, ovarian, or

TABLE 5.1 Classification of the types of Cushing's syndrome

Type	Cause	Adrenal status	Approximate frequency (%)
A. ACTH dependent			
Pituitary	With tumor	Adrenal hyperplasia	60–65
	Without tumor	Adrenal hyperplasia	5
Ectopic	ACTH or CRF production by nonendocrine tumor	Adrenal hyperplasia	15–20
Iatrogenic	Prolonged use of ACTH	Adrenal hyperplasia	
B. ACTH independent			
Adrenal	Adrenal tumor	Adenoma	7–10
		Carcinoma	8–10
		Nodular hyperplasia	
Iatrogenic	Prolonged use of glucocorticoids	Involuted	

Based on information contained in ref. 3 and 47.

adrenal origin. Subsequently, in a review of a series of 12 patients who showed a similar "polyglandular syndrome," six of whom had proven pituitary adenomas, Cushing then correctly suggested the pituitary dependency of the disease which bears his name [1]. At the same time, it was acknowledged that a similar syndrome could result from a tumor of the adrenal cortex. Further forms of Cushing's syndrome have since been recognized of which the most important is caused by the production of substances with ACTH-like activity by tumors (ectopic Cushing's). Recently, a form in which the stimulation of the adrenal cortex is caused by immunoglobulins, possibly adrenal autoantibodies functioning in a similar way to those which affect the thyroid in Graves' disease has been described. Cushing's syndrome may also have an iatrogenic cause due to administration of excess glucocorticoids or ACTH. These classes may also be divided on the basis of whether the syndrome is ACTH dependent or not (Table 5.1).

5.1.1 Etiology

Pituitary Cushing's (Cushing's Disease) Adrenal hyperplasia dependent on excess secretion of ACTH from the pituitary is the most common cause of Cushing's syndrome and is then properly called Cushing's disease. It accounts for about 70 percent of the cases, and is up to eight times more frequent in women. At surgery, small adenomas are often found, which are usually less than 10 mm in diameter [2,3]. Histologically, the majority of tumors are basophilic, but chromophobic or mixed tumors also occur. ACTH is found in adenoma tissue, together with other products of the pro-opiomelanocortin (POMC) molecule (see Fig. 3.2) including β-LPH, β-endorphin, and α-melanocyte-stimulating hormone (MSH) and occasionally, γ-MSH [3,4].

There is continuing debate as to whether the tumors are truly autonomous, or result from some disorder of the nervous system resulting in increased hypothalamic drive to the pituitary. The latter view is supported by cases where no discrete adenoma is found and in which there seems to be general corticotroph hyperplasia, and by the fact that a number of hypothalamic neurotransmitter agonists and antagonists have been found to alter ACTH secretion in Cushing's disease. For instance, cyproheptadine, a drug which blocks serotoninergic transmission in the hypothalamus, has been found to cause remission in some patients [4]. Sodium valproate which acts as an agonist to the neurotransmitter gamma-amino-butyric acid, GABA, and some opioids also have been reported to cause ACTH suppression [4,5].

The possibility that a proportion of the tumors may arise from cells of the neurointermediate lobe type has been put forward recently. This form of the disease seems to be characterized by suppression of ACTH release in response to dopaminergic drugs (such as bromocriptine) together with a higher than normal incidence of hyperprolactinemia [6]. The adult human pituitary has no distinct intermediate lobe and only limited amounts of α-MSH are secreted, nevertheless levels of this peptide have sometimes been found to be elevated in plasma of patients with Cushing's disease [7–10]. Nelson's syndrome which may occur following adrenalectomy for treatment of Cushing's syndrome, is characterized enlargement of the pituitary with excessive ACTH secretion and increased brown pigmentation of the skin.

The adrenal glands are usually moderately enlarged bilaterally (6–12 g per gland) showing simple hyperplasia with widening of the zonae fasciculata and reticularis (Fig. 5.1). In a small proportion of the cases, the adrenals also contain multiple nodules, within the hyperplastic tissue; this is termed macronodular or adenomatous hyperplasia [11,12]. Occasionally, a large unilateral nodule, suggestive of a cortical adenoma has been found and there is some evidence that such a macronodule can itself become autonomous, resulting in a transition from a pituitary-dependent to an adrenal-dependent form of Cushing's Syndrome [13].

Cushing's Syndrome (Ectopic ACTH)   This is the second most common form of the disease, accounting for approximately 15 to 20 percent of cases of Cushing's syndrome. When the "ectopic" secretion of ACTH was first demonstrated in 1962, it had been known for many years that nonendocrine tumors in a variety of sites could result in the clinical features of Cushing's syndrome [14]. A wide variety of types of neoplasm have since been found to contain or produce ACTH, of which the most common is the oat-cell

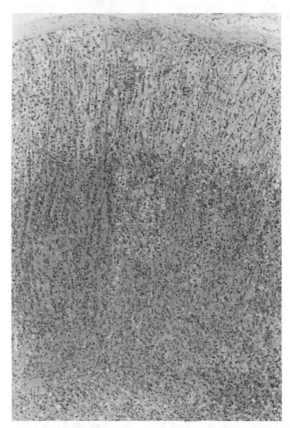

FIGURE 5.1 Light micrograph of hyperplastic adrenal from patient with pituitary Cushing's disease. Note general hyperplasia with expansion of compact cell zone (zona reticularis), C = capsule (×34).
(Micrograph generously provided by I. Doniach.)

(small-cell) carcinoma of the lung. The tumors have been found to contain ACTH both by immunohistochemistry and by radioimmunoassay after extraction; occasionally, secretion of ACTH in vivo has been also demonstrated by direct cannulation of the tumor [3]. Other products of the POMC series have been detected, abnormal processing of the molecule is common and an altered POMC-mRNA has been detected [15,16]. The adrenals in ectopic Cushing's syndrome are hyperplastic and symmetrically enlarged, often weighing more than 15 g each [11,12].

The presence in tumors of substances with CRF-like activity has been demonstrated on several occasions, although the material was not fully characterized. On the basis of recent studies performed with specific assays, it now seems that secretion of CRF-41 is rare. There is a possibility that the CRF-like activity is carried by other molecules and a medullary carcinoma of the thyroid containing bombesin (which can stimulate ACTH production) but not CRF-41 or ACTH, has been described [3].

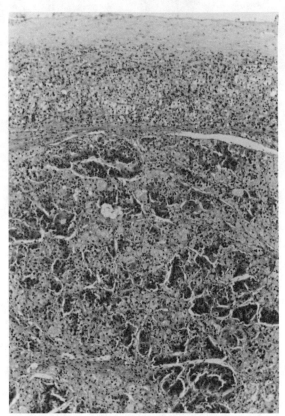

FIGURE 5.2 Light micrograph showing adrenal adenoma (A) surrounded by definitive cortex (DC), C = capsule (×34). (Micrograph generously provided by I. Doniach.)

Cushing's Syndrome (Adrenal Cushing's)   In this ACTH independent form of Cushing's syndrome, an adrenal adenoma or carcinoma is present with approximately equal frequency (7–10%). Adenomas are usually spherical, well encapsulated, and of moderate size, and the surrounding normal cortical tissue is compressed and atrophic (Fig. 5.2). They are more common in females than males. Adrenal carcinomas are usually much bigger, and large tumor weight has been considered a positive indicator of malignancy. They are the commonest cause of Cushing's syndrome in children. Steroidogenesis is frequently defective and there often appears to be a relative deficiency of 11β-hydroxylase with consequent accumulation of 11-deoxycortisol and DOC in the plasma. These can cause unusual clinical syndromes to develop, which may resemble the enzyme deficiency states described in Chap. 6 [17]. Carcinomas are often unresponsive to exogenous ACTH in vitro, indicating defects in receptor-binding or postreceptor events [18].

Cushing's syndrome (Immunoglobulin stimulation)   A number of cases of ACTH-independent hypercortisolism can be attributed to stimulation of the

TABLE 5.2 Clinical features of Cushing's Syndrome

Clinical feature	Reported incidence (%)
Centripetal obesity	79–97
Weakness/proximal myopathy	29–90
Hypertension	74–87
Skin changes	
thin skin, bruising, acne	26–84
hirsutism	64–81
plethora and striae	50–94
Pigmentation	4–16
Psychiatric changes	31–86
Disorders of sexual function	55–80
Glucose intolerance	39–90
Osteoporosis	40–50

Based on information contained in ref. 3.

adrenal cortex by immunoglobulins, possibly ACTH-receptor autoantibodies. This condition seems to be familial and inherited as an autosomal dominant characteristic, resulting in pigmented nodular hyperplasia of the adrenal cortex [19,20]. Although very few cases have been reported, it is an interesting possibility that the presence of these antibodies will be found increasingly. This may help account for the adrenal enlargement that is occasionally detected incidentally with computer scanning [21].

5.1.2 Symptoms

The principal signs and symptoms of Cushing's syndrome and their reported incidence are shown in Table 5.2. These tally well with the features described by Cushing [1] when reporting his original series; nevertheless, the pathophysiological basis for many of the symptoms remains unknown.

Central obesity with thin limbs is one of the most common features of Cushing's syndrome, with redistribution of fat to the face, trunk and cervicodorsal regions. This gives rise to the classical "moon facies" and "buffalo hump." There is evidence for an increase in total body fat in Cushing's syndrome, which may possibly be associated with hyperinsulinemia or a decreased turnover of fatty acids. The wasting of the extremities is often associated with muscular weakness and proximal myopathy. These symptoms are usually attributed to the protein catabolic actions of excess glucocorticoids (see Chap. 4). Similar mechanisms may contribute in part to the atrophy of the skin, and thinning of the stratum corneum occurs allowing the underlying capillary plexus to be seen more easily. There is loss of connective tissue, and minor trauma leads to bruising and skin ulcerations with poor healing.

Hyperpigmentation similar to that seen in Addison's disease occurs in a minority of patients with ectopic ACTH and Cushing's disease. This seems to be caused by the markedly elevated levels of ACTH and other POMC-derived peptides with melanotropic activity, rather than β-MSH as was initially thought. In Nelson's Syndrome, the pigmentation becomes particularly prominent [3]. Osteoporosis is seen in most cases of Cushing's syndrome, bone formation is decreased due to inhibition of osteoblast function and decreased collagen synthesis [22–25]. Glucocorticoids also increase bone reabsorption, this action may be mediated by parathyroid hormone [24]. The antagonism by glucocorticoids of the actions of Vitamin D may also be involved [25,26]. Glycosuria and impaired glucose tolerance occurs in approximately 80 percent of patients, although fasting blood glucose is frequently normal. In a proportion of cases, overt diabetes may be present. These symptoms are usually related to the widespread effects of glucocorticoids on carbohydrate metabolism. There is evidence of an insulin-resistant state, since hyperinsulinemia coexists with hyperglycemia [27].

Mental symptoms in Cushing's syndrome have been increasingly recognized and seem to occur in more than 50 percent of the cases. These range from simple personality changes and irritability, through agitated depression to actual psychosis. These symptoms are very closely related to the increased level of circulating cortisol, and respond very quickly when it is reduced [28].

Hirsutism and acne are most obvious in female patients and have been attributed to the secretion of excess quantities of adrenal androgens, which tends to occur most often when an adrenal tumor is present [29]. Unresponsiveness of the pituitary to gonadotropin-releasing hormone (GnRH) has been demonstrated in patients with Cushing's syndrome, and may be responsible for the menstrual irregularities in females and reduced sexual function in males [3,30].

5.1.3 Hypertension

Moderate to severe hypertension has long been known to be associated with Cushing's syndrome, the cardiovascular complications used to be a common cause of death in the disease. The incidence of hypertension varies with the different forms of Cushing's syndrome as shown in Table 5.3. Hypertension is highest when an adrenal tumor is present, and lowest in cases of excess glucocorticoid administration [31]. This difference is hard to explain, but clearly the increase in blood pressure must be related to the actions of corticosteroids, and the different patterns of secretion in the various conditions. In the case of iatrogenic Cushing's syndrome, only the effect of an individual synthetic steroid need be considered, and it seems that most of these have only a minimal effect on blood pressure in humans [32].

TABLE 5.3 Incidence of hypertension in the different types of Cushing's syndrome

Type	Incidence of hypertension (%)
Pituitary	88
Ectopic	55
Adrenal adenoma	83
Adrenal carcinoma	100
Iatrogenic	17

Reproduced with permission from ref. 31.

Hypertension in Cushing's syndrome was commonly attributed to the sodium retaining activity of cortisol when present in excess, but this approach is now acknowledged to be too simplistic. It is recognized that glucocorticoids may affect the cardiovascular system in other ways, and also that steroids without clear biological effects of their own may interact with and potentiate the actions of conventional hormones. Thus, any attempt to rationalize the differences in the incidence of hypertension, must take into account not only the levels of cortisol, but those of other steroids including DOC, corticosterone, 18-OH-DOC, 18-oxocortisol, 19-nor-DOC, and so on ([29] and see Chap. 4).

5.1.4 Diagnosis

The diagnosis of Cushing's syndrome depends in essence on demonstrating the presence of excess cortisol, and differentiation of whether the condition is dependent on ACTH secretion or not. Although simple in principle, diagnosis of Cushing's syndrome poses considerable difficulties in practice, and an appreciable proportion of patients may remain without a precise diagnosis [33]. The fact that no single test can be relied on to give a definitive answer is illustrated by the many tests, and variations upon them, that are in use. In fact, a sequential diagnostic approach using a variety of tests is usually recommended [34].

Corticosteroid excretion This was the earliest biochemical method for the detection of Cushing's syndrome based on the measurement of 17-hydroxysteroids in a 24-hour collection of urine. Better discrimination, however, is achieved by the determination of urinary free cortisol, and this is often used as a simple screening test. Urinary free cortisol is assumed to reflect the fraction of cortisol in plasma which is not bound to plasma proteins (transcortin or albumin, see Chap. 2), and which is freely filtered at the glomerulus. When plasma cortisol levels exceed about 600 nmol/l the binding is saturated and the concentration of free cortisol rises disproportionately. Such measurements, especially if corrected for body composition by creatinine excretion, give a good differentiation between Cushing's syndrome and normals, but are subject to the difficulties inherent in achieving a complete 24-hour collection of urine.

Plasma cortisol levels Simple measurement of plasma cortisol levels does not provide a suitable test for Cushing's syndrome, since isolated values are often within the normal range. The fact that patients with Cushing's syndrome often lack the normal rhythm of cortisol secretion has been exploited as an aid to diagnosis [4]. However, there has been difficulty in defining the normal circadian variation adequately. It is claimed that an effective way of detecting an abnormality is by measuring cortisol levels in plasma collected at midnight from previously asleep patients [3]. In some patients with pituitary Cushing's disease there is a rhythm of normal amplitude, which is reset at a higher level, however, the normal postprandial rise in plasma cortisol is depressed or absent, which may provide an alternative marker for the condition [35].

Low-dose dexamethasone suppression test This was introduced by Liddle in 1960, and is based on the fact that normal endocrine control mechanisms are not operative in Cushing's syndrome [36]. It represents an example of what is often called "a dynamic test," and depends on perturbing the hypothalamo-pituitary-adrenal negative feedback loop (Fig. 5.3). Thus, in normal subjects dexamethasone administration suppresses pituitary ACTH secretion and the corticosteroid concentration, and urinary excretion falls to very low or undetectable levels. Patients with Cushing's syndrome do not usually show this characteristic response. The test as originally described required administration of dexamethasone over a 48-hour period, but an "overnight" dexamethasone suppression test has been devised, and is often used as a first line outpatient screening procedure. Occasional false positives and negatives arise, however, since conditions in which the bound cortisol fraction in plasma is increased, and in which steroid metabolism is perturbed, can create ambiguities.

Patients suffering from severe endogenous depression often have high plasma cortisol levels with no circadian rhythm, and do not show cortisol suppression with low-dose dexamethasone administration. Although it is not clear why these individuals do not show any other clinical characteristics of Cushing's syndrome, this fact has been exploited as a basis of a test for the diagnosis of endogenous depression with some success [37].

Insulin tolerance test This consists of the administration of insulin at a rate sufficient to induce hypoglycemia (glucose : < 2.2 mmol/l) with the simultaneous measurement of cortisol levels. In normal subjects, this maneuver provokes ACTH secretion, and a significant increase in cortisol

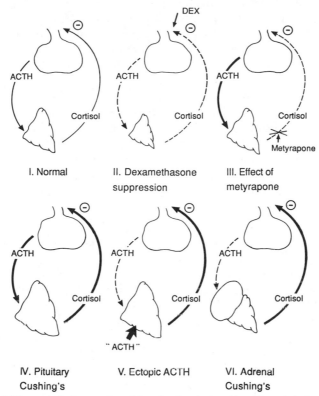

FIGURE 5.3 Normal pituitary-adrenal feedback relationships (I), and their modification by dexamethasone (II) and metyrapone (III) treatment, and in the different types of Cushing's syndrome (IV–VI). → = normal secretion; --→ = reduced secretion; → = increased secretion; ⊖ = negative feedback inhibition; × = blocked pathway.

secretion. Patients with Cushing's syndrome usually fail to show a response to this test, however patients with endogenous depression usually do. This may provide a useful method for differentiating the latter group, which can prove difficult [3,37].

5.1.5 Differential diagnosis of Cushing's syndrome

Having established that the patient's symptoms can be ascribed to hypersecretion of glucocorticoid, the next problem is to distinguish between the different causes. The first step is to differentiate between ACTH-dependent and independent forms of the disease. The possibility of transition between the two types can add considerably to the difficulty [13]. If ACTH hypersecretion is detected, it is then necessary to define whether it originates from the pituitary or from an ectopic source.

Plasma ACTH levels The wider availability of reliable immunoassays for plasma ACTH levels has made the identification of Cushing's syndrome caused by adrenal tumors much easier. In these cases normal pituitary feedback mechanisms still operate, and therefore, ACTH levels are very low (Fig. 5.3). Greater sensitivity and specificity is now provided by a two-site immunoradiometric assay (IRMA) which recognizes both the C- and N-terminals of 1-39 ACTH [38], and enables clearer discrimination between the normal and suppressed ranges.

High ACTH levels are seen both with Cushing's disease and with ectopic ACTH production, and differentiation between these two causes is proving more difficult than was originally predicted [5,39]. Typically, patients with pituitary dependent Cushing's syndrome have ACTH levels which are just above or within the upper limits of the normal range, whereas in the first descriptions of Cushing's syndrome as a result of ectopic ACTH production, the patients were found to have gross elevations of plasma ACTH (up to 2500 ng/dl; $20\times$ normal) [40,41]. The symptoms were consistent with overt, malignant tumors, and easily distinguished from Cushing's disease. However, it has now been recognized that Cushing's syndrome may also be caused by small and relatively benign tumors with no obvious symptoms, and that in these cases plasma ACTH levels are not particularly high (Fig. 5.4). Further difficulties are caused by the fact that the tumor also seems to respond the same as the pituitary in the dynamic tests to be described [39]. Information on the source of ACTH has been obtained by chromatographic fractionation of the plasma, since ectopic tumors frequently secrete unusual forms of ACTH. This arises as a result of abnormal processing of the POMC molecule in tumor cells. However, this may also occur in pituitary tumors, so that differentiation is again not absolute [3,10].

High-dose dexamethasone suppression test   This test was first described by Liddle (1960) and involves the administration of a high dose of dexamethasone (2 mg orally every 6 hours for 48 hours) and measurement of glucocorticoid production (urinary 17 hydroxycorticoids then, plasma cortisol levels more often today). Typically, in Cushing's disease there is some suppression (50% or more) of cortisol output, but with an adrenal tumor there is none. The mechanism by which a high dose of dexamethasone causes a reduction in ACTH output in Cushing's disease is not understood, but it does seem to be consistent with the view that there is an abnormally high set point for the negative feedback of cortisol, but whether this is at the level of the hypothalamus or the pituitary is not known [4].

The test has subsequently been applied to distinguishing between ectopic and pituitary production of ACTH, and has proved reasonably successful in this task. The criterion that is used is that ACTH production by ectopic tumors is also not suppressed by administration of dexamethasone. Once again, however, small, hidden tumors tend to behave differently; in one analysis it was found that as many as 33 percent of ACTH-secreting tumors showed some suppression with dexamethasone [39].

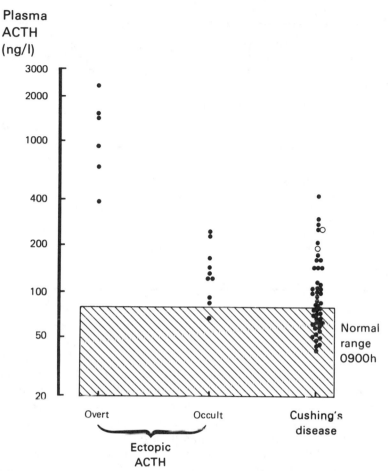

FIGURE 5.4 Plasma ACTH levels in patients with pituitary Cushing's disease, and in patients with ectopic ACTH production from 6 overt and 10 occult tumors. (Reproduced with permission from ref. 39.)

CRF stimulation test It was initially thought that the response to the administration of synthetic corticotropin releasing hormone would provide a new tool for the differential diagnosis of Cushing's syndrome. There was general agreement that in patients with Cushing's disease there was a definite and sometimes exaggerated rise in ACTH and cortisol following CRF, a response which was not seen in other forms of the syndrome [42]. Unfortunately, this initial optimism has not proved well founded and as with other tests, a number of exceptions and anomalies have been recorded [43,44]. It does, however, appear to increase the accuracy of diagnosis when used in conjunction with the dexamethasone test [33].

Metyrapone test This depends on the ability of the drug metyrapone to inhibit 11β-hydroxylation and to reduce adrenal cortisol production, with

TABLE 5.4 Tests used for the diagnosis of Cushing's syndrome, and expected results in pituitary Cushing's (PC) disease, ectopic Cushing's (EC), and adrenal Cushing's (AC) syndrome. "Increase" and "decrease" refer to adrenal steroids in plasma

Disease	PC	EC	AC
Test			
1. Basal corticosteroid excretion	high	high	high
2. Circadian rhythm of plasma cortisol	A	A	A
3. Low-dose dexamethasone	NC	NC	NC
4. Insulin tolerance	NC	NC	NC
5. Plasma-ACTH level	high	very high	low
6. High-dose dexamethasone	decrease	NC	NC
7. CRF stimulation	increase	NC	NC
8. Metyrapone	increase	NC	NC

NC = no change, A = absent

the result that ACTH levels increase as do levels of plasma 11-deoxycortisol (Fig. 5.3 and see Chap. 8). Patients with Cushing's disease respond in an exaggerated fashion in this test, whereas those with adrenal tumors show no response. However, no clear cut differences are seen between cases due to pituitary or ectopic ACTH secretion, particularly when the tumor is occult [3]. This test is no longer frequently used.

A summary of the responses to the diagnostic tests in the different types of Cushing's syndrome is shown in Table 5.4.

5.1.6 Localization of tumors

In cases where an adrenal tumor is suspected, computer aided tomography (CT) is often used to confirm the diagnosis and localize the lesion. Information may also be gained as to whether an adenoma (which tends to be small and well defined) or a carcinoma (which tends to be large and irregular) is present. Bilateral hyperplasia of the adrenals can also be visualized in this way [3,45], but confusion may occur when macronodular hyperplasia is present and the diagnosis of adenoma may be suggested.

Isotopic procedures have also been developed for the evaluation of the adrenal glands, notably scanning after the administration of ^{125}I-labeled cholesterol or related derivatives. In cases of excess ACTH secretion there is increased bilateral uptake of the tracer; autonomous adrenal function is typified by asymmetric visualization in the case of an adenoma, whereas there is no uptake in the case of a functioning carcinoma.

The localization of the source of excess ACTH secretion has been achieved by venous sampling from the great veins at a variety of sites in the body. Demonstration of a high level of ACTH compared with peripheral levels can pinpoint the source of production. Absence of a gradient of secretion tends to indicate secretion from an ectopic site which does not drain into the great veins (e.g., from lungs, pancreas, and gut) [46]. CT scanning of chest and abdomen frequently aids in the detection of small ACTH secreting tumors [3].

Radiological examination of the skull in Cushing's disease does not always detect any abnormalities, although CT scanning gives improved results. Nuclear magnetic resonance (NMR) imaging appears to be a promising technique for the evaluation of the pituitary [47]. Useful information may be obtained by bilaterally sampling the blood from the inferior petrosal sinus to determine the presence or absence of a gradient of ACTH. The procedure reliably establishes the presence of a pituitary adenoma, and often helps to localize it [48].

5.1.7 Treatment

Cushing's disease is treated increasingly successfully by surgery, and selective transsphenoidal microadenomectomy has proved a safe and effective treatment with a high-cure rate in recent years, although the rate of relapse may be appreciable [3,49]. Transient ACTH and cortisol insufficiency often occurs after surgery requiring corticosteroid replacement. Pituitary irradiation has also been used, most recently with interstitial irradiation from implanted needles containing yttrium-90 [50]. Total bilateral adrenalectomy has also been used as a treatment for ACTH dependent Cushing's syndrome, this approach obviously requires continuous corticosteroid replacement. Medical treatment using drugs which inhibit adrenal function selectively (see Chap. 8) is the alternative approach. Metyrapone and mitotane (o,p'-DDD) have been used for this purpose; other agents which have proved less successful are trilostane and aminoglutethimide [3]. Recently, the discovery that ketoconazole, an imidazole antimycotic, was able to reduce cortisol secretion through an action on cytochrome P450 enzymes has led to its use on an experimental basis in the treatment of Cushing's syndrome [51]. The potent antiglucocorticoid action of the new drug, RU 486 (see Chap. 7), has raised the possibility of its use in the treatment of hypercortisolism. However, a recent study has shown that administration of RU 486 to patients with Cushing's syndrome activated the pituitary-adrenal axis and provoked increased cortisol secretion. Thus, whether the drug proves to have any therapeutic benefit will depend on the balance between its peripheral and central actions [52]. Neurotransmitter therapy, designed to suppress ACTH, has also been advocated in Cushing's disease, bromocriptine, cyproheptadine, and sodium valproate have been found effective in some cases [3,53]. The long-acting somatostatin analogue,

SMS 201-995, has been found to be of value in treating Nelson's syndrome [54].

Management of ectopic ACTH secretion may be complicated by the fact that the tumor causing the condition is highly malignant, but obviously removal of it is attempted wherever possible. In the case of occult ACTH secreting tumors, the diagnosis procedure may extend over several years before localization is achieved and medical management will be required [33]. Adrenal adenomas can be completely cured by surgery, replacement therapy is often required for long periods of time while normal function of the pituitary and remaining adrenal is restored. Adrenal carcinomas are frequently inoperable, and aggressive drug therapy with both metyrapone and mitotane is used palliatively to correct the endocrine symptoms [3,17].

5.2 Mineralocorticoid excess states

Primary aldosteronism is a generic term for disorders in which aldosterone hypersecretion exists independently of the renin-angiotensin system. It is characterized by hypertension, hypokalemia, and low-plasma renin activity. Secondary hyperaldosteronism also occurs but in this case there is activation of the zona glomerulosa by the renin-angiotensin system, which may either be due directly to overproduction of renin or secondary to other disturbances in body fluid homeostasis. Similar symptoms to those seen in primary aldosteronism may also be found when other steroids with mineralocorticoid activity such as DOC, 18-OH-DOC, and corticosterone are secreted in excess. This may occur with adrenal tumors and with 11β- and 17α-hydroxylase deficiency (see Chap. 6).

5.2.1 Primary aldosteronism

Primary aldosteronism was first recognized by Jerome Conn (1954) as a syndrome whose symptoms could be ascribed to the overproduction of a potent mineralocorticoid and which was corrected by the removal of an adrenocortical adenoma. By 1964 it was possible to analyze the results of 145 case reports of aldosterone producing adenomas, with only extremely rare carcinomas occurring [55]. Primary aldosteronism was subsequently shown also to be caused by bilateral hyperplasia of the zona glomerulosa without the presence of a tumor; this is called idiopathic hyperaldosteronism (IHA) [56,57]. Estimates of the incidence of primary aldosteronism in the hypertensive population now vary between about 0.5 and 2 percent, with Conn's syndrome caused by a solitary aldosterone-producing adenoma (APA) as the most common cause [57,58], (Table 5.5). An extremely rare form of primary aldosteronism also exists in which zona glomerulosa hyperactivity is dependent on the pituitary and may be corrected by dexamethasone administration [59,60]. However, a case has been described in which dexamethasone-suppressible hyperaldosteronism gradually acquired the

TABLE 5.5 Frequency of the different forms of primary aldosteronism

Type	Frequency (%)
Aldosterone-producing adenoma	65–85
Aldosterone-producing carcinoma	<1
Idiopathic hyperaldosteronism	15–35
Dexamethasone-suppressible hyperaldosteronism	<1

Reproduced with permission from ref. 58.

characteristics of idiopathic hyperaldosteronism over an 8 to 12-year period. Thus, the distinction between the two conditions may not be as clear cut as was originally thought [61].

Aldosterone-producing adenoma (APA)   Solitary adrenocortical adenomas appear more often on the left side. They are usually small (less than 2 cm diameter) and have a golden-yellow color on the cut surface. Histological examination reveals a variety of cell types, some with either zona glomerulosa or zona fasciculata characteristics and some with a hybrid of the two (Fig. 5.5). Some aldosterone-producing adenomas have been found to be associated with diffuse zona glomerulosa hyperplasia in the normal portion of the gland [11,12].

In general, it has been found that the adenomatous tissue removed from patients with Conn's syndrome shows a high capacity to synthesize aldosterone and 18-hydroxycorticosterone, and there is a poor correlation between tumor size and plasma aldosterone levels [18]. Excretion of 21-deoxyaldosterone and 18-hydroxycortisol is increased in some cases of aldosterone-producing adenomas [62,63]. Aldosterone-producing carcinomas occur only very infrequently [18,57].

Idiopathic hyperaldosteronism (IHA)   The adrenal glands in patients with idiopathic hyperaldosteronism show both diffuse and focal hyperplasia of ultrastructurally normal zona glomerulosa tissue, often accompanied by nodules. These may be similar to those found in Cushing's syndrome, and to nodules sometimes found at autopsy in patients with no diagnosed adrenal disease [11,12]. Thus, in idiopathic hyperaldosteronism, the adrenals show some characteristics of a stimulated gland, and a continuing theme in this field has been the search for an agent which might be responsible [57]. Some recent studies have suggested that POMC derived peptides might be important [64,65]. Another possibility is the glycoprotein, aldosterone-stimulating factor (ASF) isolated from human urine by Sen and co-workers

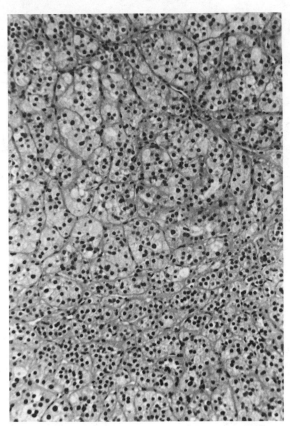

FIGURE 5.5 Micrograph showing mixed cell types in an adenoma removed from a patient with Conn's syndrome (×86).
(Micrograph generously provided by I. Doniach.)

([66] and see Chap. 3). Alternative theories link an increased sensitivity to angiotensin II to the development of idiopathic hyperaldosteronism [57], and it has been argued that idiopathic hyperaldosteronism differs only in degree from that form of essential hypertension in which relatively low-plasma renin activity and high-aldosterone levels are found together [67,68].

Dexamethasone-suppressible hyperaldosteronism This is a very rare, usually familial, form of primary aldosteronism, associated with micronodular adrenocortical hyperplasia and characterized by the relief of the symptoms by administration of a glucocorticoid [60,61,69]. This results in suppression of ACTH secretion and has led to the suggestion that a pituitary factor is responsible for the increased zona glomerulosa activity, possibly produced from POMC or a related peptide. One candidate would be ACTH itself, since there is some evidence which suggests an increased sensitivity of the zona glomerulosa to ACTH in this condition [70]. The secretion of increased amounts of 18-oxo- and 18-hydroxycortisol (which can be considered "hybrid" steroids, see Chap. 4) suggests the presence of a cell with the characteristics of both the zona glomerulosa and fasciculata, which would be consistent with chronic ACTH stimulation [70–72]. Other authors have sought to involve the MSH-peptides on the basis of their ability to stimulate aldosterone production in vitro ([73] and see Chap. 3). The loss of dopaminergic inhibition of aldosterone biosynthesis does not seem to account for the condition [70].

Symptoms and Diagnosis   Primary aldosteronism is characterized by hypertension, hypokalemia, and suppression of the renin-angiotensin system, with normal or low-cortisol production. The symptoms can be attributed to the effects of excess aldosterone secretion and are countered by a mineralocorticoid antagonist such as spironolactone [74]. Thus, there is increased sodium retention, expansion of extracellular fluid volume and increased body sodium content although this is limited by the "escape" phenomenon (see Chap. 4), to which down regulation of mineralocorticoid receptors may contribute [75,76]. Elevated plasma levels of atrial natriuretic peptide (ANP) may also play a part [77]. Increased reabsorption of sodium is most obvious in the kidney, but is seen in other sites, thus the sodium content of saliva, sweat, and feces is also reduced.

Potassium loss accompanies the sodium retention, with decreased total body and plasma potassium. The body may also be depleted of magnesium. Alkalosis also occurs, and is attributed to increased hydrogen ion (H^+) secretion in the kidney, and replacement of intracellular potassium by H^+. The increases in blood pressure seen in primary aldosteronism are most often attributed to the expansion of extracellular fluid volume consequent on retention of sodium, but other mechanisms may be involved (see Chap. 4). Suppression of the renin-angiotensin system is explained by negative feedback effects of the expanded extracellular fluid volume and increased blood pressure on the juxtaglomerular apparatus (Fig. 5.6).

The hypertension of primary aldosteronism is indistinguishable clinically from any other type, and varies considerably in severity. Hypokalemia, on the other hand, is the most consistent and obvious manifestation of primary aldosteronism, and occurs spontaneously in most patients. However, the degree of potassium loss is related in part to sodium intake, and sodium restriction leads to potassium retention and amelioration of the hypokalemia. Conversely, an increased sodium intake promotes renal potassium loss and can provoke hypokalemia where it is not already manifest. Hypokalemia associated with hypertension can also occur in patients treated with potassium-wasting diuretics, nevertheless measurement of plasma potassium provides a useful first-line screening method for the detection of primary aldosteronism in a hypertensive population [57]. Confirmation of the diagnosis is usually achieved by the demonstration of excess aldosterone production in the presence of low renin-angiotensin system activity. Measurement of either aldosterone 18-glucuronide or its tetrahydrometabolite

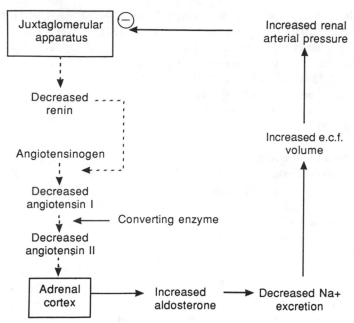

FIGURE 5.6 Diagram to illustrate the altered feedback mechanisms in primary aldosteronism. Despite reduced activity of the renin-angiotensin-system, aldosterone secretion remains high. --→ = reduced activity; ⊖ = negative feedback inhibition.

in 24-hour urine samples is used as an index of secretion. Plasma aldosterone levels can also give useful information if the conditions of collection are carefully controlled [57,74].

The basis of other tests that have been used is the lack of response of plasma aldosterone to maneuvers which expand extracellular fluid volume. These include the administration of mineralocorticoid in combination with a high-salt intake or saline infusion, both of which may not be suitable for hypertensive patients [78]. However, it has recently been suggested that a test which uses the converting enzyme inhibitor captopril to disturb the renin-angiotensin system may give useful information with less risk [79].

Differentiation between the two causes of primary aldosteronism is difficult, but is important since the symptoms of aldosterone-producing adenomas are readily relieved by surgery, whereas those of idiopathic hyperaldosteronism are usually not [57]. A number of tests have been devised to distinguish between the two conditions, usually relying on the fact that an aldosterone-producing adenoma tends to secrete autonomously, whereas the hyperplastic zona glomerulosa is at least partially responsive to changes in renin-angiotensin system activity. The results of these may be difficult to interpret and in some centers distinct subgroups of patients who show paradoxical responses have been identified [80]. The most widely used technique in the differential diagnosis of primary aldosteronism is the response of plasma aldosterone and 18-OH-corticosterone levels to change

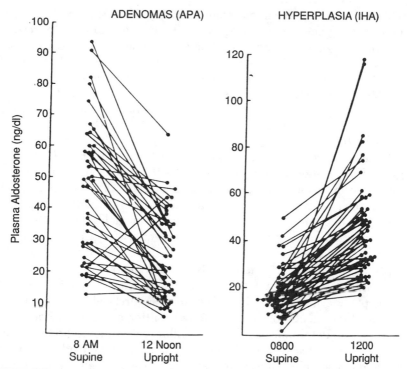

FIGURE 5.7 Plasma aldosterone levels at 8 AM and 12 noon in ambulatory patients with aldosterone producing adenomas, and with idiopathic hyperaldosteronism. (Reproduced with permission from ref. 57.)

in posture from supine to upright [57]. Plasma aldosterone levels rise between 8 AM and 12 noon following 4 hours in an upright position in idiopathic hyperaldosteronism, but in almost all patients with aldosterone-producing adenomas levels of aldosterone decline or show no change (Fig. 5.7). In addition, 18-OH-corticosterone levels tend to be higher in aldosterone-producing adenomas, and show a similar resistance to postural change [81]. Plasma aldosterone levels in idiopathic hyperaldosteronism may also respond to other disturbances of the status of the renin-angiotensin system, such as that obtained by administration of the angiotensin II antagonist, saralasin [82]. It is possible that the measurement of 18-hydroxycortisol levels in urine may prove a useful diagnostic procedure: levels of this steroid were found to be elevated in aldosterone-producing adenomas but within the normal range in idiopathic hyperaldosteronism [63,71].

If an adenoma is suspected it can be located by similar techniques to those used with Cushing's Syndrome (cannulation of the adrenal veins, adrenal imaging, CT scanning, and so on, as mentioned), although the success may be limited by the small size of the tumor (less than 1 cm in some cases) [57].

Treatment Unilateral adrenalectomy is highly effective in treating aldoste-rone-producing adenomas and potassium balance is permanently restored. However, only about 70 percent of patients are normotensive 1 year after their operation, and this declines to 50 percent after 5 years. In the case of idiopathic hyperaldosteronism, even bilateral adrenalectomy only cures the hypertension in a minority of cases, for reasons which are not understood [57]. Paradoxically, unilateral adrenalectomy has been found helpful in cases where some tests suggest the diagnosis of APA but where no tumor was detected [80].

Medical management of primary aldosteronism consists usually of correct-ing the effects of excess aldosterone, either with an antagonist at the receptor such as spironolactone, or by reversing its effects with a potassium-sparing diuretic such as amiloride or triamterene [57]. An alternative approach which has been tested is to use calcium channel antagonists, which theoretically should provide a two-pronged attack on the symptoms: by inhibition of aldosterone biosynthesis and by a direct action on the vasculature [74]. If the concept of a pituitary involvement in idiopathic hyperaldosteron-ism is substantiated then future therapy might be directed at this site; cyproheptadine has already been shown to suppress aldosterone acutely, whether other agents that alter neurotransmitter function will be helpful remains to be determined [74]. The possibility of exploiting the actions of ANP has not yet reached fruition, in fact, there is some evidence that the peptide does not affect aldosterone secretion when an adenoma is present [83,84].

5.2.2 Secondary hyperaldosteronism

The disorders in which hyperaldosteronism is secondary to increased activity of the renin-angiotensin system can be divided into two groups: those which are mediated by primary renin overproduction and which are associated with hypertension, and those which result from changes in body fluid volume distribution in which blood pressure may be normal (Fig. 5.8).

Renin secreting tumors are extremely rare, they are normally found in the kidney and contain elements of the juxtaglomerular cells. Ectopic tumors secreting renin have occasionally been reported, including one which originated in the adrenal gland and also secreted aldosterone [85,86]. Renovascular hypertension can also be associated with excess renin secretion and secondary hyperaldosteronism [78].

Bartter's syndrome is a rare, sometimes familial, disorder with various renal defects and increased renin production and elevated aldosterone levels [87,88]. Metabolic alkalosis and hypokalemia are present, but, in contrast to other forms of hyperaldosteronism, blood pressure is normal. Hyperplasia of the juxtaglomerular apparatus is characteristic, and there may be changes in the renal tubules. The syndrome is accompanied by abnormalities in the kallikrein-kinin system and urinary prostaglandin E_2 (PG E_2) excretion is increased together with vascular production of PG I_2 [88]. Treatment with

A: Primary excess renin secretion

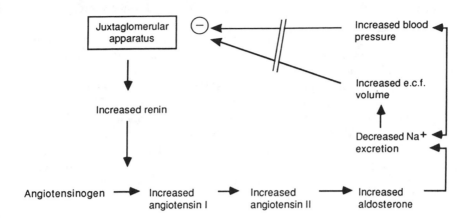

B. Loss of body fluid volume

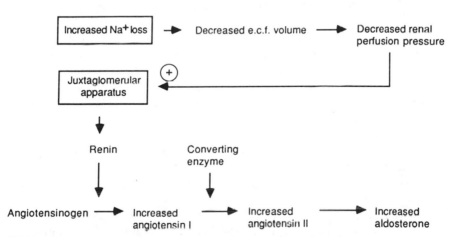

FIGURE 5.8 Diagram to illustrate the status of the renin-angiotensin system in secondary aldosteronism: (A) due to excess renin secretion, where normal negative feedback inhibition of renin is not operational, or (B) due to loss of body fluid volume, where aldosterone protects against the effects of sodium loss. ⊖ = negative feedback inhibition; ⊕ = stimulation. ‖ = interrupted feedback

indomethacin and other prostaglandin-synthesis inhibitors reverses these abnormalities and decreases plasma renin and aldosterone levels, although usually without correcting the hypokalemia [88,89]. The etiology of the condition is not fully understood, and the idea has been put forward that a number of different defects in reabsorption in the renal tubule occur, all leading to renal potassium wasting as the primary disturbance [88]. Although inappropriately high levels of atrial natriuretic peptide (ANP) have been found in some cases of Bartter's syndrome, it does not seem likely that an abnormality of ANP secretion has etiological significance [90,91].

Activation of the renin-angiotensin system leading to increased aldosterone production can also be produced by sodium or body fluid volume loss. Examples of these causes of hyperaldosteronism are seen when there is excess use of diuretics or laxatives, and in the sodium wasting caused by chronic renal insufficiency. Edematous states are also associated with secondary hyperaldosteronism, and activation of the renin-angiotensin system seems to result from changes in renal hemodynamics. However, the increase in aldosterone may then contribute to the edema [57].

In pseudohypoaldosteronism there is resistance to the effects of mineralocorticoids on tubular sodium and potassium excretion. Increased sodium and aldosterone production occur secondary to the sodium loss, however hyperkalemia is invariably present and the symptoms are those of mineralocorticoid deficit ([52] and see Chap. 6).

5.2.3 Nonaldosterone mineralocorticoid excess syndromes

Production of DOC or other mineralocorticoids in excess can occur either as a result of the activity of a tumor or as a consequence of enzyme defects which block the biosynthetic pathway at various sites leading to the accumulation of products which would normally be produced at low levels (see Chap. 2). A variety of patterns of steroid production are observed with adrenocortical carcinoma, and levels of DOC, corticosterone, 11-deoxycortisol as well as cortisol may be elevated. It is often difficult to assess what contribution the different steroids are making to the patient's symptoms ([31,92] and see Chap. 4).

In some enzyme deficiency states (see Chap. 6) there may be increased production of mineralocorticoids whose effects are sufficient to cause hypokalemia, expand extracellular fluid volume, and increase blood pressure, therefore, renin-angiotensin system activity is suppressed and aldosterone levels are low.

A rare syndrome of "apparent mineralocorticoid excess" has been recognized where severe hypertension and hypokalemia occur in the absence of increase in mineralocorticoid secretion. The disorder is correctable by the mineralocorticoid antagonist, spironolactone, and is characterized by a deficiency in the rate of oxidation of cortisol to cortisone [93]. It is suggested that the absence of this protective mechanism results in the mineralocorticoid receptor of the kidney being exposed to excessive amounts of cortisol, and consequent increased sodium retention ([94] and see Chap. 4).

References

1. Cushing, H.W. 1932. The basophil adenomas of the pituitary body and their clinical manifestations (pituitary basophilism). *Bull. Johns Hopkins Hosp.* 50:137–95.

2. Hardy, J. 1982. Cushing's disease—50 years later. *Canad. J. Neurol. Sciences.* 9:375–80.

3. Howlett, T.A., Rees, L.H., and Besser, G.M. 1985. Cushing's Syndrome. In: G.M. Besser and L.H. Rees, eds. *The Pituitary Adrenocortical Axis*, Clinics in Endocrinology and Metabolism, Vol. 14 (4). London: W.B. Saunders, pp. 911–45.

4. Krieger, D.T. 1983. Physiopathology of Cushing's disease. *Endocrine Rev.* 4:22–43.

5. Jones, M.T., Gillham B., Altaher, R., Nicholson, S.A., Campbell, E.A., and Thody, A. 1984. Clinical and experimental studies in the role of GABA in the regulation of ACTH secretion. *Psychoneuroendocrinology.* 9:107–23.

6. Lamberts, S.W.J., De Lange, S.A., and Stefanko, S.Z. 1982. Adrenocorticotropin-secreting pituitary adenomas originate from the anterior or intermediate lobe in Cushing's disease: Differences in the regulation of hormone secretion. *J. Clin. Endocrinol. Metab.* 54:286–91.

7. McNicol, A.M., Teasdale, G.M., and Beastall, G.H. 1986. A study of corticotroph adenomas in Cushing's disease: No evidence of intermediate lobe origin. *Clin. Endocrinol.* 24:715–22.

8. Coates, P.J., Doniach, I., Hale, A.C., and Rees, L.H. 1986. The distribution of immunoreactive α-melanocyte-stimulating hormone cells in the adult human pituitary gland. *J. Endocrinol.* 111:335–42.

9. Thody, A.J., Fisher, C., Kendall-Taylor, P., Jones, M.T., Price, J., and Abraham, R.R. 1985. The measurement and characterization by high pressure liquid chromatography of immunoreactive α-melanocyte stimulating hormone in human plasma. *Acta Endocrinol.* 110:313–18.

10. Hale, A.C., Besser, G.M., and Rees, L.H. 1985. Characterization of pro-opiomelanocortin derived peptides in pituitary and ectopic ACTH secreting tumours. *J. Endocrinol.* 108:49–56.

11. Neville, A.M., and O'Hare, M.J. 1985. Histopathology of the human adrenal cortex. In: G.M. Besser and L.H. Rees, eds. *The Pituitary Adrenocortical Axis*, Clinics in Endocrinology and Metabolism, Vol. 14 (4). London: W.B. Saunders, pp. 791–820.

12. Neville, A.M., and O'Hare, M.J. 1982. *The Human Adrenal Cortex. Pathology and Biology —an Integrated Approach.* Berlin: Springer-Verlag.

13. Hermus, A.R., Pieters, G.F., Smals, A.G., Pesman, G.J., Lamberts, S.W., Benraad, T.J., Van Haelst, U.J., and Kloppenborg, P.W. 1988. Transition from pituitary-dependent to adrenal-dependent Cushing's syndrome. *New Engl. J. Med.* 318:966–70.

14. Meador, C.K., Liddle, G.W., Island, D.P., Nicholson, W.E., Lucas, C.P., Nucton, J.G., and Luetscher, J.A. 1962. Cause of Cushing's syndrome in patients with tumors arising from "non-endocrine" tissue. *J. Clin. Endocrinol.* 22:693–703.

15. Pullan, P.T., Clement-Jones, V., Corder, R., Lowry, P.J., Besser, G.M., Rees, L.H. 1980. ACTH, LPH and related peptides in the ectopic ACTH syndrome. *Clin. Endocrinol.* 13:437–45.

16. De Keyzer, Y., Bertagna, X., Lenne, F., Girard, F., Luton, J.-P., and Kahn, A. 1985. Altered pro-opiomelanocortin gene expression in adrenocorticotropin-producing nonpituitary tumors. *J. Clin. Invest.* 76:1892–98.

17. Freeman, D.T. 1986. Steroid hormone-producing tumors in man. *Endocrine Rev.* 7:204–20.

18. Whitehouse, B.J., and Vinson, G.P. 1981. Hormone production in normal and abnormal human adrenocortical tissue. In: K. Fotherby and S.B. Pal, eds. *Hormones in Normal and Abnormal Human Tissues*, Vol. 1. Berlin: Walter de Gruyter, pp. 215–255.

19. Teding van Berkhout, F., Croughs, R.J.M., Wulfraat, N.M., and Drexhage, H.A. 1989. Familial Cushing's syndrome due to nodular adrenocortical dysplasia is an inherited disease of immunological origin. *Clin. Endocrinol.* 31:185–191.

20. Young, W.F., Carney, J.A., Musa, B.U., Wulfraat, N.M., Lens, J.W., and Drexhage, H.A. 1989. Familial Cushing's syndrome due to primary pigmented nodula adrenocortical disease. Reinvestigation 50 years later. *New Engl. J. Med.* 321:1659–64.

21. Findling, J.W. 1989. The Cushing syndromes. An enlarging clinical spectrum. *New Engl. J. Med.* 321:1677–78.

22. Peck, W.A., Brandt, J., and Miller, I. 1967. Hydrocortisone-induced inhibition of protein synthesis and uridine incorporation into isolated bone cells in vitro. *Proc. Natl. Acad. Sci. USA.* 57:1599–606.

23. Canalis, E. 1983. Effects of glucocorticoids on type I collagen synthesis, alkaline phosphatase activity and DNA content in cultured rat calvaria. *Endocrinology.* 112:931–939.

24. Jee, W.S.S., Park, H.Z., Roberts, W.E., and Kenner, G.H. 1970. Corticosteroid and bone. *Am. J. Anat.* 129:477–79.

25. Findling, J.W., Adams, N.D., Lemann, J. Jr., Gray, R.W., Thomas, C.J., and Tyrell, J.B. 1982. Vitamin D metabolites and parathyroid hormone in Cushing's syndrome. Relationship to calcium and phosphorus homeostasis. *J. Clin. Endocrinol. Metab.* 54:1039–44.

26. Singer, F.R. 1987 Metabolic bone disease. In: P. Felig, J.D. Baxter, A.E. Broadus and L.A. Frohman, eds. *Endocrinology and Metabolism*, 2nd ed. New York: McGraw Hill, pp. 1454–99.

27. Lenzen, S., and Bailey, C.J. 1984. Thyroid hormones, gonadal and adrenocortical steroids and the function of the islets of Langerhans. *Endocrine Rev.* 4:411–34.

28. Jeffcoate, W.J., Silverstone, J.T., Edwards, C.R.W., and Besser, G.M. 1979. Psychiatric manifestations of Cushing's syndrome: response to lowering of plasma cortisol. *Quarterly J. Med.* 48:465–72.

29. Dewis, P., and Anderson, D.C. 1985. The Adrenarche and Adrenal Hirsutism. In: D.C. Anderson and J.S.D. Winter, eds. *Adrenal Cortex*. London: Butterworths, pp. 96–119.

30. Bocuzzi, G., Angeli, A., Bisbocci, D., Fonzo, D., Gaidano, G.P., and Ceresa, F. 1975. Effect of synthetic luteinizing hormone releasing hormone (LH-RH) on the release of gonadotropins in Cushing's disease. *J. Clin. Endocrinol Metab.* 40:892–95.

31. Gomez-Sanchez, C.E. 1986. Cushing's syndrome and hypertension. *Hypertension.* 8:258–64.

32. Scoggins, B.A., Coghlan, J.P., Denton, D.A., Mills, E.H., Nelson, M.A., Spence, C.D., and Whitworth, J.A. 1984. How do adrenocortical steroid hormones produce hypertension? *Clin. Exper. Hypertension—Theory and Practice.* A6:315–84.

33. Trainer, P.J., and Besser M. 1990. Cushing's syndrome. Difficulties in diagnosis. *Trends Endocrinol. Metab.* 1:292–95.

34. Ratcliffe, J.G. 1985. Biochemical investigation of adrenocortical dysfunction. In:

D.C. Anderson and J.S.D. Winter, eds. *Adrenal Cortex*. London: Butterworths, pp. 188–207.

35. Liu, J.H., Kazer R.R., and Rasmussen D.D. 1987. Characterisation of twenty-four hour patterns of adrenocorticotropin and cortisol in normal women and patients with Cushing's disease. *J. Clin. Endocrinol. Metab.* 64:1027–35.

36. Liddle, G.W. 1960. Tests of pituitary-adrenal suppressibility in the diagnosis of Cushing's syndrome. *J. Clin. Endocrinol. Metab.* 20:1539–60.

37. Editorial, 1986. Depression and Cushing's Syndrome. *Lancet.* II 550–51.

38. Findling, J.W., Engeland, W.C., and Raff, H. 1990. The use of immunoradiometric assay for the measurement of ACTH in human plasma. *Trends Endocrinol. Metab.* 1:283–85.

39. Howlett, T.A., Drury, P.L., Perry, L., Doniach, I., Rees, L.H., and Besser, G.M. 1986. Diagnosis and management of ACTH-dependent Cushing's syndrome: Comparison of the features in ectopic and pituitary ACTH production. *Clin. Endocrinol.* 24:699–713.

40. Rees, L.H., and Ratcliffe, J.G. 1974. Ectopic hormone production by non-endocrine tumours. *Clin. Endocrinol.* 3:263–99.

41. Imura, H. 1980. Ectopic hormone syndrome. In: K. Abe, ed. *Endocrinology and Cancer*, Clinics in Endocrinology and Metabolism, Vol 9 (2). London: W.B. Saunders, pp. 235–60.

42. Muller, O., Stalla, G., and Von Werder, K. 1983. Corticotrophin releasing factor: A new tool for the differential diagnosis of Cushing's syndrome. *J. Clin. Endocrinol. Metab.* 57:227–29.

43. Gold, P.W., Loriaux, D.L., Roy, A., Calabres, J.R., Nieman, L.K., Post, R.M., Pickard, D., and Gallucci, W. 1986. Responses to corticotropin-releasing hormone in the hypercortisolism of depression and Cushing's syndrome. *New Engl. J. Med.* 314:1329–35.

44. Hermus, A.R., Peiters, G.F., Pesman, G.J., Smals, A.G., Benraad T.J., and Kloppenborg, P.W. 1986. The corticotropin-releasing-hormone test versus the high-dose dexamethasone test in the differential diagnosis of Cushing's syndrome. *Lancet.* II 540–44.

45. White, F.E., White, M.C., Drury, P.L., Kelsey-Fry, I., and Besser G.M. 1982. Value of computed tomography of the abdomen and chest in investigation of Cushing's syndrome. *Brit. Med. J.* 284:771–74.

46. Drury, P.L., Ratter, S., Tomlin, S., Williams, J., Dacie, J.E., Rees, L.H., and Besser, G.M. 1982. Experience with selective venous sampling in the diagnosis of ACTH-dependent Cushing's syndrome. *British Medical J.* 284:9–12.

47. Baxter, J.D., and Tyrell, J.B. 1987. The adrenal cortex. In: P. Felig, J.D. Baxter, A.E. Broadus, and L.A. Frohman, eds. *Endocrinology and Metabolism*. New York: McGraw Hill, pp. 511–659.

48. Oldfield, E.H., Chrousos, G.P., Schulte, H.M., Schaaf, M., McKeever, P.E., Krudy, A.G., Cutler, G.B., Loriaux, D.L., and Doppman, J.L. 1985. Preoperative lateralisation of ACTH-secreting pituitary microadenomas by bilateral and simultaneous inferior petrosal venous sinus sampling. *New Engl. J. Med.* 312:101–03.

49. Tyrell, J.B., Brooks, R.M. Fitzgerald, P.A., Cofoid, P.B., Forsham, P.H., and Wilson, C.W. 1978. Cushing's disease-selective transsphenoidal resection of pituitary microadenomas. *New Engl. J. Med.* 298:753–58.

50. Sandler, L.M., Richards, N.T., Carr, D.H., Mashiter, K., and Joplin, G.F.

1987. Long-term follow-up of patients with Cushing's disease treated by interstitial irradiation. *J. Clin. Endocrinol. Metab.* 65:441–47.

51. McCance, D.R., Hadden, D.R., Kennedy, L., Sheridan, B., and Atkinson, A.B. 1987. Clinical experience with ketoconazole as a therapy for patients with Cushing's syndrome. *Clin. Endocrin.* 7:593–99.

52. Bertagna, X., Bertagna, C., Laudat, M.-H., Husson, J.-M., Girard, F., and Luton, J.-P. 1986. Pituitary-adrenal responses to the antiglucocorticoid action of RU 486 in Cushing's syndrome. *J. Clin. Endocrinol. Metab.* 66:639–43.

53. Nussey, S.S., Price, P., Jenkins, J.S., Altaher, A.R.H., Gillham, B., and Jones, M.T. 1988. The combined use of sodium valproate and metyrapone in the treatment of Cushing's syndrome. *Clin. Endocrinol.* 28:373–80.

54. Lamberts, S.W., Uitterlinden, P., and Klijn, J.M.G. 1989. The effect of the long-acting somatostatin analogue SMS 201-995 on ACTH secretion in Nelson's syndrome and Cushing's disease. *Acta Endocrinol.* 120:760–66.

55. Conn, J.W., Knopf, R.F., and Nesbit, R.M. 1964. Clinical characteristics of primary aldosteronism from an analysis of 145 cases. *Am. J. Surgery.* 107:159–72.

56. Katz, F. 1967. Primary aldosteronism with suppressed plasma renin activity due to bilateral nodular adrenocortical hyperplasia. *Ann. Int. Med.* 67:1035–42.

57. Melby, J.C. 1984. Primary aldosteronism. *Kidney Int.* 26:769–78.

58. Conn, J.W. 1967. Diagnosis of normokalemic primary aldosteronism, a new form of curable hypertension. *Science.* 158:525–26.

59. Sutherland, D.J.A., Ruse, J.L., and Laidlaw, J.C. 1966. Hypertension, increased aldosterone secretion and low plasma renin activity relieved by dexamethasone. *Can. Med. Assoc. J.* 95:1109–19.

60. New, M.I., Seigal, E.J., and Peterson, R.E. 1973. Dexamethasone suppressible hyperaldosteronism. *J. Clin. Endocrinol. Metab.* 37:93–100.

61. Stockigt, J.R., and Scoggins, B.A. 1987. Long-term evolution of glucocorticoid-suppressible hyperaldosteronism. *J. Clin. Endocrinol. Metab.* 64:22–26.

62. Lewicka, S., Winter, J., Bige, K., Bokkenheuser, V., Vecsei, P., Abdelhamid. S., and Heinrich, U. 1987. 21-Deoxyaldosterone excretion in patients with primary aldosteronism and 21-hydroxylase deficiency. *J. Clin. Endocrinol. Metab.* 64:771–77.

63. Chu, M.D., and Ulick, S. 1982. Isolation and identification of 18-hydroxycortisol from the urine of patients with primary aldosteronism. *J. Biol. Chem.* 258:5498–502.

64. Brownie, A.C., and Pedersen, R.C. 1986. Control of aldosterone secretion by pituitary hormones. *J. Hypertension.* 4 supp. 5: S72–75.

65. Franco-Saenz, R., Mulrow, P.J., and Kitai, K. 1984. Idiopathic hyperaldosteronism: A possible disease of the intermediate lobe of the pituitary. *J. Am. Med. Assoc.* 251:2555–58.

66. Carey, R.M., Sen, S., Dolan, L.M., Malchoff, C.D., and Bumpus, F.M. 1984. Idiopathic hyperaldosteronism. A possible role for aldosterone stimulating factor. *New Engl. J. Med.* 311:94–100.

67. Davies, D.L., Beevers, D.G., Brown, J.J., Cumming, A.M.M., Fraser, R., Lever, A.F., Mason, P.A., Morton, J.J., Robertson, J.I.S., Titterington, M., and Tree, M. 1979. Aldosterone and its stimuli in normal and hypertensive man: Are essential hypertension and primary aldosteronism without tumor the same condition? *J. Endocrinol.* 81:79P–91P.

68. Melby, J.C. 1989. Endocrine hypertension. *J. Clin. Endocrinol. Metab.* 69:697–703.

69. Ganguly, A., Grim, C.E., Bergstein, J., Brown, R.D., and Weinberger, M.H. 1981. Genetic and pathophysiologic studies of a new kindred with glucocorticoid-suppressible hyperaldosteronism manifest in three generations. *J. Clin. Endocrinol. Metab.* 53:1040–46.

70. Connell, J.M.C., Kenyon, C.J., Corrie, J.E.T., Fraser, R., Watt, R., and Lever, A.F. 1986. Dexamethasone-suppressible hyperaldosteronism. Adrenal transition cell hyperplasia. *Hypertension.* 8:669–76.

71. Davis, J.R.E., Burt, D., Corrie, J.E.T., Edwards, C.R.W., and Sheppard, M.C. 1988. Dexamethasone-suppressible hyperaldosteronism: Studies of over production of cortisol in three affected family members. *Clin. Endocrinol.* 29:297–308.

72. Gomez-Sanchez, C.E., Montgomery, M., Ganguly, A., Holland, O.B., Gomez-Sanchez, E.P., Grim, C.E., and Weinberger, M.H. 1984. Elevated urinary excretion of 18-oxocortisol in glucocorticoid-suppressible hyperaldosteronism. *J. Clin. Endocrinol. Metab.* 59:1022–24.

73. Mulrow, P.J. 1981. Glucocorticoid-suppressible hyperaldosteronism: A clue to the missing hormone? *New Engl. J. Med.* 305:1012–14.

74. Hsueh, W.A. 1986. New insights into the medical management of primary aldosteronism. *Hypertension.* 8:76–82.

75. Wenting, G.J., Veld, A.J., Verhoeven, R.P., Derkx, F.H., and Schalekamp, M.A. 1977. Volume-pressure relationships during development of mineralocorticoid hypertension. *Circ. Res.* 40 Supp. I:163–70.

76. Armanini, D., Witzgall, H., Wehling, M., Kuhnle, U., and Weber, P.C. 1987. Aldosterone receptors in different types of primary hyperaldosteronism. *J. Clin. Endocrinol. Metab.* 65:101–04.

77. Nakamura, T., Ichikawa, S., Sakamaki, T., Sato, K., Kogure, M., Tajima, Y., Kato, T., and Murata, K. 1987. Role of atrial natriuretic peptide in mineralocorticoid escape phenomenon in patients with primary aldosteronism. *Proc. Soc. Exp. Biol. Med.* 185:448–54.

78. Baxter, J.D., Perloff, D., Hsueh, W., and Biglieri, E.G. 1986. The endocrinology of hypertension. In: P. Felig, J.D. Baxter, A.E. Broadus and L.A. Frohman, eds. *Endocrinology and Metabolism*, 2nd ed. New York: McGraw Hill, pp. 693–788.

79. Lyons, D.F., Kem, D.C., Brown, R.D., Hanson, C.S., and Carollo, M.L. 1983. Single dose captopril as a diagnostic test for primary aldosteronism. *J. Clin. Endocrinol. Metab.* 57:892–96.

80. Biglieri, E.G., Irony, I., and Kater, C.E. 1989. Identification and implications of new types of mineralocorticoid hypertension. *J. Steroid Biochem.* 32:199–204.

81. Kater, C.E., Biglieri, E.G., Rost, C.R., Schambelan, M., Hirai, J., Chang, B.C.F., and Brust, N. 1985. The constant plasma 18-hydroxycorticosterone to aldosterone ratio: An expression of the efficacy of corticosterone methyloxidase type II activity in disorders with variable aldosterone production. *J. Clin. Endocrinol. Metab.* 60:225–29.

82. Brown, R.D., Kem, D.C., Hogan, M.J., and Hegstad, R.L. 1984. Evaluation of a test using saralasin to differentiate primary aldosteronism due to an aldosterone-producing adenoma from idiopathic hyperaldosteronism. *Metabolism.* 33:734–38.

83. Higuchi, K., Nawata, H., Kato, K.-I., Ibayashi, H., and Matsuo, H. 1986. Lack of effect of alpha-human atrial natriuretic polypeptide on aldosteronogenesis in aldosterone producing adenoma. *J. Clin. Endocrinol. Metab.* 63:192–96.

84. Mantero, F., Rocco, S., Pertile, F., Carpene, G., Fallo, F., and Menegus, A. 1987. a-h-ANP injection in normals, low renin hypertension and primary aldosteronism. *J. Steroid Biochem.* 27:935–40.

85. Robertson, P.W., Klidjian, A., Harding, L.K., and Walters, G. 1967. Hypertension due to renin-secreting tumor. *Am. J. Med.* 43:963–67.

86. Imura, O., Shimamoto, K., Hotta, D., Nakata, T., Mito, T., Kumamoto, Y., Dempo, K., Ogihara, T., and Naruse, K. 1986. A case of adrenal tumour producing renin, aldosterone and sex hormones. *Hypertension.* 8:951–56.

87. Bartter, F.C., Pronove, J., Gill, J.R. Jr., and McCardle, R.C. 1962. Hyperplasia of the juxtaglomerular complex with hyperaldosteronism and hypokalaemic alkalosis. *Am. J. Med.* 33:811–28.

88. Stein, J.H. 1985. The pathogenetic spectrum of Bartter's syndrome. *Kidney Int.* 28:85–93.

89. Shimoyama, R. 1980. Reversal of altered vascular responsiveness in Bartter's syndrome by indomethacin treatment. *J. Clin. Endocrinol. Metab.* 51:908–11.

90. Gordon, R.D., Tunny, T.J., Klemm, S.A., and Hamlett, S.M. 1986. Elevated levels of plasma atrial natriuretic peptide in Bartter's syndrome fall to normal with indomethacin. *J. Hypertension.* 4 supp. 6:S555–58.

91. Doorenbos, C.J., Daha, M.R., Buhler, F.R., and Van Brummelen, P. 1988. Effects of posture and saline infusion on atrial natriuretic papetide and haemodynamics in patients with Bartters syndrome and healthy controls. *Europ. J. Clin. Invest.* 18:369–74.

92. Biglieri, C.E., Biglieri, E.G., Brust, N., Chang, B., Hirai, J., and Irony, I. 1989. Stimulation and suppression of the mineralocorticoid hormones in normal subjects and adrenocortical disorders. *Endocrine Rev.* 10:149–64.

93. Ulick, S., Levine, L.S., Gunczler, P., Zaconato, G., Ramierez, L.C., Rank, W., Rosler, A., Bradlow, H.L., and New, M.I. 1979. A syndrome of mineralcorticoid excess associated with defects in the peripheral metabolism of cortisol to cortisone. *J. Clin. Endocrinol. Metab.* 49:757–64.

94. Edwards, C.R.W., Burt, D., and Stewart, P.M. 1989. The specificity of the human mineralcorticoid receptor: Clinical clues to a biological conundrum. *J. Steroid Biochem.* 32:213–16.

6

Disorders of the Adrenal Cortex II: Hypofunction and Inborn Errors in Biosynthesis

6.1 Adrenal insufficiency

Reduced adrenal production of corticosteroids can result from either the destruction or malfunction of the adrenal cortex (primary adrenocortical insufficiency or Addison's disease) or can be the result of deficient ACTH production (secondary adrenocortical insufficiency). Pituitary hypofunction after cessation of glucocorticoid therapy is the most common cause of the latter. A specific deficiency in mineralocorticoid production can also occur, and may be due either to primary malfunction of the zona glomerulosa or secondary to a reduction in activity of the renin-angiotensin system.

6.1.1 Primary adrenocortical insufficiency

This is an uncommon disorder in which adrenal destruction results from a variety of causes, of which the most common today is the autoimmune or idiopathic type. Previously, destruction of the gland by a tubercular infection was the most common cause of Addison's disease and remains so in populations where the infection is poorly controlled [1]. Adrenal hemorrhage can occur, usually during extreme infections, and may lead to glucocorticoid deficiency. It has recently been recognized that the pathology of the acquired immune deficiency syndrome (AIDS) involves significant adrenal damage which may lead to actual hormone deficiency. Rare familial types of adrenal atrophy have also been described [2] (Table 6.1).

TABLE 6.1 Etiology of primary adrenocortical deficiency

Type	Approximate frequency (%)
Autoimmune/idiopathic	70–80
Tuberculosis	20
Hemorrhage	
AIDS	1
Familial adrenal hypoplasia	

Reproduced with permission from ref. 2.

Autoimmune or idiopathic Addison's disease   About 80 percent of the cases of Addison's disease are caused by autoimmune destruction of adrenal tissue, which is sometimes associated with a high incidence of other autoimmune endocrine disorders. These cases may be termed the autoimmune polyglandular syndrome (APS) [1]. Lymphocytic infiltration of the adrenal cortex is the characteristic histological feature with loss of the normal architecture of the entire cortex, although the medulla is preserved. The glands are small and atrophic and the capsule is thickened [3]. The other endocrine disorders associated with the condition are gonadal failure, thyroid dysfunction, insulin-dependent diabetes, and hypoparathyroidism; vitiligo (patchy loss of pigmentation in the skin) and pernicious anemia may also be present [4]. Two categories of polyglandular disorder have been distinguished: the first, associated with idiopathic hypoparathyroidism, is termed type-I APS, and the second, associated with diabetes and/or thyroid disease, is termed type-II APS [5]. Circulating autoantibodies to affected tissues are found, and they may also occur in the absence of overt signs of glandular failure [6,7] (Table 6.2). Adrenal autoantibodies, when present in patients without Addison's disease seem to confer a risk for the development of the disorder, which may develop gradually [8]. There is evidence that both cell surface and microsomal components of the gland are serving as antigens [9,10]. The circulating antibodies are not cytotoxic, and the adrenal damage seems to be produced by cell-mediated immunological mechanisms [4,11]. Immunoglobulins that block the stimulation of cortisol secretion by ACTH have also been isolated from the plasma of patients with Addison's disease [12,13].

There is a genetic component, in that idiopathic adrenocortical insufficiency may be familial, and the type-II APS form of the disease is associated with certain of the HLA (human leucocyte antigen) phenotypes. In contrast, the type-I APS syndrome shows no HLA linkage [14,15].

Tuberculosis   Destruction of the adrenal cortex as a result of systemic tuberculosis is now a less common cause of Addison's disease in developed

TABLE 6.2 Circulating antibodies in idiopathic Addison's disease

Antibodies directed against	Prevalance (%)
Adrenal	64
Thyroid	
cytoplasm	45
thyroglobulin	22
Parathyroid	26
Gonad	17
Islets of Langerhans	8
Stomach	
parietal cells	30
intrinsic factor	9

Reproduced with permission from ref. 2.

communities (see Table 6.1). Adrenal involvement is relatively rare, but, when it does occur, it results in the loss of both cortex and medulla, and the glands may become enlarged and are often calcified [16].

Adrenal hemorrhage This usually occurs bilaterally and can be the cause of acute adrenal failure. In adults, anticoagulant therapy (given for other causes) is responsible for approximately one third of the cases, other causes include septicemia, and adrenal metastases. In children, fulminant infections are the most common cause. Adrenal hemorrhage can also occur in a variety of severe illnesses [2]. The glands become massively enlarged and many of the cortical cells are destroyed by ischemic necrosis, leaving a thin rim of functional cells in the outer cortex [17]. In many cases the diagnosis is made at autopsy and there may be controversy as to whether acute adrenal insufficiency had contributed to mortality [18]. The term "Waterhouse-Friderichsen syndrome" may be used to describe severe septicemia with circulatory collapse and adrenal hemorrhage, but can cause confusion when there is no evidence of glucocorticoid deficiency [15]. Nevertheless, there are cases of adrenal hemorrhage where low-plasma cortisol levels have been documented, and where glucocorticoid therapy appears to have enhanced survival [2].

AIDS Acquired immune deficiency syndrome (AIDS) creates multiple defects in the immune system and makes individuals susceptible to opportunist infections and unusual neoplasms. Many of the symptoms are similar to those of adrenal insufficiency, and clinical and biochemical evidence suggestive of corticosteroid deficiency has been documented [19]. At autopsy, partial adrenal necrosis has been found in a significant proportion of patients [20]. Reduced responses to ACTH stimulation have been recorded, and there is evidence that the production of corticosterone and other 17-deoxysteroids is affected to a greater extent than is the production of cortisol. The possibility that there is also a defect at the pituitary level in AIDS patients has been considered, since the extent of adrenal destruction does not seem to be sufficient to account for the reduction in steroidogenesis [21].

Familial adrenal insufficiency (adrenal hypoplasia) This is a rare type of disorder in which familial ACTH resistance is found, with less than 100 reported cases since 1959 [22]. It is characterized by low cortisol and adrenal androgen levels coupled with raised ACTH. Usually, aldosterone levels are normal as are the responses to posture and sodium deprivation. The adrenals are atrophic, normal glomerulosa cells are present but very few fasciculata cells are found [23]. There is insensitivity to ACTH, and a defect in the adrenal receptor has been postulated [24]. It has recently been shown that human mononuclear lymphocytes possess ACTH receptors, and that high-affinity binding did not occur in cells obtained from a child with ACTH-insensitivity ([24] Fig. 6.1).

Adrenoleukodystrophy is an inherited disorder characterized by progressive neurological malfunction associated with adrenal insufficiency. There appears to be an abnormality of fatty acid metabolism leading to accumulation of very long chain, unbranched compounds. Clinical and laboratory evidence of adrenal insufficiency may precede the onset of neurological manifestations [25].

Symptoms and diagnosis The symptoms of Addison's disease are usually explained in terms of glucocorticoid deficiency which can cause muscular weakness, fatigue, hypotension, hypoglycemia, anorexia, nausea, and vomiting. Aldosterone secretion is rarely affected [18]. The clinical presentation of the disease depends on the rate at which adrenal destruction progresses, and clear cut symptoms do not occur until more than 90 percent of both glands has been lost. Until then normal regulatory mechanisms may be able to compensate during unstressed conditions. There is, however, increased secretion of ACTH other pro-opiomelanocortin (POMC)-derived peptides such as LPH and the N-terminal fragment (see Fig. 3.2). It is believed that the MSH sequences within these are responsible for the hyperpigmentation which is one of the main signs when adrenocortical insufficiency develops slowly [2]. Hypotension is common in Addison's disease but generally causes problems only on changing posture, although in severe cases recumbent hypotension or shock may appear [18]. Acute adrenal crisis can develop in patients with Addison's disease who are exposed to stressful conditions such as surgery, and alternatively, it may appear suddenly due to adrenal

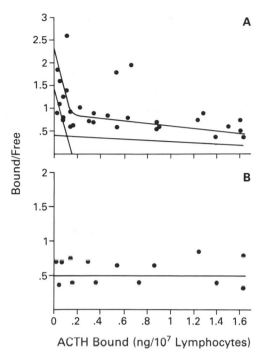

FIGURE 6.1 Scatchard analysis of the binding of ^{125}I-labeled ACTH to mononuclear leukocytes from a normal subject (A), upper panel, and from a patient with ACTH insensitivity (B), lower panel. Note the lack of high-affinity binding in the patient. (Reproduced with permission from ref. 24.)

hemorrhage. Dehydration and volume depletion are aggravated by nausea and vomiting. Shock and coma can prove fatal quickly in untreated patients [2,18,26]. Therapy includes the immediate administration of glucocorticoids (see Chap. 7), together with correction of the body fluid disturbances.

In the diagnosis of adrenocortical insufficiency, isolated measurement of steroid levels may not provide definitive information. Basal levels of adrenocortical steroids in either urine or plasma may be normal when the malfunction is only partial, so that tests of adrenocortical reserve may be necessary to establish the diagnosis. The rapid ACTH stimulation test is used initially in the assessment of adrenocortical insufficiency, and can provide the information on adrenal secretory capacity. The test does not differentiate between primary and secondary causes of adrenocortical insufficiency which can be accomplished by measurement of plasma ACTH levels, these are markedly different in the two conditions. Tests based on induction of endogenous ACTH secretion, with the use of insulin-induced hypoglycemia or the metyrapone test (see Chap. 5) may be also necessary to distinguish cases of partial secondary ACTH deficiency, although this may be considered too risky a procedure. In these cases, the adrenal may not be totally atrophied and can show a normal response to acute exogenous

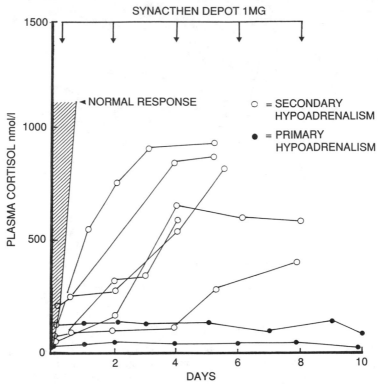

FIGURE 6.2 Effects of prolonged stimulation with ACTH 1-24 (injections shown by arrows) in patients with primary and secondary adrenocortical insufficiency. Note the progressively increasing response in the patients with secondary adrenocortical insufficiency (ACTH deficiency).
(Reproduced with permission from ref. 18.)

ACTH stimulation, however, the suppressed pituitary function will be revealed by the other tests [27] (Fig. 6.2).

6.1.2 Secondary adrenocortical insufficiency

Here deficiency of ACTH secretion is the primary event, most commonly caused by administration of excess synthetic glucocorticoids, with suppression of the hypothalamo-pituitary axis. On cessation of the treatment, pituitary activity may be suppressed for a significant length of time, and a variety of "withdrawal syndromes" may occur [28]. The inner zones of the cortex may be atrophied and cortisol levels depressed, however, the zona glomerulosa is usually unaffected and secretion of aldosterone is adequate [29]. As the recovery of ACTH secretion proceeds, basal secretion of cortisol may become normal. However, a decreased response to stress may be present for long periods of time (perhaps several months, although considerable

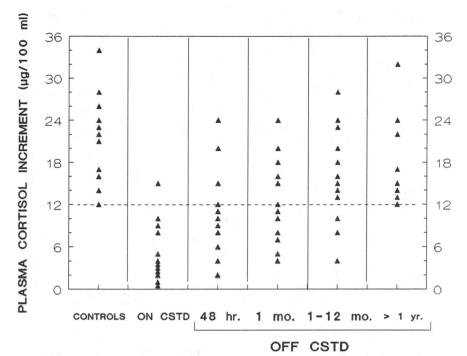

FIGURE 6.3 Recovery of the response to insulin-induced hypoglycemia in patients on corticosteroids and at various intervals following termination of therapy. CSTD = corticosteroid.
(Reproduced with permission from ref. 30.)

individual variation is found; Fig. 6.3 [30]). This needs to be taken into account if surgery is planned (see Chap. 7).

6.1.3 Mineralocorticoid deficiency

Symptoms of mineralocorticoid deficiency may also be present in those conditions where there is total destruction of the adrenal cortex, and also occasionally in familial adrenal hypoplasia (see previous text). A primary and selective deficiency in aldosterone biosynthesis can be caused either by a reduction in zona glomerulosa mass or by defects in the biosynthetic process. The former situation can occur transiently after unilateral adrenalectomy for an aldosterone producing tumor. There are also rare inherited disorders of the terminal stages of aldosterone biosynthesis which are discussed in Sect. 6.2. An acquired selective defect in aldosterone biosynthesis also has been described, and this condition may occur in patients with critical illness [31–34]. Chronic heparin treatment also has been found to lead to a deficiency of aldosterone production, possibly due to a direct action on adrenal cortex [35]. Secondary mineralocorticoid deficiency occurs when the activity of the renin-angiotensin system is reduced, which can be either as a result of a defect in production or a consequence of volume expansion. Renin deficiency is being increasingly recognized as a cause of hypoaldosteronism, and seems to be associated with chronic impairment of kidney function and diabetes in the elderly [31,36–38]. Hyperkalemia and acidosis are found, and the ability to conserve sodium is reduced. Plasma renin and aldosterone concentrations are low and unresponsive to the usual maneuvers (changes of posture and administration of diuretics). Secretion of cortisol and ACTH dependent steroids is normal. Autonomic insufficiency may also be a cause of reduced renin production [31]. Cases which may be attributed to a reduction in converting enzyme activity have also been described, possibly due to the presence of endogenous inhibitors [39]. A similar condition of hypoaldosteronism with hyperkalemia has been produced iatrogenically by treatment with the converting enzyme inhibitor, captopril [40].

Pseudohypoaldosteronism is a condition in which the symptoms of hypoaldosteronism are caused by end-organ resistance to the actions of aldosterone, and in which, paradoxically, plasma aldosterone levels are elevated. This condition seems to occur familially and is suggested when administered mineralocorticoids fail to cause sodium retention [31,41]. The defect does not seem to be restricted to the renal tubule and may appear at all sites of sodium exchange, including salivary glands, sweat glands, and colon [42]. There is evidence that the target-organ unresponsiveness to mineralocorticoids is associated with a Type I receptor defect, and the inheritance is currently being investigated [43].

Deficiency of aldosterone production may occur when excess amounts of mineralocorticoid hormones are administered, and also with the chronic ingestion of some mineralocorticoidlike substances. In these cases although endogenous aldosterone production is low, due to suppression of the renin-angiotensin system, the symptoms shown are those of mineralocorticoid excess: hypokalemia, acidosis, sodium retention, volume expansion, and hypertension. This condition can be produced by excess consumption of licorice and certain chewing tobaccos. The agents responsible appear to be glycyrrhizinic acid and its derivatives, which have been shown to cause sodium retention and increased potassium excretion, possibly by interacting with kidney mineralocorticoid receptors. There may, however, be other explanations for the phenomenon ([44,45] and see Chap. 4).

6.2 Inborn errors in corticosteroid biosynthesis: Congenital adrenal hyperplasia (CAH)

Inborn deficiencies of the enzyme systems involved in the biosynthesis of adrenocortical steroids can lead to impaired production of the end products of secretion, cortisol and/or aldosterone and corticosterone. As a consequence, secretion of ACTH may proceed unchecked by the feedback inhibition exerted by circulating steroids. This in turn leads to hyperplasia of the adrenal cortex and excessive production of inappropriate secretory products,

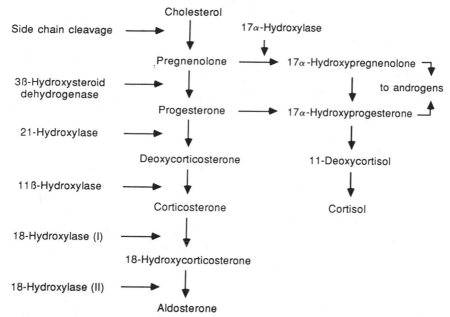

FIGURE 6.4 Summary of sites of inherited defects in corticosteroid biosynthesis in the human adrenal cortex (cf. Chap. 2).

the nature of which leads to diagnosis of the precise defect (Fig. 6.4). Clearly, these result in a variety of pathological conditions, usually corrected by hormone replacement therapy (see Chap. 7).

Histologically, the picture presented by congenital adrenal hyperplasia (CAH) is what might be predicted from our knowledge of the actions of excessive ACTH secretion [17]. By far the best studied form of this disease is the 21-hydroxylation defect where the cortex is broadened and consists largely of "compact cells" (i.e., reticularis) type as identified by light microscopy and the columns of these cells extend out from the medulla to reach the zona glomerulosa. Only a thin layer of the "clear" (zona fasciculata) cells may occur, representing the residue of a normal fasciculata. The zona glomerulosa too is hyperplastic and this is consistent with the usual finding that aldosterone as well as glucocorticoid secretion is diminished, although this is not always the case [17]. Neville and O'Hare also write that hyperplasia of the zona glomerulosa would not be expected in cases of 11β-hydroxylase deficiency since (as noted) there is no salt loss in this case [17]. However, the histological evidence does not seem to be available. Examples of other forms of congenital adrenal hyperplasia are so rare that general statements about the histological appearance of the gland are difficult to make. One exception, however, is in the case of cholesterol side-chain cleavage deficiency in which the cells have an extremely lipid laden appearance, including what may even be foci of cholesterol crystals [17].

6.2.1 21-Hydroxylase deficiency

This is by far the most common form of inherited impairment of corticosteroid biosynthesis, accounting for 90 to 95 percent of cases of congenital adrenal hyperplasia [46]. The incidence lies between one in 5000 to one in 15,000 in white populations but is reported to be much higher in certain groups, for example, an incidence of one in 700 is reported in the Yupik Eskimos of Alaska [47,48]. Impairment of 21-hydroxylation results (for the reasons stated and see Chap. 2) in impaired cortisol production and consequently in over-production of 17α-hydroxyprogesterone and 17α-hydroxypregnenolone and their precursors, pregnenolone and progesterone (Fig. 6.4). The availability of unusually high concentrations of the 17α-hydroxylated substrates for 17–20 lyase activity results in a heightened output of C_{19} steroids particularly dehydroepiandrosterone (DHEA) and androstenedione [49]. Androstenedione may be converted peripherally to testosterone, and consequently the 21-hydroxylase defect can result in virilization, to the extent that a newborn female may be incorrectly identified as male because of the masculinized genitalia. In both sexes there may be rapid somatic growth, early maturation of the epiphyses and short stature and continued excessive growth of the penis or clitoris [49,50]. The zona glomerulosa is variably affected, suggesting different forms of this enzyme in the different sites, but aldosterone biosynthesis is impaired in two thirds of the cases, and the disease may consequently result in hyponatraemia and hyperkalemia [49,50]. Consequently, this disease is also called salt-losing or salt-wasting CAH. A milder, nonclassic form of 21-hydroxylase deficiency exists in which subjects acquire symptoms of excessive androgen production during puberty. This disease, occurs in 0.3 percent of the white population and 3 percent of Jews of European origin. Maria New and her colleagues (who have studied this problem extensively) have constructed a nomogram which relates baseline to ACTH stimulated serum concentrations of circulating 17α-hydroxyprogesterone [50] and the figure is reproduced (Fig. 6.5). Cases of classic adrenal hyperplasia and the acquired adrenal hyperplastic conditions are clearly clustered in this figure, as are the heterozygotes and the normal population.

The same group has done extensive work on the genetics of inheritance of the disease using recombinant DNA techniques. Genes coding for 21-hydroxylase cytochrome P450 are found on the shorter arm of chromosome 6 where they are associated with the genes coding for the HLA major histocompatibility complex [51] (Fig. 6.6). One consequence of this degree of linkage is that siblings who both have 21-hydroxylase deficiency usually have identical HLA type, and therefore HLA typing can be used to determine the carrier status of any member of the family. Two genes code for 21-hydroxylase but from the studies of the variants shown in a number of patients, it appears that only one of these, designated OH21B, codes for the active cytochrome P450 involved in adrenal steroidogenesis, since homozygosity for a OH21A gene variant may occur in hormonally normal individuals. The two 21-hydroxylase genes are about 90 percent homologous

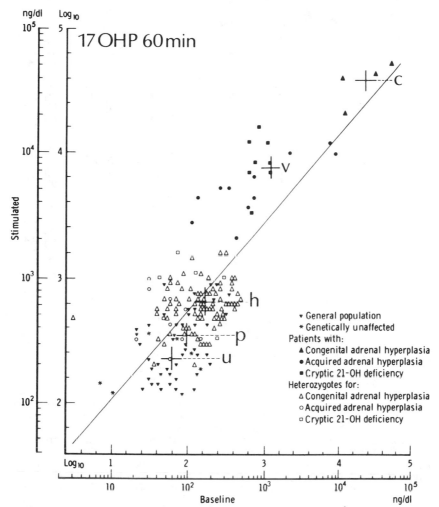

FIGURE 6.5 21-Hydroxylase deficiency: nomogram relating base line to 60-minute ACTH stimulated serum concentrations of 17α-hydroxyprogesterone. Means for these groups are shown with a large cross: C = classic 21-hydroxylase deficiency, V = variant (nonclassic type), P = general population, U = unaffected family members. The regression line is for all data points. (Reproduced with permission from ref. 49.)

with each other and some results of the study of individuals of varying genotype using genomic blot hybridization are shown in Figure 6.6, [52]. Cloned OH21A or OH21B genes will give, as expected, a single band hybridizing with the radioactive probe. A normal subject will give two bands corresponding to the A and the B genes and the absence of the A gene, clearly (as mentioned) confers no disadvantage. Absence of the B gene, however, gives the classical 21-hydroxylase deficiency disease. The heterozygote shows an unequal hybridization pattern since this subject will have two OH21A genes but only one OH21B. Interestingly, the nonclassical condition

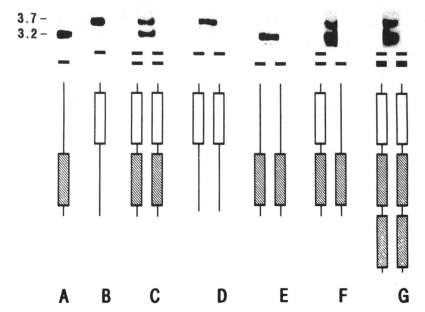

FIGURE 6.6 Analysis of 21-hydroxylase deficiency by genomic blot hybridization. Samples of cloned and uncloned human DNA were digested with Taq I endonuclease, subjected to agarose gel electrophoresis and hybridized with a radioactive probe encoding $P450_{21}$. The 3.7 kilobase pair fragment corresponds to the OH21B gene (open bars) and the 3.2 kilobase pair fragment with the 21OHA gene (hatched bar). The OH21 genes are interspersed with the C4 genes encoding the fourth component of serum complement which are not shown here. The fragment contributed by each OH21 gene is shown diagramatically. A shows the cloned OH21A gene, and B the OH21B gene, while C shows DNA from a normal subject. D shows the pattern obtained when the 21OHA gene is deleted: the absence of this pseudogene has no clinical effect. E shows the hybridization pattern in a classically affected patient, when the 21OHB gene is deleted. F shows a heterozygote for 21OHB deletion. G shows the pattern for a patient with nonclassic 21-hydroxylase deficiency, who has four A genes and two B genes. The nondeleted B gene presumably has a small mutation not detectable with these techniques. (Reproduced with permission from ref. 52.)

also shows the same hybridization pattern, apparently because of a duplication of the A gene, although this is merely a nonactive marker for the disease, which arises from a mutation in the B gene not distinguishable by these methods.

6.2.2 11β-Hydroxylase deficiency

In the classical condition, this accounts for 5 to 8 percent of congenital adrenal hyperplasia cases [53–55]. As a consequence of impaired 11β-hydroxylase activity, production of cortisol, corticosterone, and aldosterone are impaired, giving greatly increased circulating levels of 11-deoxycortisol and 11-deoxycorticosterone (Fig. 6.4). In contrast to the 21-hydroxylase deficiency syndrome, in 11β-hydroxylase deficiency there is no "salt losing" effect since, because of the mineralocorticoid activity of deoxycorticosterone, salt is retained. These subjects, therefore, develop a moderate hypertension which may become severe on occasions [53,56]. There is a good deal of heterogeneity in this group, and, in some patients, aldosterone biosynthesis is not affected [56]. In addition, a mild, ("nonclassical"), form of 11β-hydroxylase deficiency is also known [52]. The gene for cytochrome $P450_{11\beta}$ is present in the human genome in a single copy on the long arm of chromosome 8. Although Southern blot studies have been performed, so far no patients with 11β-hydroxylase deficiency have been shown to have deletions or rearrangements of the gene. Consequently the nature of the genetic disorder is not yet clear [52].

6.2.3 17α-Hydroxylase deficiency

This is a rare disorder arising from impaired 17-hydroxylation and impaired 17–20 lyase activity which occurs both in adrenals and gonads. A consequence is that males may show incomplete masculinization and females primary amenorrhea. There is an overproduction of 17-deoxysteroids, especially corticosterone, with hypertension and hyperkalemia resulting from excess deoxycorticosterone (Fig. 6.4). Aldosterone secretion remains low, apparently because of low renin activity in plasma [57,58].

The molecular biology of the $P450_{17\alpha}$ gene has been reported quite recently [59]. In the human genome, a single copy of the gene exists, and transfection studies confirm that both 17α-hydroxylase and 17,20 lyase activities are catalyzed by the gene product. The $P450_{17\alpha}$ gene appears to be the only member of a distinct family within the P450 supergene family (see Chap. 2). In studies on a patient with a mutant form, it was found that the normal and mutant genes were identical, except for a four base duplication in the mutant gene, leading to a reading-frame shift affecting the C-terminal portion of the protein, and possibly, as a consequence, impaired interaction with P450 reductase.

6.2.4 Cholesterol side-chain cleavage deficiency

This is also a very rare disease resulting presumably, in impairment of biosynthesis of all steroid hormones (see Chap. 2). In the classic case, described by Prader and Gurtner [60,61] a male pseudohermaphrodite with severe salt wasting and impaired synthesis of all adrenal steroids was described. At autopsy the subject was found to have female external genitalia and male genital duct. The adrenals were filled with lipoidal material containing cholesterol and cholesterol esters: this condition is, therefore, also known as lipoid adrenal hyperplasia [17,52].

6.2.5 18-Hydroxylase deficiency

Incomplete understanding of the terminal sequences involved in the aldosterone synthetic pathway (see Chap. 2) have led to some problems

TABLE 6.3 Clinical and laboratory features of various forms of adrenal disorders of steroidogenesis

Clinical features						Urinary excretion	
Newborn with sexual ambiguity							
Female	Male	Salt-wasting	Hyper-tension	Postnatal virilization	Enzyme deficiency	17-KS	17-OHCS
+	0	0	0	+	21-hydroxylase simple virilizing	↑↑	Normal or ↓
+	0	+	0	+	salt-wasting	↑↑	↓
+	0	0	+	+	11β-hydroxylase	↑↑	↑↑*
+	+	+	0	+	3β-OHSDH**	↑†	↓↓
0	+	0	+	0	17α-hydroxylase	↓↓	↓↓
0	+	+	0	0	Cholesterol side-chain cleavage	↓↓	↓↓
0	0	+	0	0	18-hydroxylase (methyl oxidase type I)	Normal	Normal
0	0	+	0	0	18-hydroxylase (methyl oxidase type II)	Normal	↑‡
?	+	—	—	+§	17β-hydroxysteroid dehydrogenase●	Normal or ↑	Normal
?	+	—	—	+§	C17,20-lyase	↓	Normal

* Mostly THS.
** Values presented apply to the infant and very young child.
† Mostly Δ^5-17 ketosteroids.
‡ Largely 18-hydroxy-11-dehydrocorticosterone.
§ Only in males at puberty.
Table reproduced with permission from ref. 65.

with the nomenclature of these conditions. In some subjects, 18-hydroxycorticosterone and aldosterone production is impaired, and the disease was, therefore, originally attributed to simple 18-hydroxylase deficiency (see Fig. 6.4). These very rare cases present in infancy with hyponatraemia and hyperkalemia, dehydration, poor weight-gain and intermittent fever. Another rare condition is characterized by deficient aldosterone production combined with normal (or possibly higher than normal) 18-hydroxycorticosterone. This was originally considered to be a dehydrogenase deficiency. However, in view of the presently held concepts of aldosterone biosynthesis

TABLE 6.3

Laboratory findings						
Urinary excretion		**Circulating plasma hormones**				
P'triol	Aldo	17-OHP	Δ4	DHA	T	Renin
↑↑	Normal	↑↑	↑↑	Normal or (DHA/Δ4 ↓)	↑	Normal or ↑
↑↑	↓	↑↑	↑↑	Normal or ↑	↑	↑↑
↑	↓	↑	↑↑	↑	↑	↓↓
Normal or ↓	↓	Normal or ↑	Normal or ↑	↑↑↑	‡‡	↑
↓↓	↓	↓	↓	↓	↓	↓
↓↓	↓↓	↓	↓	↓	↓	↑
Normal	↓↓	Normal	Normal	Normal	Normal	↑
Normal	↓↓	Normal	Normal	Normal	Normal	↑
Normal	Normal	Normal	↑↑	Normal or ↑	Normal or ↓ (Δ4/T ↑↑)	Normal
Normal or ↑	Normal	Normal or ↑	↓↓	Normal or ↓	↓↓	Normal

● This defect may occur only in the gonad.
‡‡ Normal or ↓ in males; normal or ↑ in females.
17-KS: 17-ketosteroids; 17-OHCS: 17-hydroxycorticosterolds. 17-OHP:
17-hydroxyprogesterone, P'triol: pregnanetriol; Aldo: aldosterone; Δ4: 4-androstenedione;
T: testosterone.

(see Chap. 2), these two conditions have been termed corticosterone methyl oxidase type I and type II deficiencies [62].

6.2.6 3β-Hydroxysteroid dehydrogenase deficiency

This condition resulting in impairment of both gonadal and adrenal steroid production is associated with excessive urinary excretion of DHEA and pregnanetriol (a metabolite of pregnenolone). There is incomplete masculinization in the male, salt-wasting symptoms, secondary to aldosterone deficiency, are associated with this condition [63,64].

Table 6.3 summarizes the effects of deficiencies in adrenal steroidogenesis.

References

1. Irvine, W.J., Stewart, A., and Scarth, L. 1967. A clinical and immunological study of adrenocortical insufficiency (Addison's disease. *Clin. Exp. Immunol.* 2:31–70.

2. Baxter, J.D., and Tyrrell, J.B. 1987. The Adrenal Cortex. In: P. Felig, J.D. Baxter, A.E. Broadus and L.A. Frohman, eds. *Endocrinology and Metabolism.* New York: McGraw Hill, pp. 511–650.

3. Neville, A.M., and O'Hare, M.J. 1985. Histopathology of the human adrenal cortex. In: G.M. Besser and L.H. Rees, eds. *The Pituitary Adrenocortical Axis,* Clinics in Endocrinology and Metabolism 14 (4). London: W.B. Saunders, pp. 791–820.

4. Irvine, W.J., Toft, A.D., and Feek, C.M. 1979. Addison's disease. In: V.H.T. James, ed. *The Adrenal Gland.* New York: Raven Press, pp. 131–64.

5. Neufeld, M., McLaren, N.K., and Blizzard, 1981. Two types of autoimmune Addison's disease associated with different polyglandular autoimmune (PGA) syndromes. *Medicine.* 60:355–62.

6. Anderson, J.R., Goudie, R.B., Gray, K.G., and Timbury, G.C. 1957. Adrenal autoantibodies in Addison's disease. *Lancet.* I:1123–24.

7. Blizzard, R.M., Chandler, R.W., Kyle, M.A., and Hung, W. 1962. Adrenal antibodies in Addison's disease. *Lancet.* II:901–03.

8. Ketchum, C.H., Riley, W.J., and MacLaren, N.K. 1984. Adrenal dysfunction in asymptomatic patients with adrenocortical autoantibodies. *J. Clin. Endocrinol. Metab.* 58:1166–70.

9. Khoury, E.L., Hammond, L., Gottazzo, G.F., and Donniach, D. 1981. Surface reactive antibodies to human adrenal cells in Addison's disease. *Clin. Exp. Immunol.* 45:48–55.

10. Bright, G.M., and Singh, I. 1990. Adrenal autoantibodies bind to adrenal subcellular fractions enriched in cytochrome-c reductase and 5'-nucleotidase. *J. Clin. Endocrinol. Metab.* 70:95–99.

11. Doniach, D., and Bottazzo, G.F. 1981. Polyendocrine autoimmunity. In: E.C. Franklin, ed. *Clinical Immunology Update.* New York: Elsevier, pp. 95–121.

12. Kendall-Taylor, P., Lambert, A., Mitchell, R., and Robertson, W.R. 1988. Antibody that blocks stimulation of cortisol secretion by adrenocorticotrophic hormone in Addison's disease. *Brit. Med. J.* 296:1489–91.

13. Wulffraat, N.M., Drexhage, H.A., Bottazzo, G.-F., Wiersinga, W.M., Jeucken, P., and Van der Gaag, R. 1989. Immunoglobulins of patients with idiopathic Addison's disease block the in vitro action of adrenocorticotropin. *J. Clin. Endocrinol. Metab.* 69:231–38.

14. Eisenbarth, G.S., Wilson, P.W., Ward, F., Buckley, C., and Lebovitz, H. 1979. The polyglandular failure syndrome: disease inheritance, HLA type and immune function. *Ann. Int. Med.* 91:528–33.

15. Maclaren, N.K., and Riley, W.J. 1986. Inherited susceptibility to autoimmune Addison's disease is linked to human leukocyte antigens-DR3 and/or DR4,

except when associated with type I autoimmune polyglandular syndrome. *J. Clin. Endocrinol. Metab.* 62:455–59.

16. Vita, J.A., Silverberg, S.J., Goland, R.S., Austin, J.H.M., and Knowlton, A.I. 1985. Clinical clues to the cause of Addison's disease. *Am. J. Med.* 78:461–66.

17. Neville, A.M., and O'Hare, M.J. 1982. *The human adrenal cortex.* Berlin: Springer-Verlag.

18. Burke, C.W. 1985. Adrenocortical insufficiency. In: G.M. Besser and L.H. Rees, eds. *The Pituitary Adrenocortical Axis*, Clinics in Endocrinology and Metabolism 14 (4). London: W.B. Saunders, pp. 947–76.

19. Guenthner, E.E., Rabinowe, S.L., Van Niel, A., Naftilan, A., and Dluhy, R. 1984. Primary Addison's disease in a patient with acquired immunodeficiency syndrome. *Ann. Int. Med.* 100:847–48.

20. Welch, K., Finkbeiner, W., Alpers, C.E., Blumenfeld, W., Davis, R.L., Smuckler, E.A., and Beckstead, J.H. 1984. Autopsy findings in the acquired immune deficiency syndrome. *J. Am. Med. Ass.* 252:1152–59.

21. Membreno, L., Irony, I., Dere, W., Klein, R., Biglieri, E.G. and Cobb, E. 1987. Adrenocortical function in acquired immunodeficiency syndrome. *J. Clin. Endocrinol. Metab.* 65:482–87.

22. Migeon, C.J., Kenny, F.M., and Kowarski, A. 1968. The syndrome of congenital adrenocortical unresponsiveness to ACTH. Report of six cases. *Pediatric Res.* 2:501–13.

23. Geffner, M.E., Lippe, B.M., Kaplan, S.A., Berquist, W.E., Bateman, J.B., Paterno, V.I., and Seegan, R. 1983. Selective ACTH insensitivity, achalasia and alacrima: A multisystem disorder presenting in childhood. *Pediatric Res.* 17:532–36.

24. Smith, E.M., Brosnan, P., Meyer, W.J. III, and Blalock, J.E. 1987. An ACTH receptor on human mononuclear leukocytes. Relation to adrenal activity. *New Engl. J. Med.* 317:1266–69.

25. O'Neill, B.P., and Moser, H.W. 1982. Adrenoleukodystrophy. *Canad. J. Neurol. Sci.* 9:449–52.

26. Irvine, W.J., and Barnes, E.W. 1972. Adrenocortical insufficiency. *Clinics in Endocrinology. Metab.* 1:549–94.

27. Tyrell, J.B., and Forsham, P.H. 1986. Glucocorticoids and adrenal androgens. In: F.S. Greenspn and P.H. Forsham, eds. *Basic and Clinical Endocrinology.* Los Altos, California: Lange Medical Publications, pp. 272–309.

28. Christy, N.P. 1971. Iatrogenic Cushing's syndrome. In: N.P. Christy, ed. *The Human Adrenal Cortex.* New York: Harper Row, pp. 395–425.

29. Tyrrell, J.B., and Baxter, J.D. 1987. Glucocorticoid therapy. In: P. Felig, J.D. Baxter, A.E. Broadus and L.A. Frohman, eds. *Endocrinology and Metabolism.* New York: McGraw Hill, pp. 788–817.

30. Livanou, T., Ferriman, D., and James, V.H.T. 1967. Recovery of hypothalamo-pituitary-adrenal function after corticosteroid therapy. *Lancet.* II:856–59.

31. Veldhuis, J.D., and Melby, J.C. 1981. Isolated aldosterone deficiency in man: Acquired and inborn errors in the biosynthesis or action of aldosterone. *Endocrine Rev.* 2:495–517.

32. Melby, J.C. 1981. Diagnosis and treatment of primary and isolated hypoaldosteronism. In: G.M. Besser and L.H. Rees, eds. *The Pituitary Adrenocortical Axis*, Clinics in Endocrinology and Metabolism 14 (4). London: W.B. Saunders, pp. 977–94.

33. Kokko, J.P. 1985. Primary acquired hypoaldosteronism. *Kidney Int.* 27:690–702.

34. Zipser, R.D., Davenport, M.W., Martin, K.L., Tuck, M.L., Warner, N.L., Swinney, R.L., Davis, C.L., and Horton, R. 1981. Hyperreninemic hypoaldosteronism in the critically ill: A new entity. *J. Clin. Endocrinol. Metab.* 53:867–73.

35. O'Kelly, R., Magee, F., and McKenna, T.J. 1983. Routine heparin therapy inhibits adrenal aldosterone production. *J. Clin. Endocrinol. Metab.* 56:108–12.

36. Hudson, J.B., Chobanian, A.V., and Relman, A.S. 1957. Hypoaldosteronism: a clinical study of a patient with an isolated mineralocorticoid deficiency, resulting in hyperkalemia and Stoke-Adams attacks. *New Engl. J. Med.* 257:529–36.

37. Phelps, K.R., Lieberman, R.L., Oh, M.S., and Carroll, H.J. 1980. Pathophysiology of the syndrome of hyporeninemic hypoaldosteronism. *Metabolism.* 29:186–99.

38. DeFronzo, R.A. 1980. Hyperkalemia and hyporeninemic hypoaldosteronism. *Kidney Int.* 17:118–34.

39. Findling, J.W., Adams, A.H., and Raff, H. 1987. Selective hypoaldosteronism due to an endogenous impairment in angiotensin II production. *New Engl. J. Med.* 316:1632–35.

40. Warren, S.E., and O'Connor, D.T. 1980. Hyperkalemia resulting from captopril administration. *J. Am. Med. Ass.* 244:2251–52.

41. Cheek, D.B., and Perry, J.W. 1958. A salt wasting syndrome in infancy. *Arch. Dis. Child.* 33:252–56.

42. Oberfield, S.E., Levine, L.S., Carey, R.M., Bejar, R., and New, M.I. 1979. Pseudohypoaldosteronism: multiple target organ responsiveness to mineralocorticoid hormones. *J. Clin. Endocrinol. Metab.* 48:228–34.

43. Armanini, D., Kuhnle, U., Srasser, T., Dorr, H., Butenandt, I., Weber, P., Stockigt, J.R., Pearce, P., and Funder, J.W. 1985. Aldosterone receptor deficiency in pseudohypoaldosteronism. *New Engl. J. Med.* 313:1178–81.

44. Mantero, F. 1981. Exogenous mineralocorticoid-like disorders. In: E.G. Biglieri and M. Schambelan, eds. *Endocrine Hypertension*, Clinics in Endocrinology and Metabolism 10 (3). London: W.B. Saunders, pp. 465–79.

45. Stewart, P.M., Wallace, A.M., Valentino, R., Burt, D., Shackleton, C.L., and Edwards, C.R.W. 1987. Mineralocorticoid activity of liquorice: 11-beta-hydroxysteroid dehydrogenase deficiency comes of age. *Lancet.* II:821–24.

46. New, M.I., Dupont, B., Grumback, K., and Levine, L.S. 1983. Congenital adrenal hyperplasia and related conditions. In: J.B. Stanbury, J.B. Wyngarden, D.S. Frederickson, J.L. Goldstein, and M.S. Brown, eds. *The Metabolic Basis of Inherited Disease.* New York: McGraw Hill, pp. 973–1000.

47. Muller, W., Prader, A., Kofler, J., Glatzl, J., and Geir, W. 1979. Zur haufigkeit des kongenitalen syndroms (AGS). *Paediatr. Paedol.* 14:151–55.

48. Pang, S., Murphey, W., Levine, L.S., Spence, D.A., Leon, A., LaFranchi, S., Surve, A.S., and New, M.I. 1982. A pilot newborn screening for congenital adrenal hyperplasia in Alaska. *J. Clin. Endocr.* 55:413–20.

49. White, P.C., New, M.I., and Dupont, B. 1987. Congenital adrenal hyperplasia. *New England J. Med.* 316:1519–24.

50. Mulaikal, R.M., Migeon, C.J., and Rock, J.A. 1987. Fertility rates in female patients with congenital adrenal hyperplasia due to 21-hydroxylase deficiency. *New England J. Med.* 316:178–82.

51. New, M.I., and Speiser, P.W. 1982. Genetics of adrenal steroid 21-hydroxylase deficiency. *Endocrine Rev.* 7:331–49.

52. White, P.C., New, M.I., and Dupont, B. 1987. Congenital adrenal hyperplasia. *New England J. Med.* 316:1580–86.

53. Zachmann, M., Tassinari, D., and Prader, A. 1983. Clinical and biochemical variability of congenital adrenal hyperplasia due to 11β-hydroxylase deficiency. A study of 25 cases. *J. Clin. Endocr.* 56:222–29.

54. Drucker, S., and New, M.I. 1987. Disorders of adrenal steroidogenesis. *Ped. Adolescent Endocr.* 34:1055–66.

55. Eberlein, W.R., and Bongiovanni, A.M. 1956. Plasma and urinary corticosteroids in the hypertensive form of congenital adrenal hyperplasia. *J. Biol. Chem.* 223:85.

56. Levine, L.S., Rauh, W., Gottesdiener, K., Chow, D., Gunczler, P., Rapaport, R., Pang, S., Schneider, B., and New, M.I. 1980. New studies on the 11β-hydroxylase and 18-hydroxylase enzymes in the hypertensive form of congenital adrenal hyperplasia. *J. Clin. Endocr.* 50:258–63.

57. Biglieri, E.G., Herron, M.A., and Brust, N. 1966. 17-hydroxylation deficiency in man. *J. Clin. Invest.* 45:1946 54.

58. Mantero, F., and Scaroni, C. 1984. Enzymatic defects of steroidogenesis. *Ped. Adolescent Endocr.* 13:83–94.

59. Kagimoto, M., Winter, J.S.D., Kagimoto, K., Simpson, E.R., and Waterman, M.R. 1988. Structural characterization of normal and mutant human steroid 17α-hydroxylase genes: Molecular basis of one example of combined 17α-hydroxylase/17,20 lyase deficiency. *Mol. Endocr.* 2:564–70.

60. Prader, A., and Gurtner, H.P. 1955. Das syndrom des pseudohermaphroditismus masculinus bei kongenitalen der nebennierenrinden hyperplasie ohne androgen-uberproduktion (adrenaler pseudohermaphroditimismus masculinus). *Helv. Paed. Acta.* 10:397.

61. Prader, A., and Anders, G.J.P.A. 1962. Zur genetik der kongenitalen lipoidhyperplasie der nebennieren. *Helv. Paed. Acta,* 17:285.

62. Ulick, S. 1976. Diagnosis and nomenclature of the disorders of the terminal portion of the aldosterone biosynthetic pathway. *J. Clin. Endocr.* 43:92–96.

63. Lobo, R.A., and Goebelsman, U. 1981. Evidence for reduced 3β-ol-hydroxysteroid dehydrogenase activity in some hirsute women thought to have polycystic ovary syndrome. *J. Clin. Endocr.* 53:394–400.

64. Pang, S., Lerner, A.J., Stoner, A., Levine, L.S., Oberfield, S.E., Engel, I., and New, M.I. 1985. Late-onset adrenal steroid 3β-hydroxysteroid dehydrogenase deficiency. I. A cause of hirsutism in pubertal and postpubertal women. *J. Clin. Endocr.* 60:428–39.

65. New, M.I. 1985. In: H.J.A. Makin, ed. *Biochemistry of steroid hormones.* Oxford, Blackwells, pp. 606–07.

7

Pharmacological Uses of Corticosteroids

The therapeutic uses of corticosteroids can be divided into two broad categories: replacement therapy in disorders of the pituitary-adrenal axis; and immunosuppression/anti-inflammatory therapy in nonadrenal disorders. Steroid replacement therapy is indicated when the adrenals are not producing adequate amounts of corticosteroid, and may include replacement of mineralocorticoid as well as glucocorticoid hormones. The glucocorticoids are the most powerful anti-inflammatory drugs available, acting on several aspects of the inflammatory response (see Chap. 4) and also having a direct action on the immune system, where they are potent immunosuppressants.

Glucocorticoids have been widely used, both systemically and topically, for a variety of disorders, and are particularly useful in the treatment of certain life-threatening conditions. There are, however, many disadvantages of using steroids therapeutically, including pituitary-adrenal suppression and a range of other side effects, as a result of which glucocorticoids may be only the third or fourth choice of drug in most chronic allergic and inflammatory diseases. However, they remain the first-line treatment for many serious conditions.

The relative ease with which steroids may be synthesized has resulted in the production of a range of both synthetic and naturally occurring glucocorticoids (Fig. 7.1). The relative potencies of these steroids is given in Table 7.1, together with some indication of the forms in which they are available for therapeutic use. Structural modifications have allowed some alteration of the ratio of sodium-retaining effect to anti-inflammatory effect

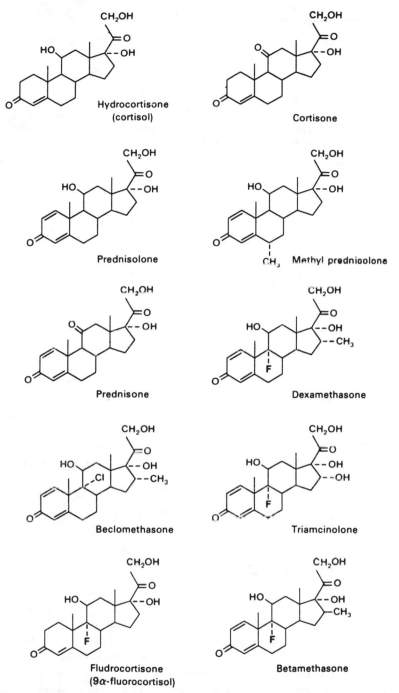

FIGURE 7.1 Chemical structures of the major corticosteroids in current therapeutic use.

TABLE 7.1 Potencies of the major corticosteroids in clinical use using hydrocortisone as a standard

Steroid	Relative affinity for glucocortical receptor[a]	Relative potency in clinical use		Duration of action after oral dose	Preparations available
		Anti-inflammatory	Sodium-retaining		
Hydrocortisone (Cortisol)	1	1	1	S	O I T
Cortisone	0.01	0.8	0.8	S	O I T
Fludrocortisone	3.5	10–15	125–150	S	O
Prednisolone	2.2	4	0.8	I	O I T
Prednisone	0.05	4	0.8	I	O
Methylprednisolone	11.9	5	0.5	I	O I T
Triamcinolone	1.9	5	0	I	O I T
Dexamethasone	7.9	25–30	0	L	O I T
Betamethasone	5.4	25–30	0	L	O I T
Beclomethasone	—	+	—	—	Inhaled

[a] Human foetal lung cells
Duration of action: S, $t_{\frac{1}{2}}$ = 8–12 h; I, $t_{\frac{1}{2}}$ = 12–36 h; L, $t_{\frac{1}{2}}$ = 36–72 h
Preparations available: O = oral, I = injectable, T = topical

Data from refs. 1 and 2.

(Table 7.1), although to date it has not been possible to isolate the anti-inflammatory action from other glucocorticoid effects which may be undesirable.

The naturally occurring steroid, cortisol, is widely used therapeutically, and in this context is generally called hydrocortisone.

7.1 Steroid replacement therapy

It is necessary to administer steroids when the adrenals are not producing adequate amounts. This hypofunction may be caused by a primary adrenal lesion or a pituitary defect (see Chap. 5). When the pituitary is the site of the lesion, it is usually only necessary to give glucocorticoid replacement since aldosterone secretion is not affected. With primary adrenal insufficiency there may be a total failure of all adrenal steroid secretion, or alternatively only one secretory product may be affected, and the replacement therapy must be arranged appropriately. Insufficient adrenal steroid secretion may appear as either chronic adrenal insufficiency or as acute adrenal crisis.

7.1.1 Acute adrenal insufficiency (Addisonian crisis)

This condition frequently results from the abrupt withdrawal of steroid therapy for a nonadrenal disorder, but is also caused by other structural or functional disturbances of the adrenal (see Chap. 6). It may also be seen in patients with untreated chronic adrenal insufficiency or on replacement therapy, where an acute Addisonian crisis is precipitated by a "stressful" incident causing an increased requirement for corticosteroid. Acute adrenal insufficiency is rarely of pituitary origin.

Characteristics of this disorder are hypotension, dehydration, weakness, vomiting, fever, and shock [3]. It is a life-threatening condition and must be treated promptly with hydrocortisone, and correction of any water and electrolyte imbalance. Initially, hydrocortisone is given as an intravenous injection of 100 mg, followed by 100 mg every 6 to 8 hours until normal oral fluid intake is restored [4]. The dose can be reduced to 25 mg given intramuscularly every 6 to 8 hours during recovery and cessation of intravenous fluid administration.

Where acute adrenal insufficiency is unconfirmed, 4 mg dexamethasone is given instead of hydrocortisone, and an adrenal function test is performed, in which 5 units per hour ACTH is infused and adrenal cortisol secretion monitored [2]. The dexamethasone provides adequate steroid replacement which is essential where adrenal insufficiency is suspected, and does not interfere with the ACTH stimulation test of adrenal function (see Chap. 6).

7.1.2 Chronic adrenal insufficiency

In chronic adrenal insufficiency, daily steroid replacement is necessary. The replacement steroid is given orally, usually as an enteric-coated tablet, or at mealtimes to minimize gastrointestinal disturbances. The dose required is best determined from the body surface area of the patient: 12 to 15 mg hydrocortisone (or equivalent) per square meter body surface area per day [5]. The dose regime is usually arranged with a large dose (two thirds of the daily dose) on awakening in the morning, and a smaller dose later in the day, mimicking the normal diurnal rhythm of cortisol secretion.

In the case of primary adrenal insufficiency (Addison's disease), mineralo-corticoid replacement is usually required, and may be achieved by the administration of 0.1 to 0.2 mg fludrocortisone acetate daily, although this is rather empirical and guided by the patient's sense of well being. It may be appropriate to check the patient's plasma potassium levels to determine whether the mineralocorticoid replacement is adequate [6]. Where the adrenal insufficiency is of pituitary origin, mineralocorticoid replacement is usually unnecessary as aldosterone secretion approximates to the normal range (see Chap. 6).

In all patients with adrenal insufficiency additional steroid cover is provided during times of physiological or psychological stress. When surgery is planned, 100 mg hydrocortisone is usually given the night before, and this dose is repeated at 6 AM on the day of surgery, with a further 100 mg hydrocortisone prior to induction of anaesthesia [7]. For lesser trauma, the daily steroid dose may simply be doubled to provide adequate cover [5].

7.2 Steroids in nonadrenal disorders

Corticosteroids have a profound effect on the inflammatory process in a range of different tissues (see Chap. 4), and so are effective in the treatment of inflammatory diseases. While corticosteroids provide rapid symptomatic relief, they are not usually curative and the underlying process in chronic progressive diseases is still present even while its manifestations are suppressed by the steroid treatment. During long term steroid therapy, there may be some degree of acclimatization to the steroid used, such that increasing doses are required to achieve the same level of control of the symptoms, resulting in a higher incidence or greater severity of side effects. It is also exceptionally difficult to withdraw patients from long-term steroid therapy.

Taking these factors into account, together with the development of nonsteroidal anti-inflammatory drugs, steroids are no longer considered to be the first-line drugs of choice for the systemic treatment of most chronic inflammatory or allergic conditions, and are generally used only when these conditions have proved resistant to other forms of treatment. Glucocorticoids are still widely used, however, to treat acute life-threatening inflammatory disorders, and also where topical application reduces the systemic side effects. Clearly the benefits from glucocorticoid therapy must be weighed carefully against the disadvantages of this form of treatment. The major clinical considerations prior to the use of glucocorticoids as pharmacologic agents are presented in Table 7.2.

7.2.1 Dose regimes

The main aim, when deciding on a dose regime, is to use the lowest possible dose which gives effective control of the symptoms, in order to minimize both the frequency and severity of side effects. Steroid dose is determined from both the amount and the relative potency of the steroid given (see Table 7.1). Where possible, local application of steroid is preferable to systemic therapy. In disorders where systemic steroid therapy is necessary, as in the treatment of life-threatening disorders, a dose of steroid, large enough to control the disease, is initially given. At first, oral steroids may be given in divided doses (typically around 1 mg prednisolone/kg/day, or equivalent), or intravenous steroids may be used. When the disease is controlled adequately on this regime, corticosteroids may be given as a single morning dose for an indefinite period until substantial improvement is seen. Steroids may then be given as alternate day therapy where possible, with a gradual reduction in dose until the minimum effective dose is

TABLE 7.2 Clinical considerations prior to the initiation of glucocorticoid therapy in nonadrenal disorders

---

A. Is glucocorticoid therapy appropriate?
1. How serious is the underlying disorder? Do the potential benefits outweigh the known hazards?
2. Have other, more conservative, therapies already been used?
3. Is there a concomitant infection (e.g., tuberculosis, *herpes simplex*) which may become serious if steroid therapy is commenced?
4. Is the patient predisposed to any of the major contraindications for glucocorticoid therapy; hypertension or cardiovascular disease, diabetes mellitus, osteoporosis, psychological disorders?

B. What form of glucocorticoid therapy is most suitable?
1. Which steroid should be used?
2. Is local application of the steroid possible?
3. How long will therapy be required?
4. Is it possible to reduce the dose of steroid or the side effects by use of a complementary form of therapy?
5. Is an alternate day dose regimen indicated?

Modified from ref. 8.

achieved, or until steroid therapy can be stopped completely [9,10]. Any dose reduction is achieved gradually, to avoid precipitating an Addisonian crisis, and adequate cover must be given during times of stress (see Sec. 7.1.2). A protocol for withdrawal of glucocorticoid therapy is given in ref. 11.

Symptomatic relief is paramount, and when dose regimes are changed, the patient must be monitored carefully, and higher doses reinstated where necessary to give continued relief of symptoms. A daily dose of 7 mg prednisolone or equivalent, is considered to be the highest dose which may be given with minimal side effects. Ideally, this dose is given on alternate days (i.e., 14 mg/2 days), as this regime allows a certain degree of recovery of pituitary function between doses, and minimizes pituitary-adrenal suppression, one of the more serious side effects of steroid therapy [12]. In all cases, however, it is desirable to employ the lowest possible dose of steroid which effectively controls the symptoms.

There has been a recent trend towards using very high doses of intravenous methylprednisolone (typically 1 g over 2 hr) either as a single treatment or repeated for up to three days, in preference to longer-term oral therapy, for a variety of disorders. The rationale for this dose regime has been the suggestion that the short duration of therapy minimizes pituitary-adrenal suppression and other side effects associated with longer term therapy. It now appears that more conservative use of steroids, notably low-dose oral steroids, is equally effective therapeutically [13]. The mechanism of action of very high doses of steroids is uncertain, although it is possible that large doses cause impaired excretion of steroid, and there is evidence to suggest that such treatment may induce the accumulation of steroids above plasma levels in some organs, particularly lymphoid tissue [14].

7.2.2 ACTH versus steroid therapy

Currently the major clinical use of ACTH is as a diagnostic agent in adrenal insufficiency (see Chap. 6). ACTH has been used to treat nonadrenal disorders, however, instead of corticosteroid therapy, the rationale being that ACTH was less likely to cause pituitary-adrenal suppression. The disadvantages of using ACTH heavily outweigh the advantages; ACTH must be given parenterally as it is inactive orally; ACTH stimulates both mineralocorticoid and glucocorticoid secretion, and so causes water and salt retention not usually associated with administration of a "pure" glucocorticoid. In the treatment of primary adrenal insufficiency, ACTH is obviously inappropriate, and there is, furthermore, no evidence to suggest that ACTH is more effective than glucocorticoid administration in the treatment of secondary adrenal insufficiency. For all these reasons, this peptide is rarely used therapeutically [2].

7.2.3 Current uses of glucocorticoids

Respiratory disease   In the treatment of acute asthmatic episodes intravenous hydrocortisone (100 mg over 8 hr) is used initially, in combination with other drugs (notably β-agonists) to reverse the status asthmaticus, followed by oral prednisolone (10 mg twice daily) for up to a week after the acute attack. Provided that remission is sustained, the oral steroid therapy may be replaced by an inhaled preparation, beclomethasone, which acts locally and is not readily absorbed into the bloodstream [15]. In the acute phase there is evidence that hydrocortisone potentiates the actions of β-agonists, in addition to its anti-inflammatory effects which reduce airway mucosal oedema [16]. The effects of corticosteroids are seen 2 to 3 hours following intravenous administration, and peak after 8 to 12 hours [17].

In the treatment of chronic bronchial asthma, oral corticosteroids are only considered when other measures have failed, and short-acting preparations are preferred, given on alternate days, to minimize side effects [18]. In cases where the disease is not effectively controlled with inhaled beclomethasone, oral prednisolone (5–10 mg/day) is frequently required [2]. In chronic obstructive pulmonary disease steroids are used to treat acute exacerbations, and are particularly useful when the sputum is eosinophilic [19]. Where the disease is primarily due to emphysema or bronchitis, the efficacy of steroid therapy has not been conclusively established.

Glucocorticoid therapy is often the only effective treatment for interstitial lung disease such as cryptogenic fibrosing alveolitis [20]. Patients are usually started on 60 mg oral prednisolone daily, reducing the dose once remission of symptoms is achieved, as described.

It has also been suggested that corticosteroids are useful in reducing the incidence of respiratory distress syndrome in infants delivered prematurely [21]: when delivery is expected before 34 weeks of gestation two doses of betamethasone, 12 mg, can be given on consecutive days. This steroid crosses the placenta to a greater extent than cortisol, as there is less binding to the maternal CBG, and acts to stimulate lung maturation [22].

Collagen-vascular diseases   The collagen-vascular diseases comprise several disorders, also known as connective tissue disorders. Steroid therapy is the first-line treatment for serious collagen-vascular diseases, including temporal arteritis and other vasculitic syndromes, polymyalgia rheumatica and polymyositis. In contrast, apart from the associated pulmonary fibrosis, scleroderma is unresponsive to corticosteroids. In most of these disorders oral prednisolone is given at an initial dose of 1 mg/kg/day in divided doses, until the disease is controlled, a period which varies between different patients and which may also vary between different clinical conditions.

Fulminating systemic lupus erythmatosus, as a life-threatening condition, must be promptly treated with a large dose of steroid, sufficient to induce rapid remission. The value of systemic steroid therapy in the long-term management of lupus erythmatosus has not been established [23].

Although rheumatoid arthritis was the first disease to be treated with corticosteroids [24], this form of therapy largely fell out of favor as the complications arising from prolonged steroid use became clear, and it was found that although many patients showed striking symptomatic relief on corticosteroid therapy, they continued to show radiographic evidence of progressive joint destruction [25]. With a number of nonsteroidal anti-inflammatory drugs available, there are now few indications for the use of systemic glucocorticoids in the management of rheumatoid arthritis. Where they are used, it is usually for a limited time period, possibly serving as a bridge between nonsteroidal anti-inflammatory agents and the slow acting agents such as gold, and the antimalarials [25].

Acute relapse of rheumatoid arthritis has been treated successfully with intravenous methylprednisolone (up to 1 g), given over a 40-minute infusion (pulsed high-dose steroid therapy) [26]. This regime gives symptomatic relief for up to 6 weeks, and it has been suggested that it may provide a safer alternative to daily oral steroid therapy. Local application of glucocorticoids has also been used to treat rheumatoid arthritis: injection of a long-acting steroid preparation (e.g., triamcinolone acetonide or methylprednisolone acetate) directly into the affected joint causes remission of symptoms [25]. In order to avoid systemic side effects, no more than three joints should be treated at the same time. Repeated intra-articular injection may cause some joint damage, but this is difficult to distinguish from damage caused by the disease process.

Systemic corticosteroids are rarely used in pediatric arthritis as they exacerbate the growth retardation caused by the disease. In some cases, however, alternate day steroid therapy may mobilize a patient who would otherwise remain bedridden [27].

Neurological disorders Myasthenia gravis and multiple sclerosis are chronic progressive neurological disorders, associated with gradual worsening of symptoms, punctuated by episodes of severe exacerbation of the disease. Corticosteroids have been used in the management of myasthenia gravis and give some improvement of symptoms in around 90 percent of patients. High dose intravenous methylprednisolone has been used to treat acute exacerbations of both myasthenia gravis and multiple sclerosis [28,29]. In severely myasthenic patients initiation of steroid therapy may cause a transient exacerbation of weakness, which may lead to respiratory failure. For this reason patients must be closely observed as corticosteroid therapy is started [30].

Although steroids are no longer used to treat acute cerebral oedema, they are used to reduce the oedema associated with tumors of the nervous system.

Steroids are without value in the management of acute neurogenic respiratory failure [30].

Skin disorders Topical steroids are used in the treatment of many skin disorders, particularly allergic conditions: A list of indications for steroid therapy is given in Table 7.3. In mild inflammatory skin disorders a 1 percent cortisol suspension is used, for more severe conditions 0.1 percent betamethasone and cortisol butyrate are both more potent. As with systemic treatments, the lowest effective dose of the least potent steroid is the preferred treatment as topical steroids can have both local and systemic side effects, depending on the strength of preparation used, the area treated and the duration of treatment. The therapeutic effects of topical steroids are enhanced by the use of an occlusive dressing, but this also promotes the uptake of the steroid into the circulation and increases the incidence of side effects [32]. The conditions which are less responsive to steroid therapy (Table 7.3) may be successfully treated with more potent steroid preparations, or by intradermal injections of steroid at the site of the lesion [31]. Systemic steroid therapy is also used for certain skin disorders, notably bullous pemphigoid.

Eye disease There are many ocular conditions, notably allergic disorders and nonpyogenic inflammation, which respond well to corticosteroid therapy [33]. A summary of these conditions is shown in Table 7.4. In many cases timely application of a glucocorticoid preparation has been responsible for the preservation of sight. In most cases it is possible to use topical steroids, particularly for diseases of the outer eye, although systemic therapy is required for diseases of the posterior segment. Retrobulbar delivery of

TABLE 7.3 Dermal conditions and corticosteroids

1. **Conditions which respond to standard concentrations of topical steroids**

   Seborrheic dermatitis
   Atopic dermatitis
   Lichen simplex chronicus (localized neurodermatitis)
   Pruritus ani (frequently a manifestation of psoriasis)
   Psoriasis
   Later phase of allergic contact dermatitis
   Later phase of irritant dermatitis
   Xerosis (inflammatory phase)

2. **Less responsive conditions, which may require a more potent steroid preparation or intralesional therapy**

   Discoid lupus erythematosus
   Necrobiosis lipoidica
   Lichen planus
   Alopecia areata
   Hypertrophic scars
   Keloids
   Pemphigus (oral therapy required)
   Granuloma annulare
   Pretibial myxedema
   Psoriasis of palms, soles, elbows, knees
   Acne cysts

From ref. 31.

steroid has been used in animal models with good effect, but is not widely used in human subjects.

Steroid therapy is ineffective in the treatment of cataracts and other degenerative disorders, and is contraindicated in most bacterial infections and some viral infections, especially the epithelial stages of herpes simplex keratitis.

Glaucoma is associated with prolonged topical therapy in ocular disease, and it has been suggested that intraocular pressure should be monitored when corticosteroids are applied for more than two weeks [2].

Allergic disorders Corticosteroids have profound effects on the allergic response (for a review see ref. 34) and may be used to treat acute allergic episodes (e.g., drug reactions). The effects of steroids require some time to develop, and therefore, the more severe reactions such as anaphylaxis and angioneurotic oedema require the immediate administration of a rapidly acting agent, notably adrenaline, in addition to intravenous steroid. For mild allergic disorders systemic steroid therapy is rarely required; serum sickness and hay fever are usually controlled adequately with antihistamines.

TABLE 7.4 Ocular conditions reported to respond to corticosteroids

Allergic blepharitis and conjunctivitis
Vernal conjunctivitis
Contact dermatitis of conjunctiva and eyelid
Immune graft reaction
Irritant conjunctivitis
Optic neuritis
Progressive thyroid exophthalmus
Phlyctenular conjunctivitis and keratitis
Pseudotumor of the orbit
Temporal arteritis
Marginal corneal ulcers
Viral ocular diseases
Herpes simplex (disciform stage)
Herpes zoster
Adenovirus
Superficial punctate keratitis
Posterior uveitis
Juvenile xanthogranuloma
Sympathetic ophthalmia
Chemical burn of cornea and conjunctiva
Scleritis and episcleritis
Acne rosacea keratitis
Ocular pemphigus
Iritis, iridocyclitis
Mucocutaneous conjunctival lesions
Interstitial keratitis
Retinal vasculitis
Infiltrative corneal disease

From ref. 33.

Glucocorticoids, applied topically, have an important role in the management of allergic disorders of the skin and eye.

Transplant rejection Patients receiving transplanted organs are routinely given oral prednisolone (50–100 mg) at the time of surgery, together with other immunosuppressants. After surgery a smaller maintenance dose of prednisolone is continued, usually 10 to 20 mg per day, although the dose is increased if rejection is suspected. Very high doses of prednisolone have been administered intravenously (1 g over 2 hr) to reverse organ rejection [35], as this regime was thought to cause maximum lymphocytic damage with fewer undesirable effects than longer term dosing with oral steroid. A study in children, however, has found that oral prednisolone (3 mg/Kg for 3 days) was just as effective as high dose intravenous steroid in preventing transplant rejection, and not associated with any side effects. Furthermore, oral therapy was cheaper, and less unpleasant for these pediatric patients [36].

Inflammatory bowel disease   In chronic ulcerative colitis which is uncontrolled by dietary management and salazopyrine, methyl prednisolone (40 mg) may be applied topically in the form of a nightly retention enema. In severe episodes of the disease systemic treatment with larger doses of steroid may be given. The disease process continues, however, despite glucocorticoid therapy, and one major problem is that the steroid treatment may mask signs of bowel perforation [37], although it should be noted that steroid therapy considerably reduces the risk of perforation.

Chemotherapy   Prednisolone is used in the treatment of lymphomas and in lymphocytic leukemias, where it is effective as a result of the suppressive effects of corticosteroids on the immune system (see Chap 4). Steroids are usually used in conjunction with other chemotherapy in the treatment of these conditions. A small proportion of patients with mammary carcinoma show tumor regression in response to 30 mg per day prednisolone [38]. This regime is only effective in estrogen receptor positive patients, in whom it is believed corticosteroids act by suppressing synthesis of adrenal androgens, which may be converted to estrogens in peripheral tissues. Other forms of treatment are usually preferred.

Renal disease   High dose pulsed intravenous steroids have been used successfully in the treatment of lupus nephritis [39], and glomerulonephritis, although glucocorticoid treatment does not alter its course [40]. In renal disease, the normal dose regime is 60 mg per day in divided doses for about four weeks. Further maintenance is given for up to a year and for this the daily steroid dose is given only for the first three days of each week [2].

Miscellaneous uses   Glucocorticoids are used in the treatment of hypercalcemia, particularly in conjunction with calcitonin, where a part of their action is to reduce the resistance to calcitonin which may develop. Corticosteroids also act directly to inhibit intestinal calcium absorption, and increase urinary excretion. Steroids usually have only a transient effect on the hypercalcemia of malignancy [3].
   Steroid therapy has been used in the treatment of septic shock, but there is no consensus as to its efficacy: in one study mortality was reduced from 34 to 10 percent by the use of corticosteroids [41], while another study found that steroid therapy *increased* the mortality rate in a dose-dependent manner [42]. A prospective study by Sprung et al. [ref. 43] found a decrease in the incidence of multiorgan system failure, but no increase in overall survival with steroid therapy. The use of steroids to treat septic shock should, therefore, still be considered experimental [44].
   In chronic active hepatitis corticosteroids are used in conjunction with

azathioprine, and a 70 percent 2-year remission rate has been reported with this therapy [45]. Prednisolone, 60 mg, is given for 7 to 14 days, and then reduced to 20 mg daily, or 10 mg plus 50 mg azathioprine [46].

7.2.4 Side effects of steroid therapy

In general side effects of steroid therapy only occur with prolonged use or when high doses are given and are not associated with appropriate replacement therapy. The side effects seen are usually exaggerations of the normal physiological and therapeutic actions of corticosteroids (see Chap. 4). Careful choice of drug preparation and dose regime minimizes side effects (see Sect. 7.2.2 of this chapter).

Iatrogenic Cushing's syndrome   When corticosteroids are given in relatively large doses for long periods of time, they may cause metabolic changes resulting in a typical appearance of Cushing's syndrome (see Chap. 5). Iatrogenic Cushing's syndrome may result from doses of cortisol in excess of 100 mg per day given for any period longer than 14 days. There is a characteristic accumulation of fat on the trunk with a typically Cushingoid moon face, and increased hair growth on the trunk and legs. With longer term therapy osteoporosis and diabetes may become clinically manifest, with frequent occurrence of insulin resistance. A summary of the major differences between natural Cushing's and iatrogenic Cushing's is shown in Table 7.5.

Pituitary-adrenal suppression   Long-term steroid therapy is associated with chronic suppression of ACTH secretion by the pituitary gland. This effect is related to both the dose of the steroid and to its biological half life (see Table 7.1). It is now clear that it is not only systemic administration of steroid that results in pituitary-adrenal suppression; steroids applied topically are absorbed percutaneously and may circulate in blood at a concentration sufficient to cause pituitary suppression [32].

Adrenal suppression can be reduced by the use of alternate day dosage regime and shorter acting steroid preparations (Table 7.1). It is important that steroid therapy is gradually reduced rather than suddenly terminated, as withdrawal of therapeutic steroids may result in acute adrenal insufficiency. Adrenal function takes around 9 months to recover from the suppressive effects of long-term steroid therapy [47], and this must be taken into account when any surgical procedure is planned, and additional steroid given if necessary, even up to 18 months after cessation of therapy.

Immunosuppression   Steroid therapy is associated with a suppressed immune response, making patients on high dose steroid therapy susceptible to potentially serious infection, which must be treated appropriately. The clinical symptoms of many infections may be masked by the anti-inflammatory actions of the corticosteroids. There may also be impaired wound healing.

TABLE 7.5 Comparison of natural with iatrogenic Cushing's syndrome

1. Symptoms almost unique to iatrogenic Cushing's syndrome:

Benign intracranial hypertension
Glaucoma
Posterior subcapsular cataract
Pancreatitis
Aseptic necrosis of bone
Panniculitis

2. Symptoms more common in natural Cushing's syndrome:

Hypertension
Acne
Hirsutism or virilization
Menstrual disturbance or impotence
Striae
Purpura
Plethora

3. Symptoms common to both syndromes:

Obesity
Psychiatric disturbances
Poor wound healing
Edema (depending on the steroid used in iatrogenic Cushing's)
Increased susceptibility to infection
Osteoporosis

Modified from ref. 30.

Central nervous system   Psychosis may occur in patients receiving large doses of steroid, even when the treatment is of short duration. Glucocorticoid therapy should, therefore, be used with caution in patients with a previous history of mental disturbance, as this is thought to predispose to steroid-induced psychosis. Euphoria has often been reported, but is generally not considered to be an undesirable side effect. Paradoxically, there are also many reports of severe depression following corticosteroid administration [48].

Pseudotumor cerebri has been reported in association with steroid therapy, mainly in children, presenting as papilloedema, and is reversible on withdrawal of steroids [49].

Dermal side effects of topical steroids   While local application of steroid reduces the systemic effects there are still some local side effects; short-term application of steroid may cause exacerbation of any local intercurrent infection and acne at the site of application. More prolonged use can cause a thinning of the skin which is only partly reversible after cessation of treatment. There may also be increased hair growth and irreversible striae atrophicae. Topical steroids can also produce allergic contact dermatitis from either sensitization by the vehicle, or by sensitivity to impurities in the preparation [50].

Ocular side effects   Ocular side effects may result from either systemic steroid therapy or from the local application of steroids to the eye. In both cases glaucoma may be precipitated, or exacerbated if previously present. The glaucoma is not always reversible following cessation of steroid therapy. Local steroid application may cause cataracts [51] but this condition is a more frequent complication of systemic steroid therapy [33]. In the presence of herpes virus infection of the eye, topical steroids can cause dendritic ulcers. Fungal infections of the eye are common in patients using topical corticosteroids for more than 3 weeks [52]. There may also be systemic effects seen from steroids applied topically to the eye, as the high degree of vascularization of the eye enhances absorption.

Oral candidosis and dysphonia   These are both complications of inhaled steroid preparations. The dysphonia is caused by a direct effect of steroid on the larynx and is reversible with reduced steroid dosage. Candidosis is reduced by rinsing the mouth with water immediately after dosing.

Complications of high-dose intravenous-pulsed methylprednisolone   There are several reports of serious cardiovascular complications of high-dose methylprednisolone treatment in recipients of renal allograft [53,54,55]. An effect of methylprednisolone on peripheral resistance and mean arterial pressure has been demonstrated (see Chap. 4), but the possibility of an acute toxic action of this steroid on the myocardium has not been excluded.

7.3 Inhibitors of the actions of adrenocortical steroids

7.3.1 Antimineralocorticoids

There are two specific aldosterone antagonists currently in clinical use, spironolactone and potassium canrenoate. These compounds are both steroid derivatives, spirolactones, and form the same metabolite, canrenone (Fig. 7.2). Spironolactone is the more commonly used aldosterone antagonist, canrenoate being used only when parenteral administration is desirable, as the high solubility of its potassium salt renders it more suitable than the almost insoluble spironolactone. Canrenoate is inactive until converted to canrenone, but while spironolactone is also converted to canrenone, the formation of this metabolite does not account for all the biological activity of spironolactone [56]. These agents are competitive inhibitors of aldosterone, acting by binding to the aldosterone receptor and preventing it from assuming an active conformation, thus blocking the biological effects of

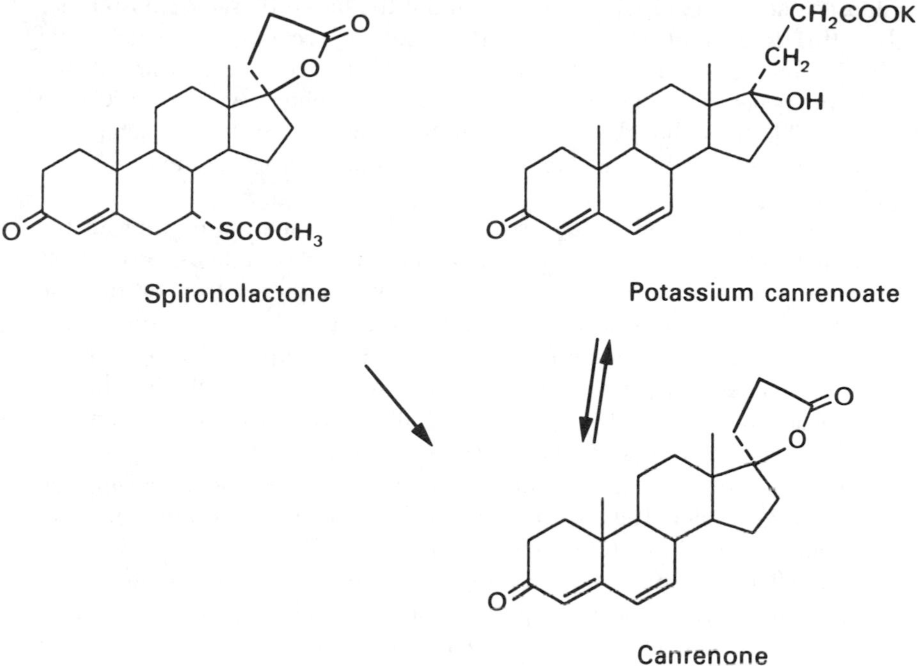

Spironolactone

Potassium canrenoate

Canrenone

Anti-mineralocortoids

FIGURE 7.2 Structures of the antimineralocorticoids in therapeutic use. Potassium canrenoate is inactive until converted to canrenone.

FIGURE 7.3 Mifepristone (RU 486): A recently developed synthetic steroid with both antiglucocorticoid and antiprogesterone properties. RU 486 has been used experimentally in the treatment of Cushing's syndrome, but is not currently in routine clinical use.

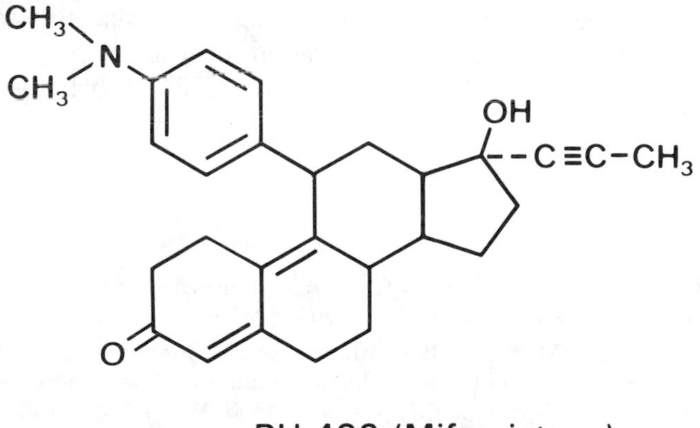

RU 486 (Mifepristone)

aldosterone. Spironolactone can also inhibit the biosynthesis of aldosterone, but this is a short-lived effect and rapidly overcome by activation of the renin-angiotensin-aldosterone system [57]. The mechanism of the antimineralocorticoid action of spironolactone has been reviewed by Corvol et al. [58]. It is thought that the inhibitory action of spironolactone on aldosterone biosynthesis is due to competition for steroid hydroxylases [see ref. 59].

Spironolactone is clinically used as a potassium-sparing diuretic in the treatment of hypertension and refractory edema or ascites. For these conditions it is often used in conjunction with either a loop or thiazide diuretic. In most cases doses of around 100 mg per day are adequate, but in edema caused by chronic liver disease, doses of up to 800 mg per day may be needed. In primary hyperaldosteronism, it is debatable whether spironolactone treatment is a reasonable alternative to surgery because of the high incidence of side effects. It is used preoperatively, however, to correct the metabolic abnormalities arising from hyperaldosteronism, and render surgery safer. The pharmacology and clinical uses of antimineralocorticoids are reviewed by McInnes and Ramsay [ref. 57].

Side effects of spirolactone therapy include hyperkalaemia, menstrual irregularities, gynacomastia, impotence, and possibly impaired renal function. These adverse effects are dose-dependent, and more common at doses above 100 mg per day [57].

7.3.2 Antiglucocorticoids

Many naturally occurring steroids, including some androgens, oestrogens and progesterone, and their derivatives have antiglucocorticoid properties. These compounds are not generally used as antiglucocorticoids, however, because of the side effects caused by their natural hormonal actions (for a review see ref. 60). A recently developed synthetic steroid, RU 486 (Fig. 7.3), has high affinity for both progesterone and glucocorticoid receptors [61,62]. Originally used as an abortifacient, RU 486 has been used experimentally to treat Cushing's syndrome [63,64]. While RU 486 effectively opposed the actions of glucocorticoids, there was a suggestion that this agent may cause symptoms of glucocorticoid deprivation, even in these patients with Cushing's syndrome [63]. At the present time, RU 486 has no routine clinical application.

References

1. Baxter, J.D., and Rousseau, G.G. 1979. "Glucocorticoid hormone action." *Monographs on Endocrinology*, 12. Berlin: Springer-Verlag.

2. Haynes, R.C., and Murad, F. 1985. Adrenocorticotrophic hormone; adrenocortical steroids and their synthetic analogues; inhibitors of adrenocortical steroid biosynthesis. In: A.G. Gilman, L.S. Goodman, T.W. Rall, and F. Murad, eds. *The Pharmacological Basis of Therapeutics*, 7th ed. New York: Macmillan.

3. Bagdade, J.D. 1986. Endocrine Emergencies. *Med. Clin. N. Am.* 70:1111–28.

4. Myles, A.B., and Daly, J.R. 1974. *Corticosteroid and ACTH treatment.* Baltimore: Williams and Wilkins.

5. Dluhy, R.C., Lauler, D.P., and Thorn, G.W. 1973. Pharmacology and chemistry of adrenal glucocorticoids. *Med. Clin. N. Am.* 57:1155–66.

6. Thompson, D.G., Mason, A.S., and Goodwin, F.J. 1979. Mineralocorticoid replacement in Addison's disease. *Clin. Endocrin.* 10:499–506.

7. Thorn, G.W., and Lauler, D.P. 1972. Clinical therapeutics of adrenal disorders. *Am. J. Med.* 53:673–84.

8. Axelrod, L. 1976. Glucocorticoid therapy. *Medicine.* 55:39–65.

9. Allen, N.B., and Studenski, S.A. 1986. Polymyalgia rheumatica and temporal arteritis. *Med. Clin. N. Am.* 70:369–84.

10. Haynes, B.F., Allen, N.B., and Fauci, A.S. 1986. Diagnostic and therapeutic approach to the patient with vasculitis. *Med. Clin. N. Am.* 70:355–84.

11. Byyny, R.L. 1976. Withdrawal from glucocorticoid therapy. *N. Eng. J. Med.* 295:30–32.

12. Harter, J.G., Reddy, W.J., and Thorn, G.W. 1963. Studies on an intermittent corticosteroid dose regimen. *N. Eng. J. Med.* 269:591–96.

13. Lancet editorial. 1983. Prednisolone pulses in collagen disease; grammes or milligrammes. *Lancet.* 1:280.

14. Durant, S., Duval, D., and Homo-Delarche, F. 1988. Effect of cortisol on the plasma and lymphoid tissue distributions of tritiated glucocorticoids in C57BL/6 mice. *J. Endocr.* 117:373–78.

15. Brown, M., Storey, G., and George, W.H.S. 1972. Beclomethasone dipropionate: A new steroid aerosol for the treatment of allergic asthma. *Br. Med. J.* 1:585–90.

16. Barnes, P.J. 1984. Drugs in respiratory disease. *Med. Int.* 2:311–16.

17. Stibolt, T.B. 1986. Asthma. *Med. Clin. N. Am.* 70:909–20.

18. O'Loughlin, J.M. 1979. Drug therapy of bronchial asthma. *Med. Clin. N. Am.* 63:391–96.

19. Rosen, R.L. 1986. Acute respiratory failure and chronic obstructive lung disease. *Med. Clin. N. Am.* 70:895–908.

20. Turner-Warwick, M., Burrows, B., and Johnson, A. 1980. Cryptogenic fibrosing alveolitis: Response to corticosteroid treatment and its influence on survival. *Thorax.* 35:593–99.

21. Ballard, P.L., and Ballard, R.A. 1979. Corticosteroids and respiratory distress syndrome: status 1979. *Pediatrics.* 63:163–65.

22. Katzung, B.G. 1987. *Basic and Clinical Pharmacology,* 3rd ed. Norwalk, Ct: Appleton-Lange.

23. Pisetsky, D.S. 1986. Systemic lupus erythmatosus. *Med. Clin. N. Am.* 70:337–54.

24. Hench, P.S., Kendall, E.C., Slocumb, C.H., and Polley, H.F. 1949. Effect of hormone of adrenal cortex (17 hydroxy-11 dehydrocorticosterone; Compound E) and of pituitary adrenocorticotropic hormone on rheumatoid arthritis: Preliminary report. *Proc. Mayo Clin.* 24:181–97.

25. St. Clair, E.W., and Polisson, R.P. 1986. Therapeutic approaches to the treatment of rheumatoid disease. *Med. Clin. N. Am.* 70:285–304.

26. Williams, I.A., Bayliss, E.M., and Shipley, M.E. 1982. A double-blind placebo-

References

controlled trial of methyl prednisolone pulse therapy in active rheumatoid disease. *Lancet.* 1:237–40.

27. Kredich, D.W. 1986. Chronic arthritis in childhood. *Med. Clin. N. Am.* 70:305–22.

28. Arsura, E., Brunner, N.G., Namba, T., and Grob, D. 1985. High dose intravenous methylprednisolone in myasthenia gravis. *Arch Neurol.* 42:1149–53.

29. Milligan, N.M., Newcombe, R., and Compston, D.A.S. 1987. A double-blind controlled trial of high dose methylprednisolone in patients with multiple sclerosis: 1 clinical effects. *J. Neurol. Neurosurg. Psych.* 50:511–16.

30. Bleck, T.P., and Klawans, H.L. 1986. Neurologic emergencies. *Med. Clin. N. Am.* 70:1167–84.

31. Maibach, H.I., and Stoughton, R.B. 1973. Topical corticosteroids. *Med. Clin. N. Am.* 57:1253–64.

32. Scoggins, R.B., and Kliman, B. 1965. Percutaneous absorption of corticosteroids: systemic effects. *N. Eng. J. Med.* 273:831–40.

33. Levine, S.B., and Leopold, I.H. 1973. Advances in ocular corticosteroid therapy. *Med. Clin. N. Am.* 57:1167–77.

34. Schleimer, R.P. 1985. The mechanisms of anti-inflammatory steroid action in allergic diseases. *Ann. Rev. Pharmacol. Toxicol.* 25:381–412.

35. Bell, P.R.F., Briggs, J.D., Calman, K.C., Paton, A.M., Wood, R.F.M., MacPherson, S.G., and Kyle, K. 1971. Reversal of acute clinical and experimental organ rejection using high doses of intravenous methylprednisolone. *Lancet.* 1:876–80.

36. Orta-Sibu, N., Chantler, C., Benick, M., and Haycock, G. 1982. Comparison of high dose intravenous methylprednisolone with low dose oral prednisolone in acute renal allograft rejection in children. *Br. Med. J.* 285:258–60.

37. ReMine, S.G., and McIlrath, D.C. 1980. Bowel perforation in steroid-treated patients. *Ann. Surg.* 192:581–86.

38. Stoll, B.A. 1972. Androgen, corticosteroid and progestin therapy. In: B.A. Stoll, ed. *Endocrine Therapy of Malignant Disease.* Philadelphia: Saunders Co., pp. 176–82.

39. Nebout, T., Sobel, A., Lagrue, G. 1977. Intravenous methylprednisolone pulses in diffuse proliferative lupus nephritis. *Lancet.* 1:1063.

40. DeTorrente, A., Popovtzer, M.M., and Guggenheim, S.J. 1975. Serious pulmonary hemorrhage, glomerulonephritis and massive steroid therapy. *Ann. Int. Med.* 83:218–19.

41. Schumer, W. 1976. Steroids in the treatment of clinical septic shock. *Ann. Surg.* 184:333–41.

42. Kreger, B.E., Craven, D.E., and McCabe, W.R. 1980. Gram-negative bacteremia. IV. Re-evaluation of clinical features and treatment in 612 patients. *Am. J. Med.* 68:344–55.

43. Sprung, C.L., Caralis, P.V., Marcial, E.H., Pierce, M., Gelbard, M.A., Long, W.M., Duncan, R.C., Tendler, M.D., and Karpf, M. 1984. The effects of high dose corticosteroids in patients with septic shock: A prospective controlled study. *N. Eng. J. Med.* 311:1137–43.

44. Karakusis, P.H. 1986. Considerations in the therapy of septic shock. *Med. Clin. N. Am.* 70:933–44.

45. Soloway, R.D., Summerskill, W.H., Baggenstoss, et al. 1972. Clinical, biochemi-

cal and histological remission of severe chronic active liver disease: A controlled study of treatments and early prognosis. *Gastroenterology.* 63:820–33.

46. Whitcomb, F.F. 1979. Chronic active liver disease. *Med. Clin. N. Am.* 63:413–22.

47. Graber, A.L., Ney, R.L., Nicholson, W.E., Island, D.P., and Liddle, G.W. 1965. Natural history of pituitary-adrenal recovery following long-term suppression with corticosteroids. *J. Clin. Endocr.* 25:11–16.

48. Carpenter, W.T., and Gruen, P.H. 1982. Cortisol's influence on human mental functioning. *J. Clin. Psychopharmacol.* 2:91–101.

49. Dujovne, C.A., and Azarnoff, D.L. 1973. Clinical complications of corticosteroid therapy: A selected review. *Med. Clin. N. Am.* 57:1331–42.

50. McMeekin, T.O., and Moschella, S.L. 1979. Iatrogenic complications of dermatologic therapy. *Med. Clin. N. Am.* 63:441–52.

51. Hewell, J.B. 1976. Eye diseases induced by topically applied steroids: The thin end of the wedge. *Arch. Dermatol.* 112:1529–30.

52. Mitsui, Y., and Hanabusa, J, 1955. Corneal infections after cortisone therapy. *Brit. J. Ophthalmol.* 39:244–50.

53. Stubbs, S.S., and Morell, R.M. 1973. Intravenous methylprednisolone sodium succinate: Adverse reactions reported in association with immunosuppressive therapy. *Transplant Proc.* 5:1145–46.

54. McDougal, B.A., Whittier, F.C., and Cross, D.E. 1976. Sudden death after bolus steroid therapy for acute rejection. *Transplant Proc.* 8:493–96.

55. Warren, D.J., and Smith, R.S. 1983. High dose prednisolone. *Lancet.* 1·594.

56. Mudge, G.H., and Weiner, I.M. 1985. Drugs affecting renal function and electrolyte metabolism. Introduction. In: A.G. Gilman, L.S. Goodman, T.W. Rall, and F. Murad, eds. *Pharmacological Basis of Therapeutics*, 7th ed. New York: Macmillan. pp. 879–86.

57. McInnes, G.T., and Ramsay, L.E. 1987. Pharmacology and clinical uses of antimineralocorticoids. In: B.J.A. Furr, and A.E. Wakeling, eds. *Pharmacology and Clinical Uses of Inhibitors of Hormone Secretion and Action.* London: Bailliere Tindall, pp. 233–54.

58. Corvol, P., Claire, M., Oblin, M.E., Geering, K., and Rossier, B. 1981. Mechanism of the antimineralocorticoid effects of spironolactone. *Kidney Int.* 20:1–6.

59. Muller, J. 1987. "Regulation of aldosterone biosynthesis." *Monographs on Endocrinology*, 29. Berlin: Springer-Verlag.

60. Sakiz, E. 1987. Survey on antiglucocorticoids. In: B.J.A. Furr, and A.E. Wakeling, eds. *Pharmacology and Clinical Uses of Inhibitors of Hormone Secretion and Action.* London: Bailliere Tindall.

61. Healey, D.L., Chrousos, G.P., Schulte, H.M., Williams, R.F., Gold, P.W., Baulieu, E.E., and Hodgen, G.D. 1983. Pituitary and adrenal responses to the anti-progesterone and anti-glucocorticoid steroid RU 486 in primates. *J. Clin. Endocrinol. Metab.* 57:863–65.

62. Bertagna, X., Bertagna, C., Luton, J-P., Husson, J-M., and Girard, F. 1984. The new steroid analog RU 486 inhibits glucocorticoid action in man. *J. Clin. Endocrinol. Metab.* 59:25–28.

63. Bertagna, X., Bertagna, C., Laudat, M-L., Husson, J-M., Girard, F., and

Luton, J-P. 1986. Pituitary-adrenal response to the antiglucocorticoid action of RU 486 in Cushing's syndrome. *J. Clin. Endocrinol. Metab.* 63:639–43.

64. Beaufrere, B., De Parscau, L., Chatelain, P., Morel, Y., Aguercif, M., and Francios, R. 1987. RU 486 administration in a child with Cushing's syndrome. *Lancet.* II:217.

8

Inhibitors of Corticosteroid Biosynthesis

A thorough knowledge of the nature of the enzymes which are involved in the biosynthetic process of steroid hormone formation has led to the development of a number of substances which are of value as pharmacologically active inhibitors of steroid biosynthesis [1]. These interact with the various enzyme systems which have been described in Chap. 2. Thus, agents are available which interfere with cytochrome P450 mediated hydroxylation, while others inhibit the Δ^5–3β-hydroxysteroid-dehydrogenase system. It is possible that other routes to inhibition of corticosteroid synthesis exist, for example, through using peptide analogues to block the actions of ACTH. However, these do not appear to have any general application and are not discussed here.

The study of the interactions of substrates (and other substances) with cytochrome P450 is facilitated through the use of spectrophotometric methods. These depend on the changing absorption spectrum of cytochrome P450 in the presence of other substances. Thus, the Type I difference spectrum is typical of substrate binding to cytochrome P450 and is characterized by an absorption maximum at 420 nm and a minimum at 392 nm. These spectra do not appear to be attributable to the binding of substrate to the heme portion of the cytochrome but may be due to the modification of an existing ligand of the heme. The Type II difference spectrum is characteristic of the reaction between a basic amine and the ferrihemoprotein giving, depending on the amine used, an absorption maximum between 425 nm and 435 nm [2]. These interactions are also reflected in spin state changes

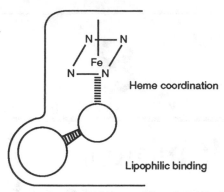

FIGURE 8.1 Interactions between inhibitors with a lipophilic binding domain and a coordinating nitrogen function, such as metyrapone, with heme in cytochrome P450. (Reproduced with permission from ref. 2.)

in the Fe^{3+} heme ion which can be studied by electron paramagnetic resonance spectroscopy. Thus the induction of a Type I difference spectrum corresponds to the transition of the Fe^{3+} heme ion from the low to the high-spin state [3,4]. The production of a Type II difference spectrum shifts the enzyme to the low spin form and this spin state change is accompanied by a change in the redox potential of the enzyme that makes its reduction by cytochrome P450 reductase more difficult. The most effective reversible inhibitors of this type interact strongly with both the protein and heme moieties of cytochrome P450 (Fig. 8.1).

8.1 Mitotane (o,p'-DDD)

The insecticide dichlorodiphenyldichloroethane or DDD (a structural ana-logue of DDT) was found, about 40 years ago, to cause cytotoxic atrophy of the adrenal cortex and liver damage in dogs, and also to decrease plasma cortisol levels [5–7]. The activity was found to be a result of the small amount of o,p'-DDD which was contained in the preparations [8-10] (Fig. 8.2). Subsequently, this drug was used in humans, as mitotane, in the treatment of Cushing's syndrome and adrenocortical carcinoma [11]. Its effects are long lasting since it appears to be stored in fatty tissue and also in the adrenal [12], at least at an early stage of treatment. However, the doses required are large and cause severe side effects although prolonged remission of the disease was claimed in some cases of adrenocortical carcinoma [13–15]. Curi-ously, the drug shows marked species specificity in that the rat is refractory to its action. In dogs, on the other hand, the effect of o,p'-DDD is character-ized by rapid inhibition of steroidogenesis and extensive disruption of adreno-cortical mitochondria. Its toxicity is specific for the adrenal cortex, as other tissues are relatively unaffected [16]. Information on its mode of action comes from studies in which animals were treated with the drug in vivo, and,

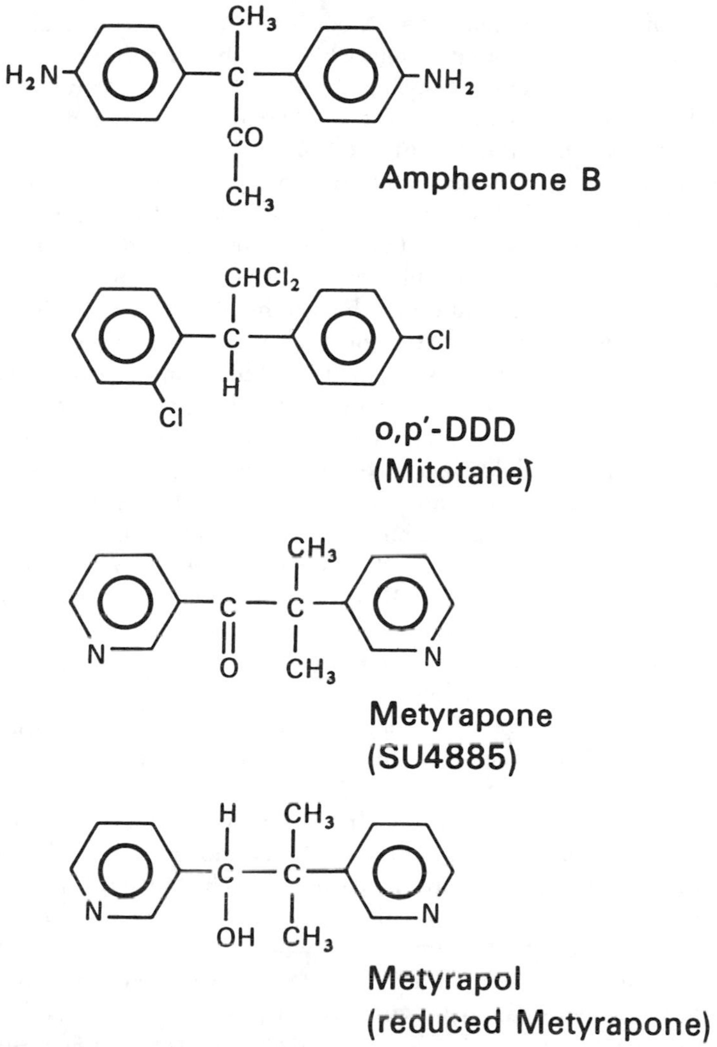

FIGURE 8.2 Structural relationships between amphenone related inhibitors.

because it has been found to be inactive or only weakly active in vitro, the possibility that one of its metabolites may be the active form cannot be excluded [17]. Another possibility is that some of the effects of o,p'-DDD may be due in part to changes in the extra-adrenal metabolism of cortisol, perhaps through the induction of hepatic enzymes [18].

8.2 Metyrapone

The findings that the insecticide, DDD, and also amphenone B inhibited adrenal function in vivo [19] led to the development of several specific inhibitors of steroid biosynthesis. Accordingly, a series of analogues of amphenone B were prepared among which 1,2-bis-(3-pyridyl)-2-methyl-1-propanone (also designated SU4885 and called metyrapone) proved to be a powerful inhibitor of 11β-hydroxylation. The relationship between all of these structures is illustrated in Figure 8.2.

This compound gave potent inhibition of corticosteroid production in dogs in vivo, and, to a lesser extent, in rat adrenal tissue in vitro [20]. In these studies it was shown that production of 11-deoxycortisol and deoxycorticosterone could be enhanced while cortisol and corticosterone were suppressed, strongly indicating that the major site of action is the 11β-hydroxylase system (although the primary site for the action of amphenone B is cholesterol side-chain cleavage). Direct studies on the site of action of metyrapone in rat adrenal preparations (in the presence of NADPH) showed that the compound inhibits conversion of deoxycorticosterone to corticosterone (and 11-dehydrocorticosterone) in a competitive fashion, and furthermore that 21-hydroxylation was unaffected [21]. In vivo treatment of experimental animals such as the rat can considerably reduce corticosterone secretion into the adrenal vein thus clearly overcoming the effect of supermaximal stimulation by ACTH resulting from anesthetics and surgical stress [22]. Circulating deoxycorticosterone concentrations are enhanced and the animals can develop, under these conditions, the symptoms of deoxycorticosterone-induced hypertension [23], (see Chap. 4). The drug shows characteristic Type II difference spectra in bovine mitochondrial cytochrome P450, and there is a direct relationship between the spectral change and the potency of inhibition [24,25].

Similar to the specificities of substrates for hydroxylases (see Chap. 2), the specificity of the action of inhibitors of steroidogenesis is only relative. This may indeed be expected, given the similarities between the hydroxylases and those between the substrates themselves. One clear additional site of inhibition of steroidogenesis by metyrapone is at C-18, and in the rat the conversion of deoxycorticosterone to 18-hydroxydeoxycorticosterone is inhibited as efficiently as 11β-hydroxylation [26]. It will be recalled that much evidence suggests that these two hydroxylations are in fact catalyzed by a single cytochrome P450 species [27]. Higher concentrations of metyrapone than are required for inhibition of 11β-hydroxylation can also inhibit cholesterol side-chain cleavage [28,29]. Metyrapone may also interact with liver cytochrome P450, and it is possible that clearance of steroid hormones may be affected in this way [30].

The major metabolite of metyrapone both in the rat and in humans is the reduced form metyrapol (see Fig. 8.2) which is also an 11β-hydroxylase inhibitor, though less active than metyrapone itself [31]. Metyrapone also inhibits placental aromatase [32] and sterol 7α-hydroxylase [33].

The most common clinical application of metyrapone is in the test for pituitary ACTH reserve [34–36] (see Chap. 5). Treatment with metyrapone should normally result in greatly increased ACTH secretion because of

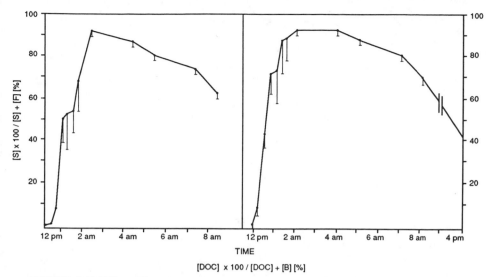

FIGURE 8.3 Effect of metyrapone treatment on circulating steroid levels in human subjects during a single dose metyrapone test. Relative serum concentrations of deoxycortisol are expressed as a percentage of total deoxycortisol [S] plus cortisol [F] and deoxycorticosterone as a percentage of total deoxycorticosterone [DOC] plus corticosterone [B]. (Figure reproduced with permission from ref. 36.)

the decrease in circulating cortisol, therefore resulting in enhanced 11-deoxysteroid output (Fig. 8.3).

8.3 Aminoglutethimide

This drug (see Fig. 8.4) was introduced originally as an anticonvulsant but was subsequently shown to have inhibitory actions on the adrenal cortex [37,38]. In rats, the secretion of corticosterone and aldosterone into the adrenal vein is reversibly inhibited, and in adrenal tissue preparations aminoglutethimide (and the related compound glutethimide) block conversion of endogenous or added cholesterol to corticosteroids. The conversion of labeled pregnenolone, progesterone or 17α-hydroxyprogesterone is, however, unaffected. This data suggests that the major action of aminoglutethimide is at the site of cholesterol side-chain cleavage. In humans and in dogs, aminoglutethimide does not affect the Δ^5–3β-hydroxysteroid-dehydrogenase system, or hydroxylation at the 11β, 17α, or 21 loci. In other species aminoglutethimide causes suppression of steroid secretion into the adrenal vein. This confirmation of its mode (and site) of action on the gland is important, because in the human species aminoglutethimide treatment rarely produces any effect on circulating levels of cortisol or on its secretory rate, apparently because of an increase in the secretion of ACTH which

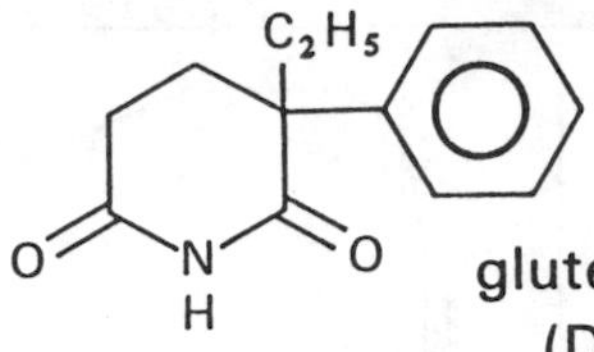

glutethimide
(Doriden)

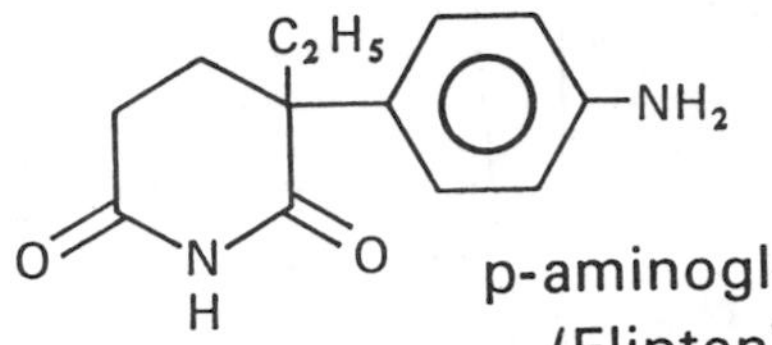

p-aminoglutethimide
(Elipten)

FIGURE 8.4 Structures of glutethimide and p-aminoglutethimide.

overcomes the initial inhibitory action of the drug [39]. The response of cortisol to further ACTH stimulation is however greatly reduced [40] (see Fig. 8.5). Aldosterone levels, on the other hand, remain suppressed during aminoglutethimide treatment, perhaps because glomerulosa function is not similarly supported by the action of ACTH [41] (see Chap. 3).

Although the cholesterol side-chain cleavage step is the major locus for aminoglutethimide action, in some circumstances other hydroxylations can be also inhibited, but the effect is less potent. 11β-hydroxylation and 18-hydroxylation have been shown to be suppressed at higher concentrations [42,43]. Aromatase, a widespread enzyme which catalyzes estrogen production from C_{19} steroids is also inhibited, and this has led to the use of aminoglutethimide treatment in cases of hormone sensitive cancer [44,45]. There have been some suggestions that the Δ^5–3β-hydroxysteroid-dehydrogenase/isomerase system is *enhanced* by aminoglutethimide', In sheep and rat adrenals, treatment with aminoglutethimide and ACTH results in increased cholesterol and cholesterol ester formation—an effect not observed with ACTH alone [40,46,47]. Aminoglutethimide interacts with cytochrome $P450_{SCC}$ to give typical Type II difference spectra, and reduction of the cytochrome is concomitantly prevented [48]. Though most samples of aminoglutethimide are racemic, the d (+) form is about three times as active as the l (−) form, both in binding to cytochrome P450 and in inhibitory action [49]. Of other analogues which have been tested, one in which the amino group was transferred to the nitrogen atom of the glutarimide ring is of particular interest. This compound was three times more active than glutethimide in inhibition of side-chain cleavage but inactive in aromatase inhibition [50].

A number of metabolites of aminoglutethimide have been described, and in the rat and in humans one major product is the N-acetyl derivative. This metabolite is itself a weak inhibitor of steroid synthesis [51]. Metabolic

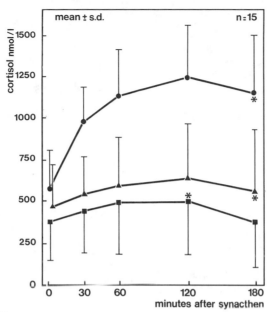

FIGURE 8.5 Effects of aminoglutethimide treatment for 4 (▲) and 8 (■) weeks on circulating cortisol and the responses to injected ACTH in patients with breast cancer, compared with responses before treatment (●). (Reproduced with permission from ref. 40.)

acetylation in humans is polymorphically inherited and the extent of metabolism of aminoglutethimide is closely paralleled by the degree of acetylation of other substrates [52].

8.4 Inhibitors of 3β-hydroxysteroid-dehydrogenase: cyanoketone and trilostane

Unlike the inhibitors of cytochrome P450s, inhibitors of the 3β-hydroxysteroid dehydrogenase/isomerase system are all clearly structural analogues of the natural substrates. A series of heterocyclic substituted steroid derivatives which inhibit 3β-hydroxysteroid-dehydrogenase activity in steroidogenic tissues have been developed. The first was cyanoketone (see Fig. 8.6). The adrenal blocking activity of this steroid was first recognized when it was shown to cause adrenal hypertrophy in rats and to block the metabolic responses to ACTH. Subsequently, it was shown to decrease corticosterone secretion into adrenal vein blood and to inhibit the production of steroids formed from pregnenolone in vitro but not from progesterone [53,54] therefore indicating its primary action on 3β-hydroxysteroid-dehydrogenase/ isomerase. Its action is relatively long lasting, and it is essentially a noncompetitive inhibitor affecting both the dehydrogenase and isomerase

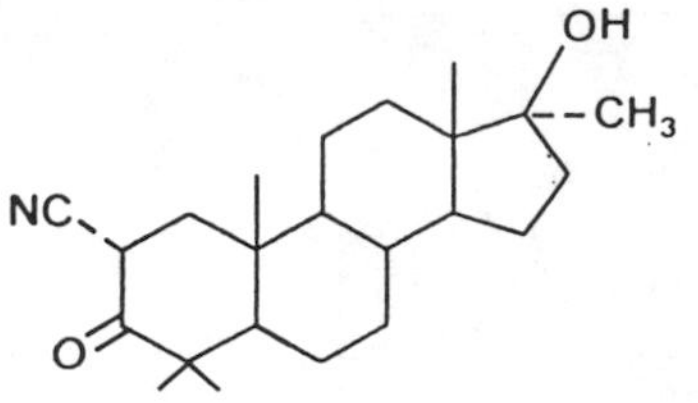

Cyanoketone
(WIN 19,578)

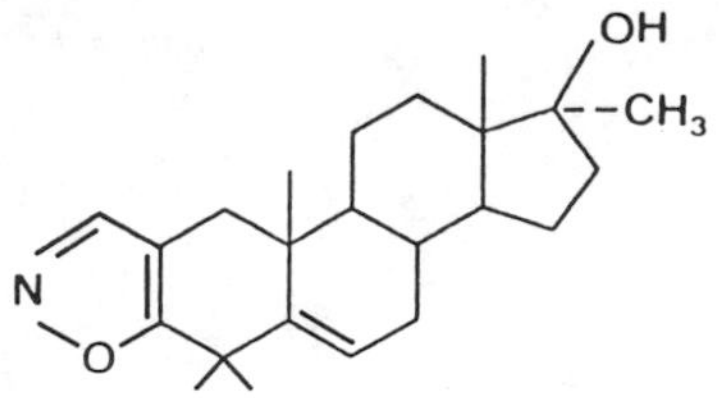

Azastene
(WIN 17,625)

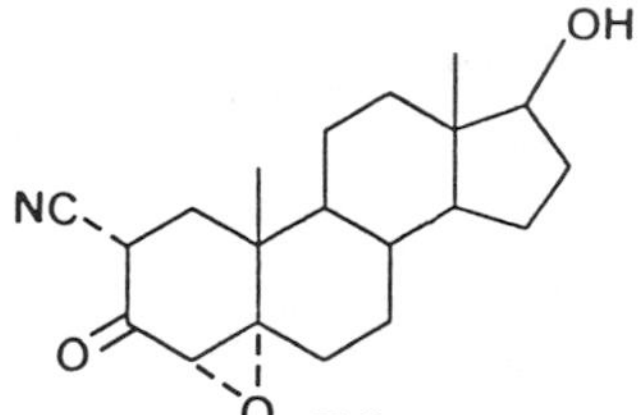

Trilostane
(WIN 24,540)

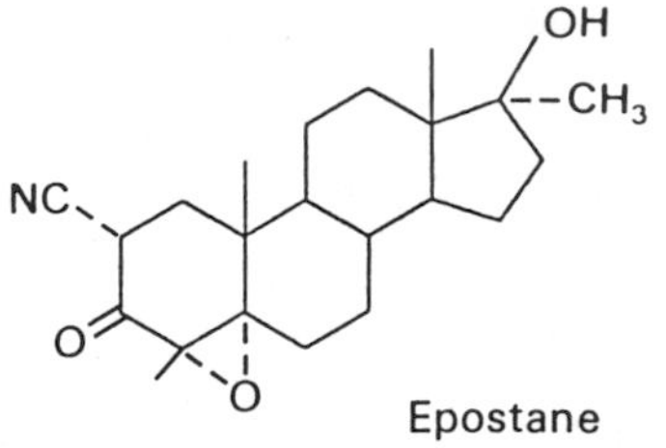

Epostane
(WIN 32,729)

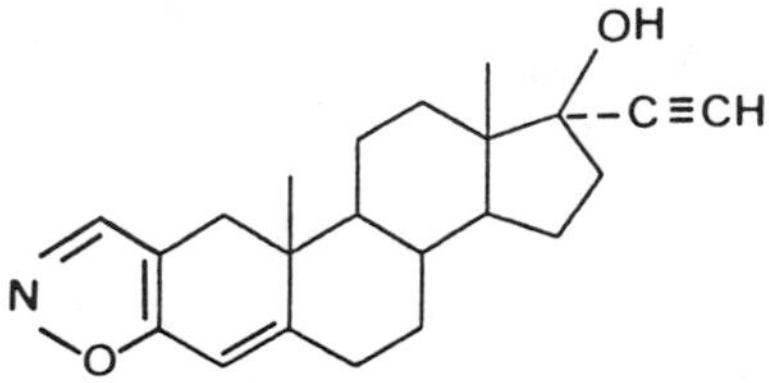

Danazol

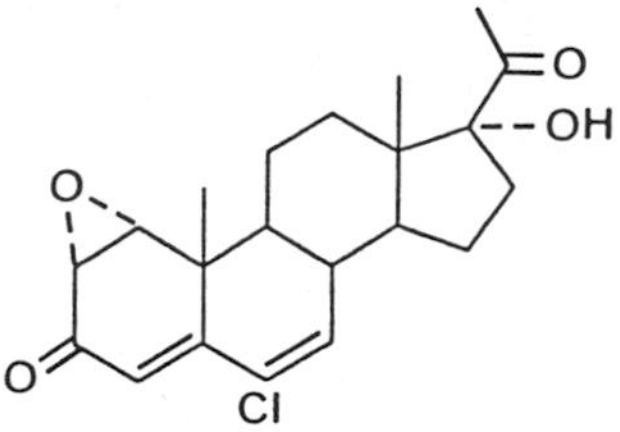

Cyproterone

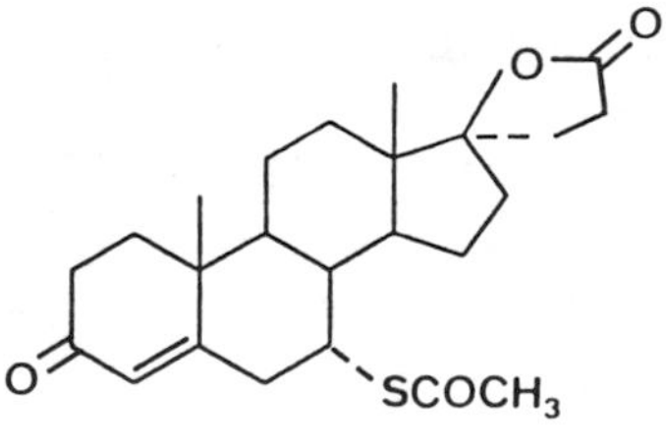

Spironolactone

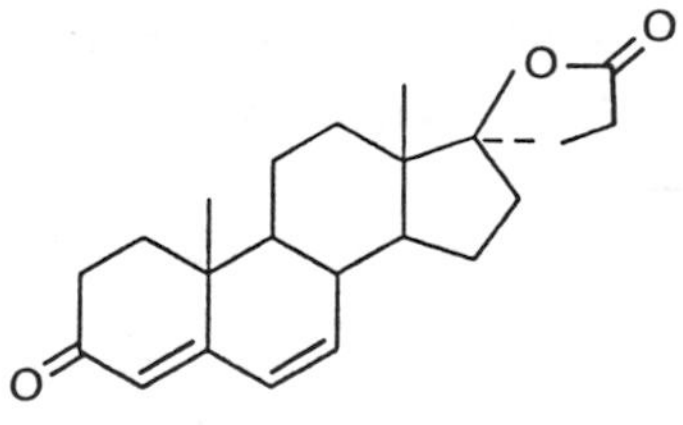

Canrenone

FIGURE 8.6 Structure of steroidal inhibitors of steroid biosynthesis.

steps [55,56]. At high levels it will also inhibit cholesterol side-chain cleavage [57]. Cyanoketone, like trilostane (see following text) has been used extensively as an experimental tool in studies of steroid biosynthesis. For example, in studies of cholesterol side-chain cleavage, the addition of either of these inhibitors to tissue preparations in vitro allows the assay of

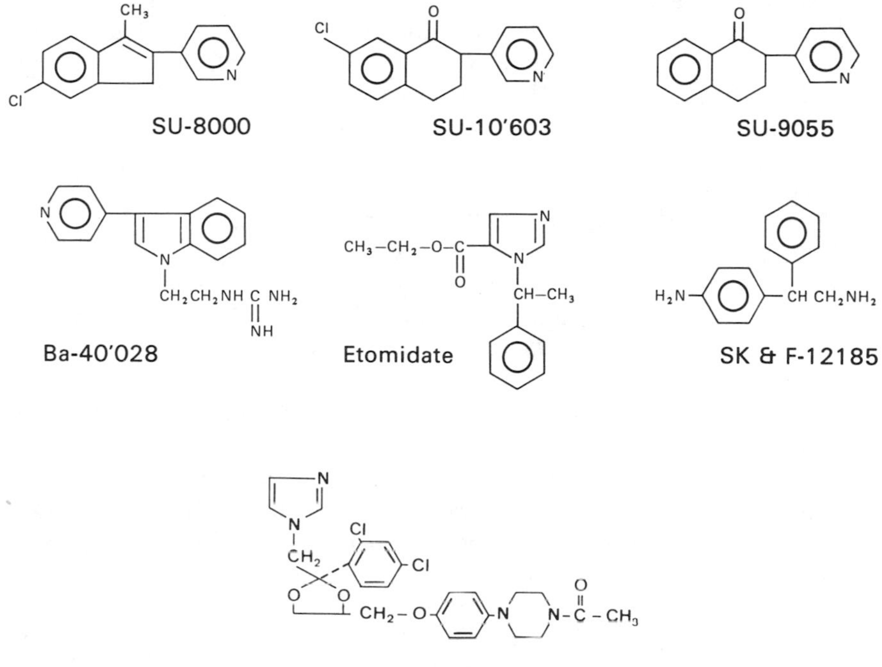

FIGURE 8.7 Miscellaneous inhibitors of steroidogenesis.

pregnenolone to be used as a direct estimate of side-chain cleavage activity [58,59].

Azastene (Fig. 8.6), in which the three ketone and two carbonitrile groups of cyanoketone are combined to form an isoxazole ring, also inhibits the 3β-hydroxysteroid-dehydrogenase system, but it is less potent than cyanoketone [55,57,60]. In trilostane, the 17α-methyl group present in azastene has been eliminated and the dimethyl group at C-4 is replaced by a 4,5-epoxy function. This compound inhibits 3β-hydroxysteroid dehydrogenase but, unlike cyanoketone, it does so in a competitive fashion [61]. It also inhibits steroid production from endogenously formed precursors (i.e., in adrenal preparations) from a number of species, but it has no action on 11β, 17α, or 21-hydroxylation [60,62,63].

Structurally, epostane (Fig. 8.6) is closely similar to trilostane and is a more effective inhibitor of 3β-hydroxysteroid dehydrogenase in the ovary and placenta than in the adrenal [64].

Cyanoketone has not been used clinically, probably because of its lack of specificity for any particular steroidogenic tissue and because its action is irreversible. Trilostane has been applied clinically, to a limited extent, to treat patients with steroid excess states. The potential advantages of inhibiting steroidogenesis at a relatively early stage in the pathway is that an inactive 3β-hydroxysteroid would accumulate rather than biologically active Δ^4,3-

ketone forms. It has been used with only moderate success in patients with Cushing's syndrome although the dosage which has been required has been found to be rather variable [65]. In pregnant women both progesterone and cortisol secretion are significantly decreased [66].

8.5 Other inhibitors of steroid hydroxylation

With the success of metyrapone, the Ciba company synthesized and tested a number of additional compounds. Some of these are illustrated in Figure 8.7. Among these, the tetralones SU 10603 and SU 9055, and the indene SU 8000 block 17α-hydroxylation [67] although actions on 18-hydroxylation and aldosterone also have been described [68]. Another compound, the 4-pyridyl indole derivative (BA40'028) was particularly interesting, since it resulted in the selective inhibition of corticosterone secretion coupled with stimulation of cortisol in bovine adrenal tissue. This was taken to indicate that the compound specifically inhibits 21-hydroxylation of 17-deoxysteroids, and suggests therefore, the existence of separate enzymes catalyzing the conversion of 17-hydroxy and 17-deoxysteroids [69]. This concept has not, however, been supported by current molecular biological studies (see Chap. 2). Another inhibitor of 11β-hydroxylation is 2-(p-aminophenyl)-2-phenyl-ethyl-amine (SKF 12185). This compound also produces classical Type II different spectra indicating a mode of action similar to that of metyrapone [70–72].

Among inhibitors discovered and described more recently, ketoconazole (Fig. 8.5) is an orally active antimycotic (used in the treatment of fungal infections) which inhibits the synthesis of ergosterol in susceptible organisms. The development of symptoms of gynecomastia in some patients led to the investigation of its effect on steroid biosynthesis. Inhibition of production of both gonadal and adrenal steroids has been described, and the major site of action appears to be 17α-hydroxylation and 17–21 lyase [73,74]. In the kidney, 24-hydroxylation is affected, and this suggests the general inhibition of cytochrome P450 enzymes [73]. Noncytochrome P450 enzymes (i.e., the dehydrogenases) are not affected. In humans, at the low-therapeutic dose of 200 mg a day, only testosterone secretion is inhibited. Higher doses also block the adrenal response to ACTH, but basal plasma cortisol levels are not usually affected. Such differential action has been exploited with the use of ketoconazole in the treatment of patients with advanced prostatic cancer [75].

Etomidate (Fig. 8.7), an anaesthetic, another imidazole (like ketoconazole) also has antiadrenal activity and has been found to lead to decreased cortisol response to ACTH, coupled with an increase in 11-deoxycortisol. Compared with other anaesthetics, levels of cortisol and aldosterone were found to be significantly lower when etomidate was used [76]. Further studies with in vitro preparations also indicate that its inhibitory activity affects 11β-hydroxylation but not 21-hydroxylation. Cholesterol side-chain cleavage is also inhibited [77]. Overall it appears to be a more powerful inhibitor of mitochondrial cytochrome P450 dependent enzymes than metyrapone.

Several further steroid analogues (Fig. 8.6) have also been found to affect steroidogenesis in adrenocortical tissue. Among these is danazol, used clinically in a variety of disorders, including endometriosis and precocious puberty. It inhibits cortisol production in guinea pig adrenal cells and human fetal adrenal preparations [78] through binding to cytochrome P450 and inhibition of 11β and 21-hydroxylation.

Cyproterone acetate, a progestagenic steroid which possesses antiandrogenic properties, is known to affect adrenal function in experimental animals, in which it causes adrenal atrophy and reduced plasma corticosteroid levels and responses to ACTH [79]. There is evidence that it acts directly on the adrenal as a noncompetitive inhibitor of 3β-hydroxysteroid dehydrogenase. It also inhibits 11β and 21-hydroxylation [80].

Spironolactone is a well-established aldosterone antagonist which is known to act by binding to renal mineralocorticoid receptors. In vitro spironolactone (and its metabolite canrenone) also inhibit the formation of aldosterone by adrenal tissue although the concentrations required are rather high [81]. In studies with bovine mitochondrial preparations, spironolactone and canrenone were shown to bind to cytochrome P450 and to inhibit 11β and 18-hydroxylation [82]. Suramin, a drug which has been used in the treatment of AIDS, has recently been shown to inhibit several adrenal steroidogenic enzymes in human adrenals, which may account for the observed side effect of adrenal insufficiency in treated patients [83].

It is likely that other drugs in use may affect steroidogenesis in a more general way. For example, drugs which inhibit cholesterol biosynthesis can also inhibit corticosteroid biosynthesis to some extent. Thus triparanol, a hypocholesterolemic drug which acts by preventing the conversion of desmosterol to cholesterol, also has been shown to affect adrenal function in some studies. Generally, the pattern is of an ACTH dependent increase in adrenal size which compensates for the reduction in corticosteroid secretion. There is little evidence for any acute effect of the drug [84].

References

1. Gaunt, R., Chart, J.J., and Renzi, A.A. 1965. Inhibitors of adrenal cortical function. *Rev. Physiol. Biochem. Exp. Pharm.* 56:114–72.

2. Schenkman, J.B., Cinti, D.L., Orrenius, S., Moldeus, P., and Kaschnitz, R. 1972. The nature of the reverse type I (modified type II) spectral change in liver microsomes. *Biochemistry.* 11:4243–50.

3. Ortiz de Montellano, P.R., and Reich, N.O. 1986. Inhibition of cytochrome P450 enzymes. In: P. Ortiz de Montellano, ed. *Cytochrome P450. Structure, mechanism, and biochemistry.* New York: Plenum Press, pp. 273–314.

4. Gower, D.B. 1984. In: H.L.J. Makin, ed. *Biochemistry of steroid hormones,* 2nd ed. Oxford: Blackwell, pp. 230–92.

5. Nelson, A.H., and Woodward, G. 1949. Severe adrenal cortical atrophy

(cytotoxic) and hepatic damage produced in dogs by feeding 2,2-bis (parachloryl-phenyl)-1, 1-dichloroethane (DDD or TDE) *Arch. Pathol.* 18:387–94.

6. Brown, J.H.V., Griffin, J.B., Smith, R.B., and Anason, A. 1956. Physiologic activity of an adrenocorticolytic drug in the adult dog. *Metabolism.* 5:594–600.

7. Cobey, F.A., Taliaferro, I., and Haag, H.B. 1956. Effect of DDD and some of its derivatives on plasma 17-OH-corticosteroids in the dog. *Science.* 123:140–41.

8. Cueto, C., and Brown, J.H.U. 1958. The chemical fractionation of an adrenocorticolytic drug. *Endocrinology.* 62:326–33.

9. Cueto, C., and Brown, J.H.U. 1958. Biological studies of an adrenocorticolytic agent and isolation of the active components. *Endocrinology.* 62:334–39.

10. Nichols, J. 1961. Studies on an adrenal cortical inhibitor. In: H.D. Moon, ed. *The Adrenal Cortex.* New York: Harper.

11. Bergenstal, D.M., Hertz, R., Lipsett, M.B., and May, R.H. 1960. Chemotherapy of adrenocortical cancer with o,p'-DDD. *Ann. Int. Med.* 53:672–82.

12. Diguglio, W., and Bejerwaltes, W.H. 1968. Tissue localization studies of a DDD analog. *J. Nuclear Med.* 9:634–47.

13. Orth, D.N., and Liddle, G.W. 1971. Results of treatment in 108 patients with Cushing's syndrome. *New England J. Med.* 285:243–47.

14. Luton, J.P., Mahondean, J.A., Bouchard, P., Thiebold, P., Hautecouverture, M., Simon, D., Laudat, M.H., Touitou, Y., and Bricaire, H. 1979. Treatment of Cushing's disease by o,p'-DDD. *New England J. Med.* 300:459–64.

15. Temple, T.E., Jones, D.J., Liddle, G.W., and Dexter, R.N. 1969. Treatment of Cushing's disease. Correction of hypercortisolism by o,p'-DDD without induction of aldosterone deficiency. *New England J. Med.* 281:801–05.

16. Hart, M.M., Reagan, R.L., and Adamson, R.H. 1973. The effects of isomers of DDD on the ACTH-induced steroid output, histology and ultrastructure of the dog adrenal cortex. *Tox. Applied Pharm.* 24:101–13.

17. Reif, V.D., and Sinsheimer, J.E. 1975. Metabolism of 1 (o-chlorophenyl)-1-(p-chlorophenyl)-2,2-dichloroethane (o,p'-DDD) in rats. *Drug Metab. Dispos.* 3:15–25.

18. Straw, J.A., and Hart, M.N. 1975. 1-(o-chlorophenyl)-1-(p-chlorophenyl)-2,2-dichloroethane (o,p'-DDD), an adrenocorticolytic agent. In: A.C. Sartorelli and D.E. John, eds. *Handbook of Experimental Pharmacology.* New York: Springer-Verlag, pp. 809–19.

19. Kielstis, J.A., and Ferguson, J.J. 1964. Inhibition in vitro of adrenal pregnenolone biosynthesis by amphenone and its analogues. *Endocrinology.* 74:567–72.

20. Chart, J.J., and Sheppard, H. 1964. Studies on adrenocortical inhibitors. In: L. Martini and A. Pecile, eds. *Hormonal Steroids, Biochemistry, Pharmacology, and Therapeutics: Proceedings of the First International Congress on Hormonal Steroids,* vol. 1. New York: Academic Press, pp. 399–411.

21. Dominguez, O.V., and Samuels, L.T. 1963. Mechanism of inhibition of adrenal steroid 11β-hydroxylase. *Endocrinology.* 73:304–09.

22. De Nicola, A.F., and Dahl, V. 1971. Acute effects of SU-4885 and its reduced derivative (SU-5236) on the adrenocortical secretion of the rat. *Endocrinology.* 89:1236–41.

23. Colby, H.D., Skelton, F.R., and Brownie, A.C. 1970. Metopirone-induced hypertension in the rat. *Endocrinology.* 86:620–28.

24. Hewick, D.S., and Young, C.K. 1973. Stereoselective inhibition of steroid 11β-

hydroxylation in ox and sheep adrenocortical mitochondria: comparison of metyrapone and the enantiomers of 2-(4-aminophenyl)-2-phenethylamine (SKF-12185). *Biochem. Pharmacol.* 26:1653–59.

25. Graves, P.E., Uzgiris, V.I., and Salhanick, H.A. 1980. Modification of enzymatic activity and difference spectra cytochrome P-450 from various sources by cholesterol side-chain cleavage inhibitors. *Steroids.* 35:543–59.

26. Sprunt, J.G., Browning, M.C.K., and Hannah, D.M. 1968. Some aspects of the pharmacology of metyrapone. *Mem. Soc. Endocr.* 17:193–203.

27. Watanuki, M., Tilley, B.E., and Hall, P.F. 1978. Cytochrome P-450 for 11β and 18-hydroxylase activities of bovine adrenocortical mitochondria: One enzyme or two? *Biochemistry.* 17:127–30.

28. Carballeira, A., Cheng, S.C., and Fishman, L.M. 1974a. Sites of metyrapone inhibition of steroid biosynthesis by rat adrenal mitochondria. *Acta Endocr.* 76:703–11.

29. Carballeira, A., Cheng, S.C., and Fishman, L.M. 1974b. Metabolism of (4-^{14}C) cholesterol by human adrenal glands in vitro and its inhibition by metyrapone. *Acta Endocr.* 76:689–702.

30. Blichert-Toft, M., Folke, K., and Lykkegard-Nielson, M. 1972. Effects of metyrapone and corticotrophin infusion on cortisol disappearance rate in man. *J. Clin. Endocr.* 35:59–62.

31. Napoli, J.L., and Counsell, R.E. 1977. New inhibitors of steroid 11β hydroxylase. Structure activity relationships of metyrapone-like compounds. *J. Med. Chem.* 20:762–66.

32. Giles C., and Griffiths, K. 1964. Inhibition of the aromatising activity of human placenta by SU 4885. *J. Endocrin.* 28:343–44.

33. Schwartz, M.A., and Margolis, S. 1983. Effects of drugs and sterols on cholesterol 7α-hydroxylase in rat liver microsomes. *J. Lip. Res.* 24:28–33.

34. Liddle, G.W., Estep, H.L., Wendall, J.W., Williams, W.C., and Towns, A.W. 1959. Clinical application of a new test of pituitary reserve. *J. Clin. Endocr.* 19:875–94.

35. Meikle, A.W., West, S.C., Weed, J.A., and Tyler, F.H. 1975. Single dose metyrapone test: 11β-hydroxylase inhibition by metyrapone and reduced metyrapone assayed by radioimmunoassay. *J. Clin. Endocr.* 40:290–95.

36. Schoneshofer, M., L'Age, M., and Oelkers, M. 1977. Short term kinetics of deoxycorticosterone, deoxycortisol, corticosterone and cortisol during single dose metyrapone test. *Acta Endocr.* 85:109–17.

37. Kahnt, F.W., and Neher, R. 1966. Uber die adrenale steroide-biosynthese in vitro. III. Selektive hemmung der nebennierenrinden-funktion. *Helv. Chim. Acta.* 49:725–32.

38. Camacho, A.M., Cash, R., Brough, A.J., and Wilroy, R.S. 1967. Inhibition of adrenal steroidogenesis by aminoglutethimide and the mechanism of action. *J. Am. Chem. Soc.* 202:114–20.

39. Vermeulen, A., Paridaens, R., and Heuson, J.C. 1983. Effects of aminoglutethimide on adrenal steroid secretion. *Clin. Endocr.* 19:637–82.

40. Bruning, P.F., Bonfrer, J.M.G., De Jong-Bakker, M., and Nooyen, W. 1984. Influence of ACTH on aminoglutethimide induced reduction of plasma steroids in postmenopausal breast cancer. *J. Steroid Biochem.* 21:293–98.

41. Mancheno-Rico, E., Kuchel, O., Nowaczynski, W., Seth, K.K., Sashki, C., Dawson, K., and Genest, J. 1973. A dissociated effect of glutethimide on the mineralocorticoid secretion in man. *Metabolism.* 22:123–32.

42. Whipple, C.A., Grodzicki, R.L., Hourihan, J., and Salhanick, H.A. 1981. Effect of cholesterol side-chain cleavage inhibitors on 11β-hydroxylation: linkage of mitochondrial oxygenase systems. *Steroids.* 37:673–79.

43. Touitou, Y., Bogdan, A., Legrand, J.C., and Desgrez, P. 1975. Aminoglutethimide and glutethime: effects on 18-hydroxycorticosterone biosynthesis by human and sheep adrenal in vitro. *Acta Endocr.* 80:517–26.

44. Santner, R.J., Rosen, H., Osawa, Y., and Santen, R.J. 1984. Additive effects of aminoglutethimide, testololactone and 4-hydroxyandrostenedione as inhibitors of aromatase. *J. Steroid Biochem.* 20:1239–42.

45. Manni, A. and Santen, R.J. 1986. Clinical uses of aromatase inhibitors. In: B.J.A. Furr and A.E. Wakeling, eds. *Pharmacology and Clinical Uses of Inhibitors of Hormone Secretion and Action.* London: Bailliere Tindall, pp. 271–87.

46. Badder, E.M., Leraan, S., and Santen, R. 1983. Aminoglutethimide stimulates extra adrenal δ-4-androstenedione production. *J. Surg. Res.* 34:380–87.

47. Moore, R.N., Penney, D.P., and Averill, K.T. 1980. Fine structural and biochemical effects of aminoglutethimide and o,p'-DDD on rat adrenocortical carcinoma 494 and adrenals. *Anat. Rec.* 198:113–24.

48. Salhanick, H.A. (1982). Basic studies on aminoglutethimide. *Cancer Res.* 42, supp. 8:3315–21s.

49. Samojlik, E., and Santen, R.J. 1980. Potency of the effect of D-stereo-isomer of aminoglutethimide on adrenal and extra-adrenal steroidogenesis. *J. Clin. Endocr.* 51:462–65.

50. Foster, A.B., Jarman, M., Leung, C.-S., Rowlands, M.G., and Taylor, G.N. 1983. Analogues of aminoglutethimide: selective inhibition of cholesterol side-chain cleavage. *J. Med. Chem.* 26:50–54.

51. Coombes, R.C., Jarman, M., Harland, S., Ratcliffe, W.A., Powles, T.J., Taylor, G.N., O'Hare, M., Nice, E., Foster, A.B., and Neville, A.M. 1980. Aminoglutethimide: metabolism and effects on steroidogenesis in vivo. *J. Endocrin.* 87:31P.

52. Adam, A.M., Roger, H.J., Amiel, S.A., and Rubens, R.D. 1984. The effect of acetylator phenotype on the disposition of aminoglutethimide. *Brit. J. Pharm.* 18:495–505.

53. Burnham, D.F., Peyler, A.L., and Potts, G.O. 1963. Adrenal blocking effects of a new steroid, cynanotrimethylandrostenolone. *Fed. Proc.* 22:270.

54. McCarthy, J.L., Reitz, C.W., and Wesson, L.K. 1966. Inhibition of adrenal corticosteroidogenesis in the rat by cyanotrimethylandrostenolone, a synthetic androstane. *Endocrinology.* 79:1123–39.

55. Goldman, A.S. 1968. Further studies of steroidal inhibitors of delta-5,3-beta-hydroxysteroid dehydrogenase and steroid isomerase by substrate analogues. *J. Clin. Endocr.* 28:49–60.

56. Neville, A.M., and Engel, L.L. 1968. Inhibition of α and β-hydroxysteroid dehydrogenase and steroid-delta-isomerase by substrate analogue. *J. Clin. Endocr.* 28:49–60.

57. Shears, S.B., and Bod, G.S. 1981. The effect of azastene, cyanoketone and trilostane upon respiration and cleavage of the cholesterol side chain in bovine adrenal mitochondria. *Eur. J. Biochem.* 117:74–80.

58. Aguilera, G., and Catt, K.J. 1979. Loci of action of regulators of aldosterone biosynthesis in isolated glomerulosa cells. *Endocrinology*, 104:1046–52.

59. Lambert, F., Lammerant, J., and Kolanowski, J. 1984. The enhancement of

pregnenolone production as the main mechanism of the prolonged stimulation of ACTH on cortisol production by guinea pig adrenal cells. *J. Steroid. Biochem.* 21:299–303.

60. Singh-Asa, P., Jenkin, G., and Thorburn, G.D. 1982. Effects of hydroxysteroid dehydrogenase inhibitors on in vitro and in vivo steroidogenesis in the ovine adrenal gland. *J. Endocrin.* 92:205–12.

61. Potts, G.O., Creange, J.E., Harding, H.R., and Schane, H.P. 1978. Trilostane, an orally active inhibitor of steroid biosynthesis. *Steroids.* 32:257–67.

62. Lambert, F., Corcelle-Cerf, F., Lammerant, J., and Kolanowski, J. 1984. On the specificity of the inhibitory effects of trilostane and aminoglutethimide on adrenocortical steroidogenesis in guinea pig. *Mol. Cell. Endocr.* 37:115–20.

63. Touitou, Y., Auzeby, A., Bodgan, A., Luton, J.P. and Galan, P. 1984. 11β-hydroxy-11-ketosteroid equilibrium, a source of misinterpretation in steroid synthesis: evidence through the effects of Trilostane on 11β-hydroxysteroid dehydrogenase in sheep and human adrenal in vitro. *J. Steroid Biochem.* 20:736–68.

64. Potts, G.O., Batzold, F.H., and Snyden B.W. 1986. Endocrinology of trilostane and epostane. In: B.J.A. Furr and A.E. Wakeling, *Pharmacology and Clinical Uses of Inhibitors of Hormone Secretion and Action.* London: Baillere Tindall, pp. 326–36.

65. Komanicky, P., Spark, R.F., and Melby, J.C. 1978. Treatment of Cushing's syndrome with trilostane (WIN 24, 540), an inhibitor of adrenal steroid biosynthesis. *J. Clin. Endocr.* 47:1042–51.

66. Van der Spuy, Z.M., Jones, D.L., Wright, C.S.W., Piura, B., Paintin, D.B., James, V.H.T., and Jacobs, H.S. 1983. Inhibition of 3-beta-OH steroid dehydrogenase activity in the first trimester of pregnancy with trilostane and WIN 32729. *Clin Endocr.* 19:521–32.

67. Kahnt, F.W., and Neher, R. 1962. On the specific inhibition of adrenal steroid biosynthesis. *Experientia.* 18:499–501.

68. Raman, P.B., Sharma, D.C., and Dorfman, R.I. 1966. Studies on aldosterone biosynthesis in vitro. *Biochemistry.* 5:1795–804.

69. Kahnt, F.W., and Neher, R. 1972. On adrenocortical steroid biosynthesis in vitro. V. Activator and inhibitors. Evidence for the presence of substrate specific 21-hydroxylases. *Acta. Endocr.* 70:315–30.

70. Hewick, D.S., and Young, C.K., 1973. Stereoselective inhibition of steroid 11β-hydroxylation in ox and sheep adrenocortical mitochondria: comparison of metyrapone and the enantionmers of 2-(4-aminophenyl)-2-phenylethylamine (SKF-12185). *Biochem. Pharmacol.* 26:1653–59.

71. Touitou, Y., Bodgan, A., Auzeby, A., and Dommergues, J.P. 1979. Glucocorticoid and mineralocorticoid pathways in two adrenal carcinomas: comparison of the effects of o,p'-dichlorophenyldichloroethane, aminoglutethimide and 2-p-aminophenyl-2-phenylethylamine in vitro. *J. Endocrin.* 82:87–94.

72. Leroux, P., Delarue, C., Leboulenger, F., Jegon, S., Tonon, M.C., Corvol, P., and Vaudry, H. 1980. Development and characterisation of a radioimmunoassay technique for aldosterone from perifused frog interrenal tissue. *J. Steroid Biochem.* 12:173–78.

73. Loose, D.S., Kan, P.B., Hirst, M.A., Marcus, R.A., and Feldman, D. 1983. Ketoconazole blocks adrenal steroidogenesis by inhibiting cytochrome P-450—dependent enzymes. *J. Clin. Invest.* 71:1495–99.

74. Couch, R.M., Muller, J., Perry, Y.S., and Winter, J.S.D. 1987. Kinetic analysis

of inhibition of human adrenal steroidogenesis by ketoconazole. *J. Clin. Endocr.* 65:551–54.

75. Trachtenberg, J., and Pont, A. 1984. Ketoconazole therapy for advanced prostate cancer. *Lancet.* II:433–35.

76. Fry, D.E., and Griffiths, H. 1984. The inhibition by etomidate of the 11β-hydroxylation of cortisol. *Clin. Endocrin.* 20:625–29.

77. Wagner, R.L., White, P.F., Kan, P.B., Rosenthal, M.H., and Feldman, D. 1984. Inhibition of steroidogenesis by the anaesthetic etomidate. *New England J. Med.* 310:1415–21.

78. Lambert, A., Mitchell, R., Frost, J., and Robertson, W.R. 1985. A simple in vitro approach to the estimation of the biopotency of drugs affecting steroidogenic tissues. *J. Steroid Biochem.* 23:235–38.

79. Barbieri, R.L., Osathanodh, R., Canick, J.A., Shilman, R.J., and Ryan, K.J. 1980. Danazol inhibits human adrenal 21 and 11β-hydroxylation in vitro. *Steroids.* 35:251–63.

80. Girard, J., and Baumann, J.B. 1976. Secondary adrenal insufficiency due to cyproterone acetate. *J. Endocrin.* 69:13P–14P.

81. Phan-Huu-Trung, M.T., de Smitter, N., Boggo, A., and Girard, F. 1984. Effects of cyproterone acetate on adrenal steroidogenesis in vitro. *Hormone Res.* 20:108–15.

82. Erbler, H.C. 1972. Stimulation of aldosterone production in vitro and its inhibition by spironolactone in rat adrenals. *Arch. Pharm.* 273:366–75.

83. Cheng, S.C., Suzuki, K., Sadee, W, and Harding, B.W. 1976. Effects of spironolactone, canrenone and canrenoate-K on cytochrome P-450, and 11β and 18-hydroxylation in bovine and human adrenal cortical mitochondria. *Endocrinology.* 99:1097–06.

84. Ashby, H., Dimattina, M., Linehan, W.N., Robertson, C.N., Queenan, J.J. and Albertson, B.D. 1989. The inhibition of human adrenal steroidogenic enzyme activities by Suramin. *J. Clin. Endocr.* 68:505–08.

9

Comparative Aspects of the Adrenal Cortex

9.1 Structure and function in nonmammalian vertebrates

Although the terms cortex and medulla are derived from the histological arrangement of these organs in mammals, the adrenocortical and medullary components have their homologues throughout most of the vertebrates, while showing structural differences. These stem from two factors. One is the variability in the extent of the association of chromaffin (medullary) tissue with the adrenal cortex. The other is that, in most groups of nonmammalian vertebrates, it is difficult to see the distinction between the adrenocortical cell types which in mammals make up the various zones of the cortex [1–3].

9.1.1 Fish

Among the Agnatha neither adrenocortical tissue nor steroids of unequivocally adrenal origin have been satisfactorily identified, although presumptive adrenal cells have been found in the pronephric region of the dorsal side of the pericardium, and along the walls of the cardinal veins. Changes in the appearance of such cells in *Petromyzon* and *Lampetra* species, after treatment with mammalian ACTH, metyrapone or cortisol, suggest that they respond in a way which is consistent with their identity as adrenocortical cells [4–6]. There is no functional evidence for hormone secretion by these cells [7]. In *Myxine*, the hagfish, adrenocortical cells have not been identified,

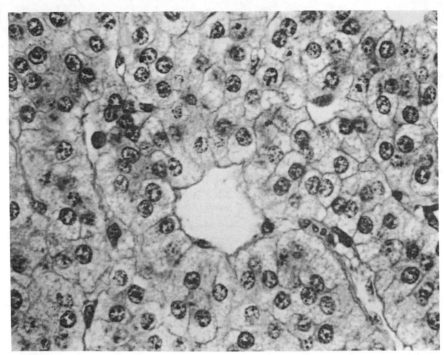

FIGURE 9.1 Dogfish interrenal tissue, × 590. (Micrograph provided by I. Chester Jones and W. Mosley.)

although there is some evidence for the presence of the glucocorticoids, cortisol, and corticosterone, in the plasma [8].

In the Elasmobranchs, however, adrenocortical tissue (usually called interrenal tissue in fish) is present as a single glandular structure lying between the kidneys. It may be in the form of a rod, or as a more compact gland, or in a horseshoe shape, and examples of these three different types are found in the sharks, rays, and in Torpedo, respectively [1,6]. In all three cases, the cortex is a discrete gland quite separate from the chromaffin cells, which appear as islets in the kidney. Histologically, interrenal cells are generally found to be homogeneous, with little zonation (Fig. 9.1).

One of the most remarkable features in some species of Elasmobranchs is the finding that a major project of the interrenal tissue both in vivo and in vitro, is 1α-hydroxycorticosterone, a compound not described in any other species of vertebrate [8–10]. In mammalian assays this compound is a potent mineralocorticoid, with approximately 70 percent of the apparent activity of DOC, and it lacks glucocorticoid activity [11]. In experiments on *Raja radiata* it was found, however, that interrenalectomy had no effect on plasma osmolarity, or concentrations of sodium, chloride, magnesium or potassium ions, or urea [12], although the animals were less able to survive the stress of reduced salinity than intact animals. Blood calcium levels were slightly, and briefly, elevated. In *Scyliorhinus canicula*, the gills have been identified as a site of Na^+/NH_4^+ and Na^+/H^+ exchange [13], and it is, therefore, particularly interesting that a binding protein for 1α-hydroxycorticosterone has been localized in *Raja ocellata* to regions of the gill rich in so called "chloride cells," thought to be the specific site of electrolyte transport [14]. Binding sites were also present in other potential target organs, including the rectal gland and a portion of the kidney tubule. Despite its lack of effect on plasma urea, it is tempting to continue to speculate whether this curious product is in some other way related to the unique mode of osmoregulation in this group, through the maintenance of high plasma concentrations of urea. This possibility has again been suggested by recent experiments in which the secretion of 1α-hydroxycorticosterone by perfused interrenal tissue from *Scyliorhinus canicula* was stimulated by increasing the urea concentration of the perfusion medium. Secretion of this steroid was also stimulated by mammalian Ile[5] angiotensin II, and, curiously, human synthetic ANP [15,16]. Obviously, several features of the endocrinology of the interrenal in this group of animals remain fairly mysterious. It is likely, for example, that the interrenal gland is under pituitary control [8,17], but renin is absent from the kidneys [16]. The significance of the actions of angiotensin II on the interrenal gland, therefore, awaits clarification.

The Teleosts are a much more widely studied group of fish. Histologically, the interrenal in this very diverse group consists of cells arranged as cords or clusters in the region of the posterior cardinal veins in the anterior part of the kidney, called the "head kidney" [1,2,6,18] (Figs. 9.1 and 9.2). Some chromaffin tissue is also found in this region, and other clusters of chromaffin cells are found elsewhere in the kidney. The interrenal cells, in many species [1,6], are quite similar in appearance to those of other vertebrates (Fig. 9.2) and even, indeed, to the glomerulosa cells in some species of carnivores such as the dog (cf. Chap. 1). Ultrastructural studies show that the interrenal cells of goldfish are packed with mitochondria with tubulovesicular cristae [19].

Cortisol is the major steroid produced by the interrenals in vitro, and it is also found in plasma in most species which have been studied [8–10]. Aldosterone is found occasionally in low amounts [20]. However it should be emphasized that the similarity between the steroid hormones of nonmammalian vertebrates and mammals should not mislead the reader into supposing that their functions are necessarily similar. In the Teleosts, for example, electrolyte exchange with the environment is controlled at a number of sites. These include not only the kidney, but also the gills, the urinary bladder, and the alimentary tract [16]. Electrolyte exchange at all of these sites is dependent to a greater or lesser extent on the presence of the interrenal, and adrenal steroid administration, in general, promotes sodium retention and potassium elimination. There is, however, no reason to suppose that aldosterone has a special role in these activities, and as noted, aldosterone has not even been clearly identified in many types [20]. Cortisol, at physiological levels, seems to have all of the activity which is

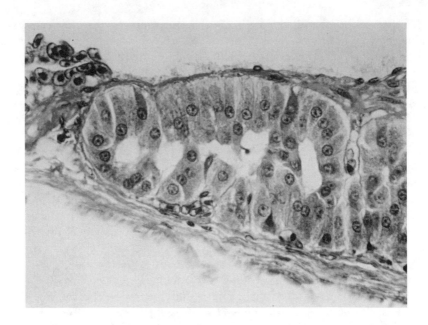

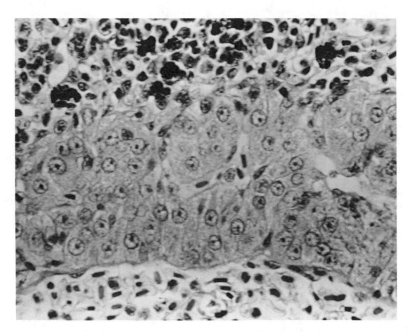

FIGURE 9.2 Interrenal tissue from (a) eel and (b) trout. (Micrographs provided by I. Chester Jones and W. Mosley.)

required to maintain electrolyte homeostasis. Cortisol also may have a role in maintaining gluconeogenesis and the balance between protein, lipid, and carbohydrate metabolism [16,21]. Accordingly, in those species of salmon which show higher corticosteroid levels during spawning [6], liver glycogen is also elevated at this time [16] and protein breakdown occurs. In these species, of course, spawning is associated with the annual migration to freshwater grounds, which makes, presumably, considerable demands on several physiological systems.

Seasonal variation is apparent in other Teleost species. In the killifish *Fundulus heteroclitus*, which undergoes a semilunar spawning cycle synchronized with the cycle of spring tides in the animal's tidemarsh habitat, serum cortisol as well as estradiol paralleled the reproductive cycle, with a peak in cortisol at the time of ovulation [19].

The structure and function of the gland is supported by ACTH, and is reduced by hypophysectomy [6] and the existence of a hypothalamo-pituitary-adrenal axis is suggested by various lines of evidence. Mammalian ACTH increases cortisol output in vitro by interrenal tissue in *Oncorhynchus kisutch* [22], and *Salmo gairdnerii* [23]. In the latter species, too, plasma cortisol was transiently raised by morphine (and by stress) and this response was reduced by the morphine antagonist naloxone. It is also known that low concentrations of morphine release CRF in this species [24], although high concentrations inhibit pituitary ACTH release. Such results suggest that both inhibitory and stimulatory opioidergic mechanisms may contribute to the mechanisms controlling hypothalamo-pituitary-interrenal activity in this trout, unlike the situation in mammals in which this type of reciprocity does not occur. Cyclic AMP has been shown to mediate the actions of ACTH in *Oncorhynchus kisutch* [25] and *Salmo gairdnerii* [26] interrenal tissue.

Maintenance of *Salmo gairdnerii* or *Anguilla anguilla* in black tanks raises plasma cortisol compared with animals kept in white tanks, suggesting either that the melanotropins may directly stimulate interrenal tissue, or that these peptides are cosecreted with ACTH [27]. In some species, histological changes in the interrenal tissue result from transfer of animals between environments of varying salinity. There appears to be a renin-angiotensin system, and although some evidence is contradictory [16,29], treatment with the angiotensin I converting enzyme inhibitor (captopril) diminished the plasma cortisol response of freshwater adapted eels to sea water transfer [28]. In *Salmo gairdnerii*, angiotensin II-Asp^1, Ile^5 had no effect on steroidogenesis in perfused interrenal tissue when added alone, but enhanced the stimulatory effects of both ACTH and forskolin [26]. Interestingly, in view of other work in amphibia these authors also found some evidence to support the concept that the angiotensin II effect was mediated by prostaglandin synthesis, at a site distal to cyclic AMP generation. Steroidogenesis was not stimulated by either the calcium ionophore, A23187, or the phorbol ester TPA, although the response to the combined effects of ACTH and angiotensin II was diminished by the calcium channel blocker verapamil,

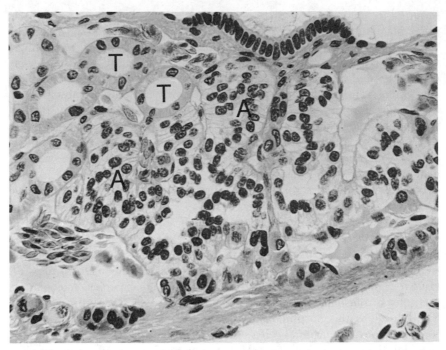

FIGURE 9.3 Adrenocortical tissue in the salamander showing cords of adrenocortical cells (A) in close proximity to kidney tubules (T). (Micrograph provided by I. Chester Jones and W. Mosley.)

suggesting that, as in mammals, calcium may have a role in mediating the actions of these stimulants.

Adrenocortical tissue also has been tentatively identified in the rarer groups of fish including the Crossopterygii, Chondrostei, Dipnoi, and Holostei [17]. In the lungfish (Dipnoi), the presumptive adrenocortical cells lie closely associated with the postcardinal veins, and their capillaries, where these pass through the kidneys, and are very similar in position to adrenocortical tissue in the urodele amphibian, *Pleurodeles waltii*. This is of particular evolutionary interest since it reinforces the view that this group appears to have many features in common with amphibia [30].

9.1.2 Amphibia

In this vertebrate class, the adrenal homologue comprises groups or clusters of cells lying on the ventral surface of the kidney, and receiving a blood supply from vessels which also supply the kidney [31]. The venous return is by vessels which enter the renal veins and thence the posterior vena cava. The cells are arranged in looped cords and are somewhat flattened in appearance, like those of teleosts and indeed of other nonmammalian vertebrates (Fig. 9.3). The cords of cells which comprise the adrenal are only loosely bound and give the gland a somewhat amorphous shape. There is no clear division into zones as seen in mammals. In the species which have been studied, mitochondria are plentiful and have cristae of the tubular form. Lipid droplets and prominent smooth endoplasmic reticulum are other features which closely resemble the appearance of these structures in mammalian adrenocortical cells. In some species of Anura, an additional cell type, the Stilling cell, is found [32]. These are of unknown origin and function. They do not seem to be concerned with steroidogenesis and they may be homologues of lymphoid cells. The histology, histochemistry, and ultrastructure of the adrenocortical cells in amphibia change after treatments with ACTH or metyrapone, or hypophysectomy. Such studies strongly support the view that a hypophyseal-adrenal axis exists in these species as it does in mammals [31–33].

In vitro, in most species, the major products of amphibian adrenals are corticosterone, aldosterone, and 18-hydroxycorticosterone [10,20]. In vitro, aldosterone and 18-hydroxycorticosterone frequently are produced in higher yields than corticosterone but in circulating plasma there is usually more corticosterone than aldosterone [30,34,35]. Cortisol is found in some species [31].

There has been a considerable amount of work on the control of steroidogenesis in vitro in the anuran model *Rana ridibunda*. In this system, perfused interrenal fragments respond sensitively to ACTH and cyclic AMP, and to frog pituitary anterior and intermediate lobe extracts [36] and the production of aldosterone under these conditions is twice that, or more, of corticosterone [37]. The system also responds sensitively to the analogue angiotensin II-Ser^2, Val^5. Of considerable interest is the finding that the response to this analogue, but not to ACTH, is inhibited by prostaglandin synthetase inhibitors, and that the angiotensin analogue was shown to stimulate prostaglandin E_2 production [38]. This is a marked point of contrast with mammalian systems, in which there is little evidence to suggest that prostaglandins act as intracellular or intraglandular messengers in the adrenal, although prostaglandins are known to be stimulatory. As in mammalian cells, intracellular calcium is also involved in the mediation of the steroidogenic response to angiotensin II, but not to ACTH, and the involvement of calcium ions occurs after the induction of prostaglandin formation [39,40]. Products of the lipoxygenase pathway, leading to the leukotrienes, are not involved [41].

Because of the close association between adrenocortical cells and chromaffin cells in amphibia, generally, it is perhaps not surprising that interrenal tissue in *Rana ridibunda* may be controlled by cholinergic influences [42] and by several neuropeptides which may be found in chromaffin tissue in this species, including the enkephalins, VIP, arginine vasopressin, oxytocin, vasotocin and mesotocin, and also serotonin [42–44].

A renin-angiotensin system clearly exists [21], but its functions are unclear, and its responses to sodium balance, as described in the literature, are contradictory [31]. However, plasma corticosterone concentrations in

Xenopus laevis are clearly affected by changes in environmental salinity, whether or not the pituitary is present [45], and in *Rana temporaria*, it has been shown that the output of aldosterone and corticosterone by perfused interrenal tissue is directly stimulated by increasing potassium ion concentrations and inhibited by increasing sodium [46]. The system is undoubtedly complex, and these authors do not exclude the possibility that other factors are involved in in vivo responses, including possibly the renin angiotensin system.

Electrolyte homeostasis is maintained at multiple sites in amphibia including the lung, kidney, skin, and cloaca. Although the toad bladder has been a classical site for the study of aldosterone action (cf. Chap. 4), it is by no means clear that this steroid has an especially important physiological role in these animals. Both corticosterone and aldosterone levels may respond to changes in sodium balance. Both steroids can also induce hyperglycemia, and hypophysectomy lowers blood sugar in all species investigated. Liver and muscle glycogen may be increased by treatment with ACTH, corticosterone, or aldosterone, but the effects may not be so readily observed as the changes in blood glucose. Adrenal steroids can also be shown experimentally to effect changes in protein and lipid metabolism. In general it can be said that the classical division of corticosteroids on the basis of their action, into mineralocorticoids and glucocorticoids, as seen in mammals, cannot necessarily be held to apply in amphibia (for references see Hanke [31]).

One remarkable aspect of adrenocortical hormone action in amphibia is that administered corticosteroids may interact with gonadotropins or gonadal steroids and elicit oocyte maturation and even ovulation [31,47]. This may be significant in that there is known to be an association between increased adrenocortical function and spawning. Increased adrenocortical activity also occurs just prior to the metamorphosis of the larval into the adult forms [31].

9.1.3 Reptiles

In the reptiles, the adrenals form discrete elongated glands lying anterior to the kidney. They thus have their own blood supply and venous drainage, although usually very close to the major blood vessels [1,48]. There is a varying degree of mixing of chromaffin and adrenocortical cells. In some species, there is a discrete dorsal mass of chromaffin tissue, as well as islets set among the cords of adrenocortical cells. In as far as generalizations are possible, the adrenocortical cells are flattened and stacked in cords which are thrown into somewhat amorphous loops and lobes (Fig. 9.4). The cells show, in ultrastructure, a profusion of mitochondria characteristic of adrenocortical cells, with tubulovesicular cristae, numerous lipid droplets and extensive smooth endoplasmic reticulum [49]. Adrenocortical zonation on the mammalian pattern does not occur, although differences between cells in the subcapsular region and those of the inner cortical region have

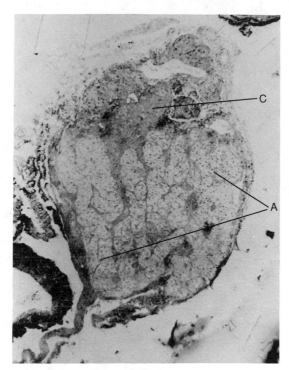

FIGURE 9.4 Adrenocortical tissue (A) from *Sphenodon* showing relationship with chromaffin tissue (C). (Micrograph provided by I. Chester Jones and W. Mosley.)

been shown in the cobra [50]. In a number of reptiles, the gland undergoes seasonal changes in morphology and there may be a direct correlation between reproductive activity and adrenocortical secretion [49,51].

There is a functional hyphophyseal-adrenocortical axis, and hypophysectomy brings about a reduction in adrenal weight, while treatment with mammalian ACTH can enhance steroid output [49,52]. Changes following treatments with metyrapone (which increases adrenocortical size) or with adrenal steroids (which produce involution of the gland) are also consistent with this view [49,52]. There are differences from mammals in the function of the hypothalamo-pituitary-adrenal system, however, since both aldosterone and corticosterone exert negative feedback control at the hypothalamus. After hypophysectomy, circulating corticosterone concentrations are about 30 percent of normal values, suggesting a degree of adrenal independence from pituitary control [53]. The major steroids produced in vitro are corticosterone, aldosterone, and 18-hydroxycorticosterone and it is likely that these steroids are present in the plasma of most species [20,52].

Components of the renin angiotensin system are present, where these have been sought, but stimulation of steroidogenesis by angiotensin II is equivocal [52]. In the turtle *Pseudemys scripta*, infusion of Asp[1], Val[5] angiotensin II gave increased circulating levels of corticosterone but not aldosterone, and in this respect its actions resembled that of porcine ACTH [54]. In perfused adrenal tissue from *Lacerta vivipara*, on the other hand, Sar^1 Val^5 angiotensin II stimulated both corticosterone and aldosterone outputs similarly, whereas human ACTH$_{1-39}$ had a more marked effect on corticosterone [55]. The question arises, as in other nonmammalian vertebrates, whether the distinction between mineralocorticoids and glucocorticoids, on the mammalian pattern, is valid in reptiles, and some authors are certainly of the opinion that it is not [53]. Accordingly, interpretation of the role of the renin-angiotensin system in mammalian terms may not be appropriate.

Sites of control of electrolyte flux in reptiles include the bladder and the salt (nasal) gland, in addition to the kidney, and steroids affect ion transport at all of these sites. Aldosterone promotes sodium reabsorption in the kidney and the bladder and diminishes sodium excretion via the salt gland. In contrast, corticosterone is generally hyponatremic and promotes sodium excretion by the salt gland. Corticosterone can also bring about changes in water distribution and mobilization. It is not always clear how circulating levels of steroids respond to changes in salinity, although increases in circulating corticosterone have been noted when certain lizards are saline loaded. Corticosteroids also affect gluconeogenesis and can bring about increases in blood sugar and hepatic glycogen deposition (for review see ref. [52]).

9.1.4 Birds

In most birds, the adrenals are paired organs lying at the anterior end of the kidneys close to the major vessels [1,56]. The gland is supplied by blood coming directly from the aorta, or from branches of the renal artery, and the veins drain into the posterior vena cava at about the level of the bifurcation. Chromaffin tissue is present in islets, bigger clumps, or strands, throughout the cortex which generally consists of cords of flattened cells thrown into loops and lobules much as seen in reptiles (Fig. 9.5). Ultrastructurally, the adrenocortical cell is characterized by the profusion of mitochondria and lipid droplets and development of the smooth endoplasmic reticulum [56,57] the characteristics, as we have seen, of adrenocortical cells of other classes. In some species, evidence for rudimentary zonation of the gland is apparent and there are features of the subcapsular zone of the gland which resemble the zona glomerulosa of mammals [58]. In particular, this tissue responds to changes in sodium balance in a manner consistent with the changes seen in the mammalian zona glomerulosa in similar situations [59]. It is also less affected by hypophysectomy (Fig. 9.5). However, other evidence suggests that the stem cells for the adrenocortical cords lie in the innermost part of the gland [60] and this would be a point of contrast since the stem cells in the mammalian adrenal cortex lie in or just below the zona glomerulosa (see Chap. 1).

The major steroids produced in vitro and present in circulating plasma in birds are of the 17-deoxy conformation. Corticosterone is the major product in vitro with smaller amounts of aldosterone and 18-hydroxycorticosterone [20,56]. It is also possible that 17-hydroxysteroids, including cortisol, are produced in some species [20]. As in mammals, there is, in some species, a distinct diurnal rhythm in the circulating corticosterone level, with a peak early in the morning and a nadir at dusk. Experiments using mammalian ACTH and cortisol treatments in vivo in intact and in hypophysectomized animals, strongly indicate the existence of a classical hypophyseal-pituitary-adrenocortical axis [56]. Studies on the subcellular events following ACTH stimulation provide some evidence that its mode of action on the adrenal is similar to that in mammals. Data obtained with superfused duck adrenal tissue suggests that the steroidogenic response is dependent on the "rapidly turned-over" protein familiar from mammalian studies [61] (see Chap. 3). This is associated with changes in mitochondrial morphology, as the cristae are transformed from a shelflike form into the tubular configuration—a feature of sustained ACTH treatment in mammals, but which has not been described in conditions of acute stimulation (cf. Chap. 1). Chicken adrenal cells respond to cyclic AMP stimulation [62], and cyclic AMP is generated in response to ACTH [63]. Studies on the metabolism and clearance of steroids in ducks suggest that the hypophyseal secretion of ACTH is also involved in the control of these processes [56]. It is possible that the control of adrenocortical function is complex, since stimulation has also been achieved with ovine prolactin [64] human, bovine, and avian parathyroid hormone [65], serotonin [66], and human growth hormone [67]. There is evidence that a degree of adrenocortical function may be independent of hypophyseal control however, and an extra-hypophyseal source of corticotropin has been suggested, although it is not clear what this source might be [56,68,69].

The elements of the renin angiotensin system appear to be present in birds, but the relationship between the function of this system and adrenocortical function is again not clear [56]. In particular, high dietary intake of sodium ions depresses circulating aldosterone concentrations in chicken, but are without effect on corticosterone [70] and low sodium intake enhances circulating aldosterone [71]. However, neither Ile^5 nor Val^5 angiotensin II nor potassium ions affected either steroid in dispersed chick adrenal cells in vitro. Rat ANP, however, both stimulated cyclic GMP production, and inhibited aldosterone [72]. In studies on intracellular mechanisms, neither the calcium ionophore A23187 nor calcium blockers, nor the phorbol ester TPA had any effect on steroidogenesis, although they are known to affect aldosterone in mammals (see Chap. 3). Accordingly, the mode of control of these steroids is incompletely understood at present.

Similar to other nonmammalian groups, it is evident that in birds electrolyte flux may be controlled at a number of sites, including not only the kidney, but also, and most importantly in some species, the nasal (or salt) gland in the orbit [56,73]. As in reptiles, the function of this gland is to eliminate unwanted sodium ingested when drinking salt water. In ducks,

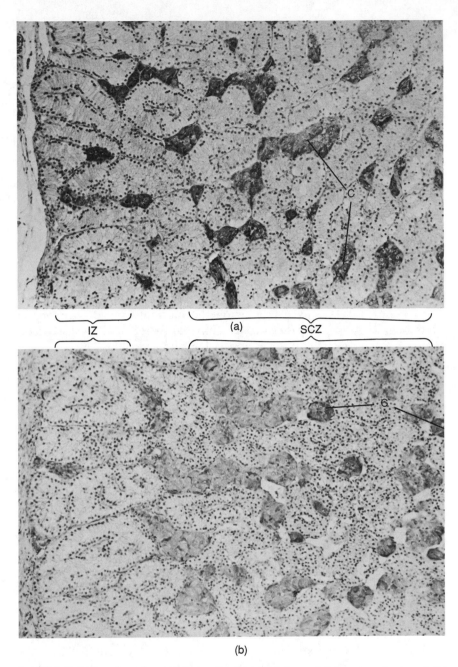

(a)

(b)

FIGURE 9.5 Outer region of duck adrenal gland from (a) intact animal and (b) hypophysectomized for 14 days showing subcapsular zone (SCZ) of interrenal tissue and inner zone (IZ). This contrast is not apparent in intact animals. C = chromaffin tissue. × 144 (Micrographs provided by J. Cronshaw), (cf. ref. [57]).

aldosterone causes a significant decrease of the kidney excretion of both sodium and potassium. Nasal gland activity, however, is also dependent on the adrenal cortex and administered corticosteroids promote the flow of nasal gland salt secretion [74]. This is also true even of aldosterone, although, in the intact animal, administered aldosterone essentially has an overall sodium retaining effect, because of its potent actions on the kidney. It is not actually clear whether these corticosteroids are a major controlling factor or merely permissive in nasal gland function, since the responses of plasma steroid levels to changes in osmolarity of drinking water are transitory and slight [75]. Sodium retention is stimulated equally by mineralocorticoids and glucocorticoids in the coprodeum (lower intestine) in hens [71].

Administered corticosteroids, in particular corticosterone, cause an overall loss in body weight even though food intake may increase. One feature of the actions of cortisol and corticosterone is that the distribution of protein, glycogen and lipid changes, and it may be that metabolism is changed in favor of fat deposition. Chronic administration can produce hyperglycemia and deposition of glycogen in the liver. Overall, liver weight changes may also reflect increases in the protein and fat content of the liver [56].

9.1.5 Monotremes and marsupials

The two monotreme species have differences in adrenocortical function. In the platypus, *Ornithorhynchus anatinus*, cortisol is the major adrenocortical product in circulating plasma, (at a concentration similar to that found in placental mammals) and when adrenal tissue is incubated in vitro [76]. In the echidna, *Tachyglossus aculeatus*, on the other hand, although cortisol is present, corticosterone is the major adrenocortical product in vivo and in vitro. Concentrations of these hormones in plasma were only one tenth of those in placental mammals [77,78]. The echidna adrenal responds to ACTH and stress in a similar way to other mammals, but clearly the level of activity is much lower [79]. As yet, the relationship between these secretory patterns and the metabolic requirements of the animals is not known.

Considerably more is known about marsupial adrenocortical function. Generally, the structure of the gland appears to be similar in all respects to those of Eutherian mammals, with the cortex, showing apparently normal zonation (although functional data to support this is generally lacking), surrounding a typical Eutherian-like medulla [80]. The single major exception to this arrangement is in the brush-tailed possum, *Trichosurus vulpecula*, with its special zone expressed only in adult females (cf. Chap. 1, and following text.)

In every species which has been examined, cortisol is the major secretory product, with smaller amounts of corticosterone, when it is detected, and aldosterone, in proportions and concentrations largely comparable to those seen in the Eutheria [77,81]. Insofar as the evidence is available, there is no reason to suppose that control of these secretions by stress, the pituitary, or, in the case of aldosterone, by changes in electrolyte availability is grossly different, in physiological terms, from the situation in Eutheria [71,81–83].

There is considerable variation in the metabolic response to glucocorticoids in the marsupials. In *Trichosurus vulpecula* and *Antechinus stuartii* there is a highly sensitive response to the diabetogenic nitrogen-mobilizing actions of cortisol, but some Macropidae, including *Macropus rufa* and *Setonix brachyurus* (but not *Macropus eugenii*) [84] are highly resistant to these effects, and it has been suggested that such resistance may be of adaptive value to macropodid marsupials on a low plane of nutrition in arid environments. Another adaptation, also beneficial in such circumstances, is the capacity to recycle urea, so that nitrogen mobilized as a consequence of glucocorticoid action is not necessarily lost [85,86].

In certain small shrewlike dasyurid marsupials, notably *Antechinus swainsonii*, all the males in natural populations die after the short intensive annual mating period. At this time, plasma cortisol levels are extremely high, and CBG binding falls, and the animals are thought to die as a consequence of the suppression of the immune system. There are several causes of the increase in circulating free cortisol, and transcortin is reduced by the actions of testosterone and ACTH. At this time, too, ACTH is no longer suppressed by dexamethasone injection, indicating impairment of this feedback loop [87].

9.2 Functional aspects of adrenocortical zonation in mammals

The observation that the mammalian adrenal cortex comprises concentric shells or zones of cells with different morphological appearance leads (cf. Chap. 1) to the hypothesis that each is associated with a different function. It was established some 50 years ago that the zona glomerulosa is associated with electrolyte status and is at least partially independent of the pituitary, whereas the inner adrenal cortical zones, the fasciculata and reticularis, depend on the presence of the pituitary [1,2,88]. The appearance of these zones under different conditions of stimulation led to the view that the inner zones are primarily concerned with glucocorticoid secretion.

Initially, much of the evidence for functional differences in the zones was indirect. However, with the advance of techniques of in vitro incubation and, subsequently, for cell purification, it became possible to achieve more direct evidence. Some 30 years ago, it was found by incubating slices of bovine adrenal cortex, that tissue from the outer part of the gland had the capacity to produce aldosterone, whereas cortisol was more readily formed in the inner zones [89]. In the rat, which does not have the capacity to secrete cortisol in significant amounts, the major steroidogenic difference between the glomerulosa and the inner zones lies in the capacity of the glomerulosa to produce aldosterone, and its companion 18-hydroxycorticosterone [90]. No aldosterone and very little 18-hydroxycorticosterone is produced by inner zones. In addition, in the rat gland, the glomerulosa (but not the fasciculata/reticularis), has the ability to sequester stores of 18-hydroxydeoxycorticosterone in the form of a lipid protein complex apparently in the cell membrane. This stored pool of steroid may be utilized in the formation of aldosterone and 18-hydroxycorticosterone [91].

The separate functions of the fasciculata and reticularis zones have not been clearly resolved. There is evidence to support the contention in earlier texts that the reticularis is a source of adrenal androgens [92]. Several studies have shown that the ratio of cortisol to androgens produced in vitro is lower in the reticularis than in the fasciculata. In microdissected dispersed cells from human adrenal tissue maintained in monolayer culture, it was found that reticularis cells produced more androstenedione and 11β-hydroxyandrostenedione than fasciculata cells from both added ^{3}H-pregnenolone and from endogenous precursors, and less corticosterone, although cortisol production by the two cell types was similar. Reticularis cells also formed twice as much sulfoconjugated steroid (principally DHEAS) as fasciculata cells [93].

Differences between the fasciculata and reticularis ratios of C_{19} to C_{21} steroids were also apparent in some related studies using guinea pig adrenal tissue, although the findings varied from the human data in detail. In guinea pigs, it may be concluded that the reticularis contribution to cortisol production is negligible [94–96]. The relative contribution of the two cell types to overall androgen production is less clear. Using purified dispersed reticularis cells, conversion of ^{14}C-pregnenolone to both 11β-hydroxyandrostenedione and androstenedione was only about 10 percent that obtained with similarly purified fasciculata cells. However, in other studies in which microdissected guinea pig adrenal was maintained in organ culture, a range of C_{19} steroids, including particularly androstenedione and DHEAS, with smaller amounts of testosterone and Δ^5-androstenediol, were produced under basal conditions from endogenous precursors predominantly (60–70%) by the reticularis [95]. When these authors later used dispersed (though not purified) cell preparations, androstenedione production was similar in the two cell types. These discrepancies may depend on differences in the techniques used. They emphasize the difficulties inherent in drawing general conclusions about this topic, which is only accessible to study in vitro, and is not easy to verify in other ways.

In the rat, 11β-hydroxylation occurs slightly less readily in the reticularis, and hence the corticosterone/deoxycorticosterone ratio is lower in this zone than in the fasciculata [97]. Production of androstenedione from endogenous precursors was similar, and very low, for the two cell types in these experiments with purified cells. Accordingly, these authors were unable to establish that the reticularis had any significant role in this species [98].

Further data, in the guinea pig and human adrenal, suggests that sulfoconjugation (a major route for steroid metabolism in these species), occurs more readily in the reticularis [93,98].

These seem to be the extent of the differences in steroidogenesis between the zones of the adult adrenal cortex. Further differences between these zones are, however, found in their responses to stimulation. As can be seen from Chap. 3, the secretion of aldosterone is under multifactorial control, whereas the secretion of glucocorticoids is highly correlated with ACTH alone. Similarly, it has been found that zona glomerulosa cells respond in vitro to a wide variety of effectors (extensively described in Chap. 3), whereas only ACTH can clearly be shown to affect the inner zones. Some qualification to this assertion is necessary, however, since angiotensin II receptors may occur throughout the adrenals in some species, notably the bovine [99]. In the inner zones of the rat adrenal cortex, angiotensin II stimulates phospholipid metabolism [100] but it evokes no response in steroid output. Another possibility is that peptides from the N-terminal portion of the POMC molecule, which may include gamma MSH, can potentiate inner zone responses to ACTH [101]. Once again, however, it is important to draw attention to the somewhat different interpretations and conclusions which different groups of workers have reached, and which may depend on variations in the precise methods employed. Using tissue from guinea pig and rat adrenals, corticosteroid output by reticularis cells is relatively refractory to ACTH stimulation [94–96,98]. Details vary: in guinea pig reticularis cells purified by unit gravity sedimentation on an albumin gradient, cortisol secretion responds reasonably well to ACTH, whereas androstenedione does not, leading to a sharp decrease in the androstenedione/cortisol ratio. Fractionation on Percoll gradients gave different results, perhaps a result of more effective cell purification, and the fasciculata cell population gave cortisol responses to ACTH two to five times greater than the reticularis cells. The decrease in androstenedione/cortisol ratio in reticularis cells was nevertheless similar to that seen in albumin gradient purified cells [96]. However, in dispersed cells from microdissected guinea pig tissue, others were unable to find any cortisol response at all to ACTH stimulation in reticularis cells [94], or to dibutyryl cyclicAMP [94,102], while concluding that androstenedione is stimulated by ACTH [102]. Thus, the reticularis may [102] or may not [96] be a significant source of androgen.

Other differences between the functions of these cell types have been explored. Although possibly refractory in terms of cortisol production, cyclic AMP production in the guinea pig adrenal reticularis and outer zones responds equally well to ACTH stimulation [103], and the increase in HMG Co-A reductase after ACTH treatment is indeed greater in the reticularis [104], while LDL receptor activity response is greater in the outer zones. After prolonged dexamethasone treatment, fasciculata volume decreases, but the reticularis does not, changes which are paralleled by cholesterol side-chain cleavage activity in these fractions [105], and since addition of later precursors (pregnenolone onwards) to reticularis cells gave healthy production of corticosteroids, including cortisol [102], the conclusion must be drawn that the low activity of this enzyme accounts for the poor production of cortisol by the reticularis in basal or stimulated conditions.

A further function for the reticularis is, however, suggested by work which shows that Δ^4-hydrogenase activity in guinea pig adrenals resides almost exclusively in the reticularis, giving 5α or 5β dihydro derivatives of corresponding Δ^4-3-ketones, such as cortisol [106]. The physiological significance of the adrenal production of these reduced compounds is unclear, although some are known to be central nervous depressants [107]. This field merits further investigation.

The possibility still exists that stimulants other than ACTH may support the reticularis. One which has been discussed in the literature in this context, as well as in connection with the human fetal zone, see following text, is prolactin. In vitro studies in which prolactin was used acutely have failed to produce evidence of its involvement [93], but in vivo evidence exists. In particular, chronic prolactin treatment led in orchiectomized rats to atrophy of the fasciculata, and a conspicuous hypertrophy of the reticularis. At the same time, plasma corticosterone was slightly, and testosterone was greatly increased [108]. Clearly this possibility too merits further study.

9.2.1 Functional interaction between zones

Another interpretation of adrenocortical function stems from the fact that the zones are arranged, as noted in Chap. 1, in concentric shells while the blood supply flows centripetally. This leads to the possibility that products of the outer zones may affect cells lying nearer the center of the gland. Examples show that aldosterone may inhibit steroid production by fasciculata and/or reticularis [109] and that increasing concentrations of cortisol towards the innermost part of the cortex results in suppression of specific enzyme activities, thus giving rise to the changes in glucocorticoid/androgen ratios seen in the innermost zones as noted [110]. In this connection, it is well understood that the flow of corticosteroids to the medulla is important in the full development of medullary function, and the final step in the pathway for epinephrine formation, phenylethanolamine N-methyl transferase depends on high local concentrations of corticosteroids (for general review see ref. 111). The reverse interaction may also occur: regulation of cortical function by the medulla is possible through paracrine mechanisms which may involve intra-adrenal nerves [112]. There is no evidence for the theory from the earlier literature, sometimes quoted in clinical textbooks, that, in the human gland, the reticularis is the major site of secretion and the zona fasciculata merely a site of cholesterol storage. In contrast, where direct comparisons between the fasciculata and reticularis have been made in experimental animals, the reverse is true (see previous text).

9.2.2 Additional adrenocortical zones

Variants on the adult system of zones are seen in adrenals of other species. Most commonly these occur as additional zones in the inner part of the gland. Examples are the X zone of the mouse and special zone in the possum adrenal cortex [1,80]. No function has as yet been ascribed to the former although it is known to be exceptionally sensitive to androgen and consequently disappears, leaving a residue of connective tissue surrounding the medulla at puberty in the male or at the first pregnancy in the female [1,80,92]. This zone is at least partially independent of ACTH, and may be supported by gonadotrophins. The same is true of the special zone in the cortex of the brush-tailed possum *Trichosurus vulpecula*, referred to previously (and Chap. 1). The capacity of this adrenal of this animal to form C_{19} steroids, including androstenedione, and testosterone [113] may be marked under some circumstances, possibly because of gonadotrophin stimulation [113,114], and this may be focused in the special zone, although the zone does not vary in size with the stages of the estrous cycle [115]. The special zone also has the capacity to form reduced ring A products, and although the secretion of testosterone or androstenedione into the adrenal vein has been doubted [116], certainly it contains all of the enzymes necessary for the formation of C_{19} products [117].

The human fetal zone which comprises the major part of the fetal cortex (see Chap. 1), has a relative lack of the Δ^5-3β-hydroxysteroid-dehydrogenase/isomerase system. As a consequence it produces large amounts of Δ^5,3β-hydroxysteroids, including pregnenolone, and its sulphate, and also, dehydroepiandrosterone sulfate (DHEAS). It is this latter compound which, when secreted in fetal life, is utilized by the placenta as a substrate for formation of 16α-hydroxyDHEAS, and is a substrate for the production of the increasingly larger amounts of estriol by the placenta [118]. The fetal gland is certainly under ACTH control, presumably secreted by the fetal pituitary [119]. De novo synthesis of cholesterol and LDL binding are stimulated by ACTH in both fetal and definitive cortex, while HMG Co-A reductase and adenylate cyclase were preferentially stimulated in the definitive cortex [120,121]. Anencephaly provides further insight into the control of the fetal adrenal. In these infants pituitary function is grossly impaired, and the adrenal gland is small, mostly because the fetal zone is absent. As a result, DHEAS production is low, and maternal estrogens are also low [122]. Anencephalic fetuses are hypercholesterolemic as a result of reduced LDL uptake by the adrenal [123]. However, not all observations are compatible with the interpretation that ACTH is the sole support for fetal adrenal function, and the period of greatest increase in adrenal weight is early in gestation, when plasma ACTH concentrations are low. Conversely, ACTH concentrations are more or less sustained during the perinatal period, when the fetal zone undergoes involution [119]. There is no evidence that other ACTH related peptides may be involved in fetal zone maintenance [121], but, despite much effort in this field, there is still little resolution about what additional trophic agent might be involved. Chorionic gonadotropin (hCG), which was once considered as a promising candidate in this role, seems on balance to be ineffective [124]. Prolactin, however, may be involved [125], a view also supported by recent studies in the baboon *Papio anubis*, in which it was found that prolactin may have a role in stimulating DHEA (but not DHEAS or cortisol) production in the midterm fetal adrenal cortex [126].

It will be clear from this account that in many ways the functions of the inner adrenocortical zones are enigmatic. However, there appear to be features of these tissues which are characteristic, and held in common. Thus the zona reticularis, the mouse X zone, the possum special zone, and the human fetal zone all have been thought to be at least partly independent of ACTH secretion, although the pituitary is important. In each case, there is some basis for supposing that either luteinizing hormone or prolactin may be the additional factor required. Furthermore (with the exception of the X zone, for which data is difficult to obtain, given that the whole mouse adrenal gland weighs only 2–3 mg), all have been shown to be particularly implicated in C_{19} steroid production. It may be, in summary, that all of these seemingly different tissues reflect somewhat varying manifestations of what is fundamentally the same organ and cell type. These are prominent tissues, with well formed healthy cells. Although the inner zones may well be a site for adrenal cell death (see following text), there are strong grounds for supposing that this secretory activity must play an important role in the metabolic (or reproductive) function of the animal.

9.2.3 Development of the adrenal cortex

Another interpretation of zonation is that the different cell types represent cells at different stages in their life cycle. This concept, originally suggested by Gottschau's work almost a hundred years ago, holds that cell division is more common in the outer part of the cortex and cell death is more common in the reticularis [127]. Although subsequently other authors suggested that other types of interchange between the different cell types may occur, this cell migration theory is undoubtedly valid. Not only have most subsequent findings shown that cell division occurs more frequently in the zona glomerulos (or just below) and that cell division is very rare in the reticularis [128–130], a cell division is also stimulated by ACTH treatment in vivo. Indeed it is possible that ACTH stimulates the whole sequence of this life cycle. Certainly, ACTH causes a transformation of the glomerulosa cell to a fasciculata (or intermediary) cell type in which aldosterone secretion is suppressed, as is the responsiveness of the tissue to the range of effectors characteristic of the glomerulosa cells (see Chaps. 1 and 3). It has recently been proposed that other peptides derived from the POMC molecule, including nongamma MSH containing N-terminal peptides may be involved in the stimulation of cell division [131] (see Chap. 1). The actions of these peptides on zonal transformations remains to be seen.

References

1. Chester Jones, I. 1957. *The Adrenal Cortex*. Cambridge: The University Press.
2. Deane, H.W. 1962. The anatomy, chemistry, and physiology of adrenocortical tissue. *Handb. Exp. Pharm.* 14:1–185.

3. Idelman, S. 1978. The structure of the mammalian adrenal cortex. In: I. Chester Jones and I.W. Henderson, eds. *General, Comparative and Clinical Endocrinology of the Adrenal Cortex*. Vol. II, New York and London: Academic Press, pp. 1–199.

4. Seiler, K., Seiler, R., and Sterba, G. 1970. Histochemische untersuchungen am interrenalsystem des bachneunauges (*Lampetra planeri*) Bloch. *iActa biol. medica Germanica*, 24:553–54.

5. Seiler, K., Seiler, R., and Hoheisel, G. 1973. Zur cytologie des interrenalsystems beim bachneunauge (*Lampetra planeri*) Bloch. *Gegenbaurs morphologisches jahrbuch, Leipzig*, 119:823–56.

6. Chester Jones, I., and Henderson, I.W. 1980. The interrenal gland in Pisces. Part 1. Structure. In: I. Chester Jones and I.W. Henderson, eds. *General, Comparative and Clinical Endocrinology of the Adrenal Cortex*. Vol. III. New York and London: Academic Press, pp. 395–472.

7. Weisbart, M., Youson, J.H., and Weibe, J.P. 1978. Biochemical, histochemical and ultrastructural analyses of presumed steroid-producing tissues in the sexually mature sea lamprey, (*Petromyzon marinus*) L. *Gen. Comp. Endocr.* 34:26–37.

8. Idler, D.R., and Truscott, B. 1972. Corticosteroids in fish. In: D.R. Idler, ed. *Steroids in Nonmammalian Vertebrates*. New York and London: Academic Press, pp. 127–252.

9. Idler, D.R., and Truscott, B. 1α-hydroxycorticosterone and testosterone in body fluids of a cartilaginous fish (*Raja radiata*). *J. Endocr.* 42:165–66.

10. Vinson, G.P., and Whitehouse, B.J. 1970. Comparative aspects of adrenocortical function. In: M.H. Briggs, ed. *Advances in Steroid Biochemistry and Pharmacology*. New York and London: Academic Press, pp. 163–342.

11. Idler, D.R., Freeman, H.C., and Truscott, B. 1967. Biological activity and protein-binding of 1α-hydroxycorticosterone: An interrenal steroid in elasmobranch fish. *Gen. Comp. Endocr.* 9:207–13.

12. Idler, D.R., and Szeplaki, B.J. 1968. Interrenalectomy and stress in relation to some blood components of an Elasmobranch (*Raja radiata*). *J. Fish. Res. Bd. Canada.* 25:2549–60.

13. Payan, P., and Maetz, J. 1973. Branchial sodium transport mechanisms in *Scyliorhinus canicula*. Evidence for Na^+/NH_4^+ and Na^+/H^+ exchanges and for a role of carbonic anhydrase. *J. Exp. Biol.* 58:487–502.

14. Burton, M., and Idler, D.R. 1986. The cellular location of 1α-hydroxycorticosterone binding protein in skate. *Gen. Comp. Endocr.* 64:260–66.

15. Hazon, N., Decourt, C., O'Toole, L.B., Lahlou, B., and Henderson, I.W. 1987. Vascular and steroidogenic effects of ANP and angiotensin II in elasmobranch fish. *J. Endocr.* 115, supp. abst. 161.

16. Henderson, I.W., and Garland, I.O. 1980. The interrenal gland in pisces. Part 2 physiology. In: I. Chester Jones and I.W. Henderson, eds. *General, Comparative and Clinical Endocrinology of the Adrenal Cortex*. vol. III. New York and London: Academic Press, pp. 473–523.

17. O'Toole, L.B., Hazon, N., Leonard, R.A., and Henderson, I.W. 1987. In-vitro perifusion of dogfish interrenal gland. *J. Endocr.* 115, supp., abst. 160.

18. Nandi, J. 1962. The structure of the interrenal gland in teleost fishes. *University of California publications in zoology.* 65:129–212.

19. Ogawa, M. 1967. Fine structure of the corpuscles of Stannius and the interrenal

tissue in the goldfish (*Carassius auratus*). *Z. Zellforsch. u. Mikrosk. Anat.* 81:174–89.

20. Vinson, G.P., Whitehouse, B.J., Goddard, C., and C.P. Sibley. 1979. Comparative and evolutionary aspects of aldosterone secretion and zona glomerulosa function. *J. Endocr.* 81:5P–24P.

21. Chan, D.K.O., and Woo, N.Y.S. 1978. Effect of hypophysectomy on the chemical composition and intermediary metabolism of the Japanese eel (*Anguilla anguilla*). *Gen. Comp. Endocr.* 35:169–78.

22. Patino, R., and Schreck, C.B. 1988. Spontaneous and ACTH-induced interrenal steroidogenesis in juvenile Coho salmon (*Oncorhynchus kisutch*): Effects of monovalent ions and osmolarity in vitro. *Gen. Comp. Endocr.* 69:416–23.

23. Mukherjee, S., Baker, B.I., Bird, D.J., and Buckingham, J.C. 1987. Hypothalamo-pituitary-interrenal responses to opioid substances in the trout. *Gen. Comp. Endocr.* 68:40–48.

24. Bird, D.J., Buckingham, J.C., Baker, B.I., and Mukherjee, S. 1987. Hypothalamo-pituitary-interrenal responses to opioid substances in the trout. I Effects of morphine on the release in vitro of corticotrophin-releasing activity from the hypothalamus and corticotrophin from the pituitary gland. *Gen. Comp. Endocr.* 68:33–39.

25. Patino, R., Bradford, C.S., and Schreck, C.B. 1986. Adenylate cyclase activators and inhibitors, cyclic nucleotide analogs, and phosphatidylinositol: Effects on interrenal function of Coho salmon (*Oncorhynchus kisutch*) in vitro. *Gen. Comp. Endocr.* 63:230–35.

26. Decourt, C., and Lahlou, B. 1987. Evidence for the direct intervention of antiotensin II in the release of cortisol in teleost fishes. *Life Sci.* 41:1517–24.

27. Baker, B.I., and Rance, T. 1981. Differences in concentrations of plasma cortisol in the trout and the eel following adaptation to black or white backgrounds. *J. Endocr.* 89:135–40.

28. Kenyon, C.J., McKeever, A., Oliver, J.A., and Henderson, I.W. 1985. Control of renal and adrenocortical function by the renin-angiotensin system in two euryhaline Teleost fishes. *Gen. Comp. Endocr.* 58:93–100.

29. Patino, R., Redding, J.M., and Schreck, C.B. 1987. Interrenal secretion of corticosteroids and plasma cortisol and cortisone concentrations after acute stress and during seawater acclimation in juvenile Coho salmon (*Oncorhynchus kisutch*). *Gen. Comp. Endocr.* 68:431–39.

30. Call, R.N., and Janssens, P.A. 1975. Histochemical studies of the adrenocortical homologue in the kidney of the Australian lungfish, (*Neoceratodus forsteri*). *Cell Tiss. Res.* 156:533–38.

31. Hanke, W. 1978. The adrenal cortex of amphibia. In: I. Chester Jones and I.W. Henderson, eds. *General, Comparative and Clinical Endocrinology of the Adrenal Cortex.* vol. II. New York and London: Academic Press, pp. 419–95.

32. Volk, T.L. 1972. Morphological observations on the summer cell of Stilling in the interrenal gland of the American bullfrog (*Rana catesbiana*). *Z. Zellforsch. u. Mikrosk. Anat.* 130:1–11.

33. Volk, T.L. 1972. Ultrastructure of the adrenal cortical cells of the interrenal gland of the American bullfrog (*Rana catesbiana*). *Z. Zellforsch. u. Mikrosk. Anat.* 123:470–85.

34. Johnston, C.I., Davis, J.O., Wright, F.S. and Howards, S.S. 1967. Effects of renin and ACTH on adrenal steroid secretion in the American bullfrog. *Am. J. Physiol.* 213:393–99.

35. Dupont, W., Leboulenger, F., Vaudry, H., and Vaillant, R. 1976. Regulation of the aldosterone secretion in the frog (*Rana esculenta*) L. *Gen. Comp. Endocr.* 29:51–60.

36. Leboulenger, F., Delarue, A., Tonon, M.C., Jegou, S., and Vaudry, H. 1978. In vitro study of frog (*Rana ridibunda* Pallas) interrenal function by use of a simple perifusion system. I. Influence of adrenocorticotrophin upon corticosterone release. *Gen. Comp. Endocr.* 36:327–38.

37. Delarue, C., Tonon, M.C., Leboulenger, F., Fegou, S., Leroux, P., and Vaudry, H. 1979. In vitro study of frog (*Rana ridibunda* Pallas) interrenal function by use of a simple perifusion system. II. Influence of adrenocorticotrophin upon aldosterone production. *Gen. Comp. Endocr.* 38:399–409.

38. Perroteau, I., Netchitailo, P., Homo-Delarche, F., Delarue, C., Lihrmann, I., Leboulenger, F., and Vaudry, H. 1984. Role of exogenous and endogenous prostaglandins in steroidogenesis by isolated frog interrenal gland: evidence for dissociation in adrenocorticotropin and angiotensin action. *Endocrinology.* 115:1765–84.

39. Lihrmann, I., Delarue, C., Homo-Delarche, F., Netchitailo, P., Leboulenger, F. and Vaudry, H. 1986. Role of prostaglandins in calcium-induced corticosteroid secretion by isolated frog interrenal gland. *Prostaglandins.* 32:127–31.

40. Lihrmann, I., Delarue, C., Homo-delarche, F., Feuilloley, M., Belanger, A., and Vaudry, H. 1987. Effects of TMB-8 and dantrolene on ACTH- and angiotensin-induced steroidogenesis by frog interrenal gland: evidence for a role of intracellular calcium in angiotensin action. *Cell calcium.* 8:269–82.

41. Delarue, C., Lihrmann, I., Feuilloley, M., Netchitailo, P., Idres, S., Leboulenger, F., Belanger, A. Perroteau, I., and Vaudry, H. 1988. In vitro study of frog adrenal function, IX. Evidence against the involvement of lipoxygenase metabolites in the control of steroid production. *J. Steroid Biochem.* 30:461–64.

42. Benyamina, M., Leboulenger, F., Lihrmann, I., Delarue, C., Feuilloley, M., and Vaudry, H. 1987. Acetylcholine stimulates steroidogenesis in isolated from adrenal gland through muscarinic receptors: evidence for a desensitization mechanism. *J. Endocr.* 113:339–48.

43. Leboulenger, F., Banyamina, M., Delarue, C., Netchitailo, P., Saint-Pierre, S., and Vaudry, H. 1988. Neuronal and paracrine regulation of adrenal steroidogenesis: interactions between acetylcholine, serotonin and vasoactive intestinal peptide (VIP) on corticosteroid production by frog interrenal tissue. *Brain res.* 453:103–09.

44. Larcher, A., Delarue, C., Idres, S., Leboulenger, F., Vandesande, F., and Vaudry, H. 1988. Vasotocin as a local regulator of adrenal function in the frog adrenal gland. *J. Endocr.* 119, supp., abst. 93.

45. Maser, C., Hittler, K., and Hanke, W. 1980. External induction of changes of corticosterone level in amphibia. *Gen. Comp. Endocr.* 40:335.

46. Maser, C., Janssens, P.A., and Hanke, W. 1982. Stimulation of interrenal secretion in amphibia 1. Direct effects of electrolyte concentration on steroid release. *Gen. Comp. Endocr.* 47:458–66.

47. Smith, L.D., Ecker, R.E., and Subtelny, S. 1968. In vitro induction of physiological maturation in *Rana pipiens* oocytes removed from their ovarian follicles. *Developmental Biol.* 17:627–43.

48. Ray, P.P., De, T.K., and Maiti, B.R. 1987. A comparative histomorphological study of the adrenal gland in seven species of turtles. *Z. mikrosk.-anat. forsch.* 101:50–60.

49. Lofts, B. 1978. The adrenal gland in reptilia. Part 1 Structure. In: I. Chester Jones and I.W. Henderson, eds. *General Comparative and Clinical Endocrinology of the Adrenal Cortex*. vol. II. New York and London: Academic press, pp. 292–369.

50. Lofts, B., and Phillips, J.G. 1965. Some aspects of the structure of the adrenal gland in snakes. *J. Endocr.* 33:327–28.

51. Lofts, B., Phillips, J.G., and Tam, W.H. 1971. Seasonal changes in the histology of the adrenal gland of the cobra (*Naja naja*). *Gen. Comp. Endocr.* 16:121–31.

52. Callard I.P., and Callard, G.V. 1978. The adrenal gland in reptilia. Part 2. Physiology. In: I. Chester Jones and I.W. Henderson, eds. *General, Comparative and Clinical Endocrinology of the Adrenal Cortex*. vol. II. New York and London: Academic Press, pp. 370–418.

53. Callard, I.P., Chan, S.W.C., and Callard, G.V. 1973. Hypothalamic-pituitary-adrenal relationships in reptiles. In: *Brain-Pituitary-Adrenal Interrelationships*. Karger, Basel, pp. 270–92.

54. Sanford, B., and Stephens, G.A. 1988. The effects of adrenocorticotropin hormone and angiotensin II on adrenal corticosteroid secretions in the freshwater turtle (*Pseudemys scripta*). *Gen. Comp. Endocr.* 72:107–14.

55. Dauphin-Villemant, C., Leboulenger, F., Xavier, F., and Vaudry, H. 1988. Interrenal activity in the female lizard Lacerta vivipera J.: In vitro response to ACTH 1-39 and to [Sar^1, Val^5] angiotensin II (ang II). *J. Steroid Biochem.* 30:457–60.

56. Holmes, W.N., and Phillips, J.G. 1976. The adrenal cortex of birds. In: I. Chester Jones and I.W. Henderson, eds. *General, Comparative and Clinical Endocrinology of the Adrenal Cortex*. vol. I. New York and London: Academic Press, pp. 293–420.

57. Pearce, R.B., Cronshaw, J., and Holmes, W.N. 1978. Evidence for the zonation of interrenal tissue in the adrenal gland of the duck (*Anas platyrhynchos*). *Cell Tiss. Res.* 192:363–79.

58. Cronshaw, J., Ely, J.A., and Holmes, W.N. 1985. Functional differences between two structurally distinct types of steroidogenic cell in the avian adrenal gland. *Cell Tiss. Res.* 240:561–67.

59. Taylor, A.A., Davis, J.O., Breitenbach, R.P., and Hartroft, P.M. 1970. Adrenal steroid secretion and a renal-pressor system in the chicken (*Gallus domesticus*). *Gen. Comp. Endocr.* 14:321–25.

60. Haack, D.W., Abel, J.H., and Rhees, R.W. 1972. Zonation in the adrenal of the duck: Effect of osmotic stress. *Cytobiologie.* 5:247–64.

61. Pearce, R.B., Cronshaw, J., and Holmes, W.N. 1981. Changes in corticotropic responsiveness and mitochondrial ultrastructure of adrenocortical cells from the inner zone of the duck (*Anas platyrhynchos*) adrenal gland: The effects of cycloheximide, puromycin, and chloramphenicol. *Cell Tiss. Res.* 221:45–57.

62. Carsia, R.V., Scanes, C.G., and Malamed, S. 1985. Isolated adrenocortical cells of the domestic fowl (*Gallus domesticus*): Steroidogenic and ultrastructural properties. *J. Steroid Biochem.* 22:273–79.

63. Rosenberg, J., Pines, M., and Hurwitz, S. 1987. Response of adrenal cells to parathyroid hormone stimulation. *J. Endocr.* 112:431–37.

64. Carsia, R.V., Scanes. S.G., and Malamed, S. 1987. Polyhormonal regulation of avian and mammalian corticosteroidogenesis in vitro. *Comp. Biochem. Physiol.* 88A:131 40.

65. Rosenberg, J., Pines, M., and Hurwitz, S. 1987. Response of adrenal cells to parathyroid hormone stimulation. *J. Endocr.* 112:431–37.

66. Cheung, A., Hall, T.R., and Harvey, S. 1987. Serotoninergic regulation of corticosterone secretion in domestic fowl. *J. Endocr.* 113:159–65.

67. Cheung, A., Hall, T.R., and Harvey, S. 1988. Stimulation of corticosterone release in the fowl by recombinant DNA-derived chicken growth hormone. *Gen. Comp. Endocr.* 69:128–32.

68. Frankel, A.I., Graber, J.W., Cook, B., and Nalbandov, A.V. 1967. The duration and control of adrenal function in adenohypophysectomised cockerels. *Steroids.* 10:699–707.

69. Bradley, E.L., and Holmes, W.N. 1971. The effects of hypophysectomy on adrenocortical function in the duck (*Anas platyrhynchos*). *J. Endocr.* 49:437–57.

70. Rosenberg, J., and Hurwitz, S. 1987. Concentration of adrenocortical hormones in relation to cation homeostasis in birds. *Am. J. Physiol.* 253:R20–R24.

71. Clauss, W., Durr, J.E., Guth, D., and Skadhauge, E. 1987. Effects of adrenal steroids on Na transport in the lower intestine (*coprodeum*) of the hen. *J. Membrane Biol.* 96:141–52.

72. Rosenberg, J., Pines, M., and Hurwitz, S. 1988. Regulation of aldosterone secretion by avian adrenocortical cells. *J. Endocr.* 118:447–53.

73. Holmes, W.N., Phillips, J.G., and Chester Jones, I. 1963. Adrenocortical factors associated with adaptation of vertebrates to marine environments. *Recent Progr. Hormone Res.* 19:619–72.

74. Holmes, W.N., and Phillips, J.G. 1964. Adrenocortical hormones and electrolyte metabolism in birds. Proceedings of the second international congress of endocrinology. *Excerpta Med., Int. Congr. Ser.* 83, part 1, pp. 158–61.

75. Harvey, S., and Phillips, J.G. 1980. Growth, growth hormone and corticosterone secretion in freshwater and saline adapted ducklings (*Anas platyrhynchos*). *Gen. Comp. Endocr.* 42:334–44.

76. Weiss, M. 1973. Biosynthesis of adrenocortical steroids by monotremes: echidna (*Tachyglossus aculeatus*) and platypus (*Ornithorhynchos anatinus*). *J. Endocr.* 58:251–62.

77. Oddie, C.J., Blaine, E., Bradshaw, S.D., Coghlan, J.P., Denton, D.A., Nelson, J.F., and Scoggins, B.A. 1976. Blood corticosteroids in Australian marsupial and placental mammals and one monotreme. *J. Endocr.* 69:341–48.

78. Weiss, M., Oddie, C.J., and McCance, I. 1979. The effects of ACTH on adrenal steroidogenesis and blood corticosteroid levels in the echidna (*Tachyglossus aculeatus*). *Comp. Biochem. Physiol.* 64B:65–70.

79. Sernia, C., and McDonald, I.R. 1977. Adrenocortical function in a prototherian mammal, (*Tachyglossus aculeatus*) (Shaw). *J. Endocr.* 72:41–52.

80. Bourne, G.H. 1949. *The Mammalian Adrenal Gland.* London: Oxford University Press.

81. McDonald, I.R. 1977. Adrenocortical functions in marsupials. In: B. Stonehouse and D. Gilmore, eds. *Biology of Marsupials.* London: Macmillan, pp. 345–77.

82. McDonald, I.R., and Bradshaw, S.D. 1977. Plasm corticosteroids and the effect of adrenocorticotrophin in a macropodid marsupial (*Setonix brachyurus*), Quoy and Gaimard. *J. Endocr.* 75:409–18.

83. Miller, T., and Bradshaw, S.D. 1979. Adrenocortical function in a field population of a macropodid marsupial (*Setonix brachyurus*), Quoy and Gaimard. *J. Endocr.* 82:159–70.

84. Janssens, P.A., and Hinds, C.J. 1981. Long-term effects of corticosteroid administration in the tammar wallaby, (*Macropus eugenii*). *Gen. Comp. Endocr.* 47:221–25.

85. Martin, I.K., and McDonald, I.R. 1986. Adrenocortical functions in a macropodid marsupial (*Thylogale billardierii*). *J. Endocr.* 110:471–80.

86. Martin, I.K., and McDonald, I.R. 1988. Metabolic actions of cortisol in a macropodid marsupial (*Thylogale billardierii*). *J. Endocr.* 116:71–79.

87. McDonald, I.R., Lee, A.K., Than, K.A., and Martin, R.W. 1986. Failure of glucocorticoid feedback in males of a population of small marsupials (*Antechinus swainsonii*) during the period of mating. *J. Endocr.* 108:63–68.

88. Long, J.A. 1975. Zonation of the mammalian adrenal cortex. Handbook of Physiology, Sec. 7: *Endocrinology*. vol. VI. Adrenal gland. American physiological society, Washington, pp. 13–24.

89. Ayres, P.J., Gould, R.P., Tait, S.A.S., and Tait, J.F. 1956. The in-vitro demonstration of differential corticosteroid production within the ox adrenal gland. *Biochem. J.* 63:19P.

90. Vinson, C.P., Whitehouse, B.J., Goddard, C., and Sibley, C.P. 1979. Comparative and evolutionary aspects of aldosterone secretion and zona glomerulosa function. *J. Endocr.* 81:5P–24P.

91. Laird, S., Vinson, G.P., and Whitehouse, B.J. 1988. Steroid sequestration within the rat adrenal zona glomerulosa. *J. Endocr.* 117:191–96.

92. Vinson, G.P., and Kenyon, C. 1978. Steroidogenesis in the zones of the mammalian adrenal cortex. In: I. Chester Jones and I.W. Henderson, eds. *General, Comparative and Clinical Endocrinology of the Adrenal Cortex.* vol. II. New York and London: Academic press, pp. 201–264.

93. O'Hare, M.J., Nice, E.C., and Neville, A.M. 1980. Regulation of androgen secretion and sulfoconjugation in the adult human adrenal cortex: Studies with primary monolayer cultures. In: A.R. Genazzini, J.H.H. Thissen, and P.K. Siiteri, eds. *Andrenal androgens.* New York: Raven Press, pp. 7–25.

94. Nishikawa, T., and Strott, C.A. 1984. Cortisol production by cells isolated from the outer and inner zones of the adrenal cortex of the guinea pig. *Endocrinology.* 114:486–91.

95. Davison, B., Large, D.M., Anderson, D.C., and Robertson, W.R. 1983. Basal steroid production by the zona reticularis of the guinea pig adrenal cortex. *J. Steroid Biochem.* 18:285–90.

96. Hyatt, P.J., Bell, J.B.G., Bhatt, K., and Tait, J.F. 1983. Preparation and steroidogenic properties of purified zona fasciculata and zona reticularis cells from the guinea-pig gland. *J. Endocr.* 96:1–14.

97. Bell, J.B.G., Bhatt, K., Hyatt, P.J., Tait, J.F., and Tait, S.A.S. 1980. Properties of adrenal zona reticularis cells. In: A.R. Genazzani, J.H.H. Thissen, and P.K. Siiteri, eds. *Adrenal androgens.* New York: Raven Press, pp. 1–6.

98. Jones, T., and Griffiths, K. 1968. Ultrachemical studies of the site of formation of dehydroepiandrosterone sulphate in the adrenal cortex of the guinea pig. *J. Endocr.* 42:559–65.

99. Maurer, R. and Reubi, J.-C. 1986. Distribution and coregulation of three peptide receptors in adrenals. *Eur. J. Pharmacol.* 125:241–47.

100. Whitley, G. St.J., Bell, J.B.G., Chu, F.W., Tait, J.F., and Tait, S.A.S. 1984. The effects of ACTH, serotonin, K^+ and angiotensin analogues on ^{32}P incorporation into phospholipids of the rat adrenal cortex: Basis for an assay method using zona glomerulosa cells. *Proc. Roy. Soc. B.* 222:273–94.

101. Pedersen, R.C., and Brownie, A.C. 1987. Gamma$_3$-melanotropin promotes mitochondrial cholesterol accumulation in the rat adrenal cortex. *Mol. Cell. Endocr.* 50:149–56.

102. Robertson, W.R., Davison, B., Anderson, D.C., Frost, J., and Lambert, A. 1986. Effect of adrenocorticotrophin on cortisol and androstenedione secretion from dispersed cells of guinea-pig adrenal zonae fasciculata and reticularis. *J. Endocr.* 109:300–404.

103. Mikami, K., Nishikawa, T., and Strott, C.A. 1985. Adenylate cyclase activity and cyclic AMP production in the outer and inner zones of the adrenal cortex. *Biochem. Biophys. Res. Comm.* 129:664–70.

104. Kubo, M., and Strott, C.A. 1987. Differential activity of 3-hydroxy-3-methylglutaryl coenzyme A reductase in zones of the adrenal cortex. *Endocrinology.* 120:214–21.

105. Obara, T., Mikami, K., and Strott, C.A. 1984. Differential suppression of the outer and inner zones of the adrenal cortex of the guinea pig. *Endocrinology.* 115:1838–41.

106. Martin, K., and Black, V.H. 1982. Δ^4-Hydrogenase in guinea pig: Evidence of localization in zona reticularis and age-related change. *Endocrinology.* 110:1749–57.

107. Holzbauer, M., Birmingham, M.K., de Nicola, A.F., and Oliver, J.T. 1985. In vivo secretion of 3α-hydroxy-5α-pregnan-20-one, a potent anaesthetic steroid by the adrenal gland of the rat. *J. Steroid Biochem.* 22:97–102.

108. Robba, C., Rebuffat, P., Mazzocchi, G., and Nussdorfer, G.G. 1985. Opposed effects of chronic prolactin administration on the zona fasciculata and zona reticularis of the rat adrenal cortex: an ultrastructural stereological study. *J. Submicrosc. Cytol.* 17:255–61.

109. Vinson, G.P., and Whitehouse, B.J. 1976. In vitro modification of rat adrenal zona fasciculata/reticularis function by the zona glomerulosa. *Acta Endocr.* 81:340–50.

110. Hornsby, P.J. 1985. The regulation of adrenocortical function by control of growth and structure. In: D.C. Anderson and J.S.D. Winter, eds. *Adrenal Cortex.* London: Butterworths, pp. 1–31.

111. Weinkove, C., and Anderson, D.C. 1985. Interactions between adrenal cortex and medulla. In: D.C. Anderson and J.S.D. Winter, eds. *Adrenal Cortex.* London: Butterworths, pp. 208–34.

112. Hinson, J.P. 1990. Paracrine control of adrenocortical function. *J. Endocr.* 124:7–9.

113. Vinson, G.P. 1974. The control of the adrenocortical secretion in the brush-tailed possum (*Trichosurus vulpecula*). *Gen. Comp. Endocr.* 22:268–76.

114. Weiss, M. 1984. Gonadotrophin induced development of the "special zone" in the adrenal cortex of immature female possums (*Trichosurus vulpecula*) with concomitant activation of steroid reductases. *Comp. Biochem. Physiol.* 79B:173–79.

115. Call, R.N., and Janssens, P.A. 1984. Hypertrophied adrenocortical tissue of the Australian brush-tailed possum (*Trichosurus vulpecula*): Uniformity during reproduction. *J. Endocr.* 101:263–67.

116. Curlewis, J.D., Axelson, M., and Stone, G.M. 1985. Identification of the major steroids in ovarian and adrenal venous plasma of the brush-tail possum (*Trichosurus vulpecula*) and changes in the peripheral plasma levels of oestradiol and progesterone during the reproductive cycle. *J. Endocr.* 105:53–62.

117. Weiss, M., and Ford, V.L. 1977. Changes in steroid biosynthesis by adrenal homogenates of the possum (*Trichosurus vulpecula*) at various stages of sexual maturation with special reference to 5B-reductase activity. *Comp. Biochem. Physiol.* 57B:15–18.

118. Neville, A.M., and O'Hare, M.J. 1981. *The Human Adrenal Cortex.* Berlin: Springer Verlag.

119. Winters, A.J., Oliver, C., Colston, C., MacDonald, P.C., and Porter, J.C. 1974. Plasma ACTH levels in the human fetus and neonate as related to age and parturition. *J. Clin. Endocr.* 39:269–73.

120. Carr, B.R., Ohashi, M., and Simpson, E.R. 1984. Low density lipoprotein binding and de novo synthesis of cholesterol in the neocortex and fetal zones of the human fetal adrenal gland. *Endocrinology.* 110:1994–98.

121. Carr, B.R., Milburn, J., Wright, E.E., and Simpson, E.R. 1985. Adenylate cyclase activity in neocortex and fetal zone membrane fractions on the human fetal adrenal gland. *J. Clin. Endocrin..* 60:718–22.

122. MacDonald, P.C., and Siiteri, P.K. 1965. Origin of estrogen in women pregnant with an anencephalic fetus. *J. Clin. Invest.* 44:465–74.

123. Parker, C.R., Carr, B.R., Winkel, C.A., Casey, M.L., Simpson, E.R., and MacDonald, P.C. 1983. Hypercholesterolaemia due to elevated low density lipoprotein-cholesterol in newborns with anencephaly and adrenal atrophy. *J. Clin Endocr.* 57:37–43.

124. Abu-Hakima, M., Branchaud, C.L., Goodyer, C.G., and Murphy, B.E.P. 1987. The effects of human chorionic gonadotrophin on growth and steroidogenesis of the human fetal adrenal gland in vitro. *Am. J. Obstet. Gynec.* 3:681–87.

125. Taga, M., Tanaka, K., Liu, T.-I., Minaguchi, H., and Sakamoto, S. 1981. Effect of prolactin on the secretion of dehydroepiandrosterone (DHEA), its sulfate (DHEA-S), and cortisol by the human adrenal fetal zone in vitro. *Endocrinologica japonica.* 28:231–37.

126. Pepe, G.J., Waddell, B.J., and Albrecht, E.D. 1988. The effects of adrenocorticotropin and prolactin on adrenal dehydroepiandrosterone secretion in the baboon fetus. *Endocrinology.* 122:646–50.

127. Gottschau, M. 1833. Struktur und embryonale entwicklung der nebennieren bei saugethieren. *Arch. Anat. u. Physiol.* pp. 412–58.

128. Wright, N.A., Voncina, D., and Morley, A.R. 1973. An attempt to demonstrate cell migration from the zona gomerulosa in the prepubertal male rat adrenal cortex. *J. Endocr.* 59:451–59.

129. Idelman, S. 1978. The structure of the mammalian adrenal cortex. In: I. Chester Jones and I.W. Henderson, eds. *General, Comparative and Clinical Endocrinology of the Adrenal Cortex.* vol. II. New York and London: Academic Press, pp. 1–199.

130. Zajicek, G., Ariel, I., and Arber, N. 1986. The streaming adrenal cortex: direct evidence of centripetal migration of adrenocytes by estimation of cell turnover rate. *J. Endocr.* 111:477–82.

131. Estivariz, F.E., Carino, M., Lowry, P., and Jackson, S. 1988. Further evidence that N-terminal pro-opiomelanocortin peptides are involved in adrenal mitogenesis. *J. Endocr.* 116:201–06.

Abbreviations

ACAT = acyl CoA : cholesterol acyltransferase

ACTH = adrenocorticotropic hormone, corticotropin

AIDS = acquired immune deficiency syndrome

ANP = atrial natriuretic peptides

APA = aldosterone-producing adenoma

APS = autoimmune polyglandular syndrome

ASF = aldosterone-stimulating factor

ATP = adenosine triphosphate

CAH = congenital adrenal hyperplasia

cAMP = cyclic 3'5' adenosine monophosphate

cDNA = complimentary deoxyribonucleic acid

CE = cholesterol esters

CEH = cholesterol ester hydrolase

CLIP = corticotropin-like intermediate lobe peptide

CoA = coenzyme A

CoQ = coenzyme Q

CRF = corticotropin releasing factor

CT = computer aided tomography

Cyt = cytochrome

DDD = dichlorodiphenyldichloroethane

DEAE = diethylaminoethyl

DNA = deoxyribonucleic acid

ECF = extracellular fluid

EDLF = endogenous-digitalis-like factor

EGF = epithelial growth -factor

EM = electron microscopy

FAD = flavine adenine dinucleotide

FMN = flavine mononucleotide

Fp = flavoprotein

FSH = follicle stimulating hormone

GABA = gamma-aminobutyric acid
GH = growth hormone
GHRH = growth hormone-releasing hormone
GNRH = gonadotropin-releasing hormone
HDL = high density lipoprotein
HLA = human lymphocyte antigen
HMG CoA = hydroxymethylglutaryl CoA
IHA = idiopathic hyperaldosteronism
IRMA = immunoradiometric assay
kDa = kiloDalton
LDL = low density lipoprotein
LH = luteinizing hormone
LPH = lipotropic hormone
mRNA = messenger ribonucleic acid
MSH = melanocyte-stimulating hormone
Na^+/K^+ ATPase = sodium-potassium ATPase
NAD^+ = nicotinamide adenine dinucleotide
NADH = nicotinamide adenine dinucleotide, reduced
$NADP^+$ = nicotinamide adenine dinucleotide phosphate
NADPH = nicotinamide adenine dinucleotide phosphate, reduced
NHFe = nonheme iron protein
PDGF = platelet derived growth factor

PG E_2 = prostaglandin E_2
POMC = proopiomelanocortin
RER = rough endoplasmic reticulum
RIA = radioimmunoassay
RNA = ribonucleic acid
SDS = sodium dodecyl sulfate
SER = smooth endoplasmic reticulum
TPA = tumor promoting agent (phorbol ester)
UDP = uridine diphosphate
VIP = vasoactive intestinal polypeptide

Steroids:

B = corticosterone
DHEA = dehydroepiandrosterone
DHEAS = dehydroepiandrosterone sulphate
DOC = deoxycorticosterone
DOCA = deoxycorticosterone acetate
E = cortisone
F = cortisol
18-OH-B = 18-hydroxycorticosterone
18-OH-DOC = 18-hydroxydeoxycorticosterone
19-OH-DOC = 19-hydroxydeoxycorticosterone
S = 11-deoxycortisol

Index